MathCAD
for Introductory Physics

Denis Donnelly
Siena College

Addison-Wesley Publishing Company

*Reading, Massachusetts • Menlo Park, California • New York • Don Mills,
Ontario • Wokingham, England • Amsterdam • Bonn • Sydney
Singapore • Tokyo • Madrid • San Juan • Milan • Paris*

MathCAD is a registered trademark of MathSoft, Inc.

Library of Congress Cataloging-in-Publication Data

Donnelly, Denis P., 1937–
    MathCAD for introductory physics / Denis Donnelly.
        p.   cm.
    Includes index.
    ISBN 0-201-54736-8
    1. Physics–Data processing. 2. MathCAD. I. Title.
QC52.D66   1992
530'.0285'5369–dc20   92-7768
                                                                        CIP

3 4 5 6 7 8 9 10 MA 96959493

Preface

Physics is the most basic, the most fundamental of the sciences. It is a study of matter and energy and their interactions. To a beginner, that may sound deep and powerful (which it is) and perhaps even a bit frightening (which it needn't be) but surely something you want to be in on (which is right on). Beginner or not, you can get in on the action right away. Our goal is this; we want you to get a feeling for what physics is about, how it works, what it can do. Our approach to attaining that goal requires three things: an active learner, this book, and a computer running MathCAD. With these tools available, we can attack a wide range of problems and think about lots of different phenomena, get a sense of physical behavior, and let the computer do most of the work. We have at our fingertips a means to explore aspects of physics that would be very cumbersome (and in a number of cases virtually impossible) to do any other way. With MathCAD, it is easy to calculate, check units, plot, change, redo, try this, try that, see what happens, get a feel for the process. That's what you need to do. That's what I hope you will be doing.

You should use this book in a different way than you would a typical text. A textbook may be loaded with information but it's not interactive the way a computer is with you at the controls. This book is not just meant to be read; it is meant to be alongside you as you sit at your computer running MathCAD. When you access one of the files on the disk, it is in an active state, ready for you to change, process, and/or extend as the need requires.

The code for a number of documents is included in the text (a number of the documents are in two parts on separate pages). This material is to be read just as is the text. Follow the development; first imitate and then create your own documents. Once you get used to moving regions, it is fun to arrange a document so that the regions are in a form that you enjoy. Some of the documents near the beginning are demonstrations, designed for you to enter values and observe outcomes (e.g., the summation of vectors). There may be some advanced features in these documents that are necessary to create the figure(s) or to solve some equations. You are not expected to write things this complex, although by the time you have finished the book, you will have gained enough experience to do so.

The book is designed primarily for students taking college-level physics who have access to either a PC, a PC clone, or a Macintosh computer that runs MathCAD. An adept high school physics student could make good use of this book as well. Of course, individuals not officially enrolled in a course but interested in learning physics on their own would find this book's approach to physics very useful, as well.

This book is meant to accompany any general physics text. Topics from both semesters of a typical course are included. Some of the discussion makes use of the calculus but the actual computations are performed by MathCAD. So don't run when you see an integral or a deriative. You only need to know what to take the derivative or integral of, you do not need to know how to take the derivative or how to perform the integration.

The phenomena described in general physics are so rich and so varied, that it is possible to direct one's attention to almost any problem or phenomenon and discover something surprising. On several occasions in the text, we explore the transition between a discrete and a continous system. For example, we consider the motion of a rocket as the fuel is expelled in different states; at one extreme, the fuel is ejected in one large discrete chunk, at the other extreme the fuel is a gas ejected over a period of time. Or we compare the pressure in a wall constructed from a series of blocks. Such a simple model hardly seems like a model of either compressible or incompressible fluids.

For such problems, to appreciate the outcomes, first you must understand the model, then see how the model is expressed in equations, and finally you must explore a number of different cases. As you vary parameters or change models and consider a variety of cases, then you begin to understand the system. One-answer calculator problems appear rather limited from this perspective.

I'm willing to make you a bet. If you spend a sufficient amount of time with MathCAD to get some facility with the language, enough so that you are comfortable with it, you will change. The more you work with the computer, the less you will want to do by hand. A typewriter is a terrific device but once you understand how to use a word processor, a typewriter has a somewhat archaic aura. A pocket scientific calculator is essential for modest computations but if you want to explore a problem in more detail and with greater generality, a more powerful approach is essential. Learn, explore, and gain insight. The opportunity is yours for the taking.

Acknowledgments

Many colleagues and students have generously taken time from their busy schedules to comment on preliminary drafts of sections of the manuscript. I particularly wish to thank Shamshad Ahmad, Burt Brody, Josh Diamond, Harvey Gould, Leonard Merrill, and Ed Rogers. Clearly, the errors, omissions, and limitations are my own.

Thanks, too, are due to members of Siena College's administration for their support of this project. In particular, I thank David Rice, former Vice President for Academic Affairs; Ken Wittig, Dean, Division of Science; and Josh Diamond, Head, Department of Physics. I also thank Stuart Johnson, the physics editor of Addison-Wesley for his interest and support, Lorraine Ferrier for the careful copy editing, and Mona Zeftel for her production work. The text was produced using TeXtures. The figures showing the documents were produced directly from MathCAD.

To   Jacqueline

Jane
Philip
Stephen
Peter

and

Lulu

Contents

Contents

List of Programs

MathCAD Fundamentals

MathCAD, an equation-solving software package, provides a convenient direct method for handling numbers, equations, graphs, and text. A unique feature of MathCAD is that it allows the computer to be used like a notepad. Equations appear on the screen as equations do when written on a pad of paper. Such equations or any other information entered into the computer are contained within regions. These regions can be arranged on the screen in any desired nonoverlapping form. The only constraint in placement is that processing proceeds from left to right and from top to bottom.

Any information entered into a MathCAD document is in one of three types of regions: equation, plot, or text. Equation is the default mode. That is, unless a text or plot region is specified, any entry will appear in an equation region. The symbols to initiate text and plot regions are " and @, respectively.

MathCAD has many built-in functions so that any calculations done on a calculator are easily performed with MathCAD. MathCAD's advantage is that, once information is on the screen, it is easy to review and modify what has been done, to extend the calculation, to explore ideas that come to mind as the process unfolds, to save the work for recall at a later time, and to print out the contents of the document. Computational possibilities with MathCAD are extensive. They include solving equations, plotting graphs, curve fitting, performing numerical integration and differentiation, and matrix operations. MathCAD provides a rich environment in which to deal with numerical questions and models.

"Central Services . . . . We do the work, you do the pleasure."

Brazil

1.1   Prelude: A Glance at MathCAD

An example gives tooth to the words. In Fig. 1.1, a series of calculations demonstrate various MathCAD capabilities and give an indication of the kinds of things that can be done quite readily. There are regions of text, equations, and plots. The equations, whether or not they are familiar, at least appear in a familiar form.

Follow the document from left to right and from top to bottom. There is text. There are examples of arithmetic operations, with units and the use of variable names. Note the different equality signs used when defining a quantity and when evaluating a quantity. Observe the use of the place marker to express a result in terms of a particular quantity (the π in asin (1)). We then see trigonometric and arctrigonometric functions; a user-defined function, $y(x)$; a range variable, x; a plot region; subscripted variables, xx_i, yy_i; derivative and integral operations; finding a zero of a function using the root function; using subscripted variables in an iterative procedure; and finally solving simultaneous equations using a solve block.

The point of this example is to see the general layout of a document and to see the kinds of things that can be done with only a few steps. The details of MathCAD and many additional examples follow. In short order, you will be performing all of these operations and others as well.

1.2   Getting Started

The most sensible way to learn to use MathCAD is to sit down at the console and perform some operations. Start out by imitating examples that you know work. As you gain experience, the overall process as well

Some examples of basic Mathcad operations.

Note that when values are assigned, the sign is `:=`; when values are calculated the sign is `=`.

Define a unit of length $m := 1L$

The circumference of a circle $2 \cdot \pi \cdot 3 \cdot m = 18.85 \cdot length$ $2 \cdot \pi \cdot 3 \cdot m = 18.85 \cdot m$

or with variable names $r := 3 \cdot m$ $c := 2 \cdot \pi \cdot r$ $c = 18.85 \cdot m$

The advantage of using variable names is that they can be used again.

$A := \pi \cdot r^2$ $A = 28.274 \cdot m^2$ $ratio := \dfrac{A}{c}$ $ratio = 1.5 \cdot length$

The use of trigonometric functions is straight forward. Angles in radians.

$\sin\left[\dfrac{\pi}{4}\right] = 0.707$ $\cos(2 \cdot \pi) = 1$ $asin(1) = 0.5 \cdot \pi$

To plot a function, there are two possible methods, a user defined function or a vector.

A parabola $y(x) := x^2$ $x := -2, -1.9 \ .. 2$

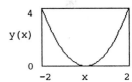

$n := 20$

$i := 0 \ .. n$ $xx_i := \dfrac{i}{n} \cdot 4 - 2$ $yy_i := xx_i^2$

Each method has its advantages.

Some more advanced functions.

Differentiation. $y'(x) := \dfrac{d}{dx} y(x)$

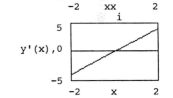

Integration $I := \displaystyle\int_0^2 y(x) \ dx$

$I = 2.667$

Find the zero, the root, of the y curve.

guess value $z := 1$ $rt := root(y(z), z)$ $rt = 0.029$

The value for y is within the default tolerance. $y(rt) = 8.661 \cdot 10^{-4}$

Solve two simultaneous equations.

$c1 := 2$ $c2 := 3$ $g := 1$ $f1(x) := c1 \cdot x$

$f2(x) := c2 \cdot x^2$

given $f1(g) \approx f2(g)$ $g := find(g)$ $g = 0.667$

check results $f1(g) = 1.333$ $f2(g) = 1.333$

Figure 1.1 MathCAD examples.

as the details will become clear and, generally, straightforward. When you see a double bullet, ••, that is a sign for you to do something. The more active you are, the more quickly you will learn.

There are implementations of MathCAD on PC and on Macintosh computers. The computer-specific comments that follow apply to the PC version; Macintosh users will have little difficulty adapting the concepts.

To begin, you need a computer, a monitor, DOS, and some implementation of MathCAD. There is more space to work if the computer has 640K of memory. A coprocessor will speed the calculations. Of course, use the fastest machines available.

Make a backup copy of MathCAD. Save the original copy, and work with the backup. See your manual for instructions.

Getting the Machine Ready.   These directions are for a hard disk system in which MathCAD has already been stored in a separate directory titled MCAD.

Turn on the power for the computer and monitor.

After DOS is loaded, change the directory to the MCAD directory. To change directories, type, following the prompt:
C:\ > cd\ mcad

On the screen will appear the new prompt: C:\ MCAD >

To load MathCAD, type following the prompt: mcad/m. (Note, in one case a backslash was used and in the other a slash.) The /m directs MathCAD to load in manual mode. This means that processing takes place only when requested. Automatic mode is a nuisance because reprocessing takes place every time a change is made. This is a waste of time.

•• If you haven't already done so, turn on the power and load MathCAD into the machine.

On occasion, the machine will crash. To restart, hold down [Ctrl][Alt] and press [Del]. Then release all three buttons. This will bring the system back up under DOS control. You must reload MathCAD.

It is a good idea to keep a notebook of MathCAD information. When you discover or are told something useful, write it down. Later, when you are alone with the machine, and that question — Now, how did I do that? — occurs, you will be prepared.

The startup screen, a space of 24 lines and 80 columns, includes a flashing underscore cursor, the message line at the very top of the screen, and the Mathsoft logo. The logo disappears with the first operation. Toward the right of the message line, there are two zeros. These numbers provide the line and column location of the cursor. If a file were loaded, the file name would appear at the left of the message line. Commands are also entered on this line.

The information about MathCAD in the rest of this chapter is intended as a brief introduction to MathCAD. It is not intended to tell you everything there is to know about MathCAD. The point is to get you started quickly. Keep the MathCAD manual handy and refer to it as necessary. The professional manual provides a more extensive coverage of the MathCAD language than does the student manual. However, the student manual in combination with this text should cover all the essentials and then some. For on-line help press [F1].

Before starting, it is useful to know **how to quit**. It is always best to exit from a program by the proper means; don't just turn off the machine (although nothing will blow up if you do). The command to quit is [Ctrl]Q; hold down the [Ctrl] key and press Q. This command tells MathCAD to end the session. If there is information on the screen that has not been saved, the following statement appears on the command line: "Changes not saved. OK to discard?" If you press return without typing anything, you are returned to your existing MathCAD document. If you type y (for yes, it is OK to discard the current document) and then press return, the document is discarded, MathCAD is removed from RAM, and control is returned to DOS. If you do want to save the document, you must do so before quitting. Press [F6] and provide a name for the file where the document is to be saved; see the section on files, below, for more details. After saving the document, press [Ctrl]Q.

1.3   The MathCAD Language

The information in the rest of this chapter falls roughly into two main categories, general operations and coding. The general operations, which include processing and printing documents and loading and saving files, will be treated last. We do this in order to start right out going over the basic MathCAD moves. (Of course, those who feel more comfortable proceeding in the other order can jump ahead to the next section, and then return.)

Beginning any language is a bit awkward. The thing to keep in mind as you go is the final result. Even if things get frustrating and you get angry, remember

"Never hate your enemies, it clouds your judgment."

The Godfather III

Many features of the MathCAD language are described in the following sections. They provide the basis for the work that follows. No document is being designed. Rather, the features are described one by one, and brief exercises are suggested. There is probably too much for one session. Get as far as is reasonable for you, stop, and come back later. Working with a friend is a good idea.

Regions.   MathCAD permits the user to enter information essentially anywhere on the screen. All information is one of three types: text, equation, or plot. Equation is the default mode; that is, unless text or plot is specified, entries are considered as equations.

To create a **text** region, type a single quotation mark. A pair of quotation marks will appear. As long as the cursor is within the text region, these quotation marks will remain visible. When the cursor exits the region, the quotation marks disappear. To exit a text region use the arrow keys, not the return or enter key.

To create a **plot** region, type the at sign @, shift 2. A rectangle will appear bordered with six place markers. These place markers are to identify what is to be plotted and the ranges for each. The text and plot regions will be discussed more fully later in this chapter. However, we see examples in Fig. 1.2.

• •   Type a quotation mark and then enter some text. The backspace can be use to delete characters to the left of the cursor. The delete key deletes characters at the cursor.

• •   Now create a plot region. There is nothing to plot yet, but the region and the markers can be observed. Move the cursor within the plot rectangle and press the letter f, for format. Observe the possible features that can be changed. Press [Esc] to return.

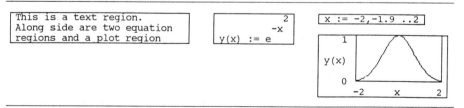

Figure 1.2 Text, equation, and plot regions, outlined.

• •   Enter something in equation mode. Type 2+3= and press [F9]. Do not attempt to enter spaces; MathCAD takes care of spacing automatically.

When creating a document, keep some space between regions. Often, it is necessary to go back and change things. This may result in regions expanding and overlapping if new material is included.

Avoid putting all regions in a vertical pattern. Get used to placing regions horizontally.

Viewing Regions/Outline/Refresh.   To see the space assigned to each region, press [Ctrl]V (hold down [Ctrl], next, press V, and then release both). This is a toggling operation. For each [Ctrl]V operation, the outlines are either drawn or removed.

• •   Press [Ctrl]V several times. In general, it is preferable to omit the outlines as it slows the screen refresh.

When a MathCAD document is processed, the processing moves from left to right and from top to bottom. Regions with the top edges aligned will be processed from left to right. For processing order, the top of the region is what counts; do not go by alignment of equality signs.

If the screen display becomes corrupted for any reason, it can be refreshed or rewritten by pressing [Ctrl]R.

• •   Press [Ctrl]R.

Cut, Copy, and Paste.   Any region can be cut (that is, removed from the document), copied, and if desired pasted elsewhere in the document. When a region is cut or copied, it is stored in a buffer. There is only one buffer; only the last region cut or copied is saved.

To cut a region, move the cursor to that region and press [F3]. To copy a region, move the cursor to that region and press [F2]; note, on the command line, the statement, region copied. To paste a region, move the cursor to the desired location of the upper left corner of the region and press [F4].

If you cut or copy a region, move to a new empty location and press [F3], for example, instead of [F4], the region in the buffer is not lost. Press [F4] and it will be pasted.

• •   Copy the regions on the screen and paste them elsewhere. Cut a region and paste it elsewhere. (You could, of course, paste it right back where it was, which is useful when you cut the wrong region.) Attempt to paste one region on top of another.

There are also commands referred to as incut, incopy, and inpaste. These will be discussed later.

Insert or Delete Blank Lines.   As a document is developed, frequent changes in the layout of the document may be very useful. In addition to the cut, copy, and paste operations, blank lines can be inserted or deleted.

The command to add a blank line is [Ctrl][F9]. The command to remove a blank line is [Ctrl][F10]. The operations can be repeated by holding down the [Ctrl] key and pressing [F9] or [F10] as many times as required. The screen takes a moment to refresh.

• •   Add and delete lines between the regions on the screen.

Reset.   To clear the screen without saving what is present, press [Esc], type reset, and press return. If the document has been changed since it was last saved, the comment "Changes not saved, OK to discard?" will appear. Type y for yes and press return.

• •   Clear the screen.

Separate.   If regions overlap, one region may obscure another. Regions can be separated. Press [Esc], type sep(arate), and press return. Curious errors can occur because regions were overlapped and one is invisible or only partly visible.

• •   Move the cursor five spaces to the right using the arrow key. Type x:2. Move the cursor left five spaces so that it is back where it started (keep the cursor on the same line as the $x := 2$ region). Type y:3.141592654. The y region now completely covers the x region. Even though it is not visible, the x region is still there and still active. Move several spaces to the right, type $x =$, and evaluate. Now separate the groups using the above procedure. If you keep the regions separated, there will be no surprises.

1.3.1   *Equation Regions*

Equality Signs.   MathCAD has four different kinds of equality. Ultimately, we will need them all. The two types that we need immediately are for definitions or assignments and for calculations. The other two kinds of equality are for global assignment (as opposed to local) and for relational equality. We will encounter the global assignment when discussing units and for convenience in processing some documents. We will encounter the relational equality when using solve blocks.

The local assignment uses the colon, as in x:2. The value 2 is assigned to x. Whenever x is used, it carries with it that value, unless it is reassigned a different value.

The calculational equality, as in $x=$, when processed, returns the value assigned to x.

The global equality, as in $k\tilde{\ }2$, which appears as $k \equiv 2$, is an assignment equality, as is the colon. However, in processing a document, MathCAD takes two passes through the document, the first time processing all global equalities and the second time everything else. This provides a mechanism whereby the code can be placed at the end of a document and plot regions, for example, that depend on the code can be placed first.

The relational equality is used in given-find and given-minerr solve blocks. It is called by typing [Alt]=, as in $2 \cdot x \approx 3 \cdot y$.

• • Assign 2 to k with a global equality. Evaluate k by typing $k=$ and pressing [F9]. Reset the screen.

Processing or Evaluating. We make some brief comments on processing now. More will be given in the general features section.

To process a document — that is, to evaluate the various quantities within the document — press [F9]. MathCAD will process the document to the bottom of the present screen. If more regions exist below the bottom of the screen they will not be processed. To process everything, press [Esc], type pro(cess), and press return.

Variable Name. A value can be assigned to a variable name. The local assignment symbol, as we have already noted, is the colon. To assign the value 4 to the name x, one would type x:4. Leave no spaces; they are handled automatically.

Variable names must start with a letter (Roman or Greek) and can include numerals, the underscore, and the percent sign. A variable name can include up to 64 characters. Let the name describe the quantity it is to represent.

Warning. Never use the same name for two different things in the same document. Two common errors are using m for meters and also for mass, and using N for Newtons and also for normal force or the upper limit of an index.

• • Enter hip : 13 and be_{bop} : hip; evaluate by entering $be_{bop} =$ and pressing [F9].

Variable names may include a literal subscript. A literal subscript appears as a subscript when the cursor exits the region. Literal subscripts are implemented by typing variable_name.literal_subscript. The specification for the literal subscript comes after the end of the variable name and is separated by a period. It is strongly suggested that literal subscripts not be integers. If they are integers, they may become confused with active subscripts.

• •   Enter $x.o:e$. Move the cursor in and out of the region. Observe the o.

The PC version of MathCAD can display 17 Greek letters, 14 lowercase and 3 uppercase. The general format for displaying these letters is [Alt]letter. For example, θ is [Alt]Q. The Roman letters that, with [Alt], produce the indicated Greek letters are: a (α), b (β), d (δ), e (ϵ), f (ϕ), g (Γ), h (Φ), l (λ), n (ν), o (Ω), p (π), q (θ), r (ρ), s (σ), t (τ), u (μ), and w (ω).

Arithmetic Operations.   The arithmetic operations use the symbols +, –, *, and / for addition, subtraction, multiplication, and division, respectively. The caret, ˆ , is used to raise something to a power, as in xˆ 2, for x^2. The backslash, \, is the symbol for a square root, as in \(4+5).

The order of priority in arithmetic operations is standard. Make sure that parentheses are used to group terms that need to be grouped. A good rule is to use parentheses for grouping whenever there is any question; the MathCAD editor will behave less schizophrenically if you do. Think ahead: Start with a left-hand parenthesis when necessary, rather than saying, "Ah yes," and then having to go back and try to insert one. In an equation region, an apostrophe creates a pair of parentheses. However, where the parentheses go depends on the construction already present. Just try it and see what happens; you can always delete what you don't like.

• •   Reset MathCAD so that you have a blank screen. Type the expression $a + b + c$. Remember, spacing is automatic in equation regions. Move the cursor to the first plus sign (half-surrounding from the right); type an apostrophe. Note the location of the parentheses. Delete the righthand parenthesis. Move the cursor to the second plus sign; type an apostrophe. Again, note the location of the parentheses. The rules are not easily summarized but, in any case, what you see is what you get.

• •   Divide the quantity 17 plus 29 by 2.54. Do this twice, once explicitly using the parentheses and once using the apostrophe. (Ans. 18.11)

• •   Take the square root of the sum of any two or more quantities.

• •   Enter 4+5. Now try to put parentheses around this quantity. Do the right-hand one first; that should be no problem. Now try the left one. Did you get a message "can't edit blank space"?

Note that the cursor in an equation region appears as ⌡, not as an underscore. This is the append cursor. Characters entered are appended to what has been entered. The insert key toggles the cursor from ⌋ to ⌊ and back. When in ⌊ mode, the cursor is referred to as the insert cursor. This is a necessary editing feature for approaching from the left. Press the insert key. The cursor should now appear as ⌊. Now insert the parenthesis from the left. Note that the cursor must be snuggled up to and half surrounding the character, not a space away. In general, keep the cursor in append mode rather than in insert mode.

● ●   Enter 6+4=. Now go back and put a minus sign in front of the 6. Evaluate.

● ●   Try 87 plus the quantity 16 times 4, the whole thing divided by the quantity 77 minus 29. (Ans. 3.146)

Format.   The answer to the previous calculation appears with three figures after the decimal point. That can be changed, as can a number of other features. To see what controls are available, press [Esc], type format, and press return.

Six controls are listed: (1) radix — decimal, octal, or hexadecimal; (2) precision — default 3, possible range 0–15; (3) exponential threshold — default 3, numbers outside the range 10 to plus or minus the threshold value are shown in exponential notation; (4) imaginary symbol i or j; (5) zero tolerance — default 15; numbers less than 10 to the minus default value are shown as zero (this is frequently not adequate; change as needed, range 0–308); and (6) complex tolerance — indicating that when the ratio between larger and smaller values is outside of range, consider the smaller one to be zero. Of these, the two of most frequent concern are (2) and (5). Whenever a document is opened, the default values are in place.

Place Marker.   The place marker, an active and useful device, is the solid box at the end of a region with a standard equality sign (=). See Fig. 1.3. The place marker is visible only if the cursor is in the region. For example, an evaluated quantity might appear as

$$x = 1024 \ \blacksquare.$$

● ●   Enter x:1024 and evaluate x. Move the cursor to the place marker, enter 2, and press [F9]. Change the 2 to 2^4 and evaluate.

This property is a general feature. Any number can be represented as a multiple of anything. When dealing with angles, frequently useful quantities to substitute are π or deg (see the section on built-in functions below).

Examples using the place marker. The result is expressed as a multiple of the quantity in the place marker.

```
r := 2      c := 2·π·r      c = 12.566

d := 2·r                    c = 6.283·r        r in place marker

                            c = 4·π             π in place marker

                            c = 3.142·d        d in place marker.
```

Figure 1.3 Example of the functionality of the place marker.

Incutting, Incopying, Inpasting.   These operations refer to cutting or copying a portion of a region and pasting that portion either into an existing region or as a new region. There are occasions when an expression needs to be repeated or moved to a different location within an existing region. The in- operations are helpful under such circumstances.

In an exercise in the section on arithmetic operations, we entered parentheses in the expression $a+b+c$ using the apostrophe. The placement of the parentheses is helpful in knowing which portions will be cut or copied.

• •   Type $a + b + c$. Place the cursor at the first plus sign and perform an incopy operation. Move to the side and inpaste into a new region. Go back to the original region, place the cursor at the second plus sign, and perform an incut operation. Type a colon. Inpaste the incut information at the right-hand place marker. Try a few examples of your own; include parentheses and other operations.

Built-in Functions.   First, let's consider two built-in symbols with values. The symbol π, expressed by pressing [Alt]P, has associated with it the value for π. The letter e has associated with it the value of the base of the natural logarithm.

• •   Evaluate π and e. What is the percent error in the approximation $\pi^2 \approx 10$?

MathCAD has built in all the functions one would expect to find on a scientific calculator plus a number of others. (There are approximately 75 built-in functions.) These include the standard trigonometric functions; the angle function, which is often preferable to the arctangent; hyperbolic functions; log and exponential functions; Bessel functions; complex numbers; interpolation; and the fast Fourier transform. Procedures for vector and matrix operations, regression, and statistical, correlation, equation-solving, and other functions are included as well.

The standard format is function_name(argument, argument, ...). Many functions, like the trig functions, have only one argument. Regression and sorting functions, for example, have two; the number of arguments with Find or Minerr depends on the number of unknowns in a solve process.

The arguments for the standard trigonometric functions are in radians. To convert from degrees to radians, the following procedure is suggested. Define the radian as equal to one; then define the degree in terms of the radian:

$$\text{rad} := 1 \qquad \text{deg} := \frac{\pi}{180} \cdot \text{rad}.$$

If you attach the name deg as a multiplier to any value in degrees, Math-CAD will make all necessary conversions, for example, $45 \cdot \text{deg}$.

• • Enter the statements for rad and deg. Type 45*deg and evaluate.

Evaluate the sine of 90° and of $\pi/2$ radians.

Take the asin of 1. (Ans. 1.571)

Take the tangent of 135° and 225°. Take the arctangent of 1 and −1. This can cause problems if one is not careful. Instead, try the following:

The coordinates of a point at 135° could be (−1,1) and at 225° (−1,−1). Evaluate angle(−1,1) and angle(−1,−1). Enter deg at the place marker and press [F9].

The general form of the angle function is angle(δx, δy). The function returns a value between 0 and 2π. The variables δx and δy are the differences in x and y between the point of interest (x, y) and the reference point (x_r, y_r); $\delta x = x - x_r$ and $\delta y = y - y_r$. If the reference point is the origin, then δx and δy are just the x and y coordinates of the point for which the angle is to be determined. The angle is measured counter-clockwise from a line originating at (x_r, y_r) and parallel to the positive x-axis.

• • The base of the natural log system, e, is available either directly as e or as exp(). Evaluate e, $e\hat{\ } 1$, and exp(1). Use the exponential function in a simple radioactive decay problem. If

$$N = N_o \cdot e^{-\lambda \cdot t}$$

calculate the number of nuclei N at time t where $\lambda = 0.2$, $t = 30$, and the number of radioactive nuclei at time $t = 0$ is 10^{12}. Name each of the quantities. Write N in terms of other names, not in terms of numbers. Evaluate N.

Moving the Cursor.   Several useful cursor commands make getting around the screen and around a file quick and easy. They are presented in list form.

The four arrow keys move the cursor one space in the direction of the arrow for each operation of the key. If the key is held down, the stepping repeats automatically.

[Ctrl][Home] moves the cursor to the beginning of the first region in the document.

[Ctrl][End] moves the cursor to the end of the last region in the document.

[Home] moves the cursor to the beginning of the previous region.

[End] moves the cursor to the end of the present region or the beginning of the next. If you are at the last region, the cursor will toggle back and forth between beginning and end with each successive operation.

[PgDn] moves the cursor down five lines.

[PgUp] moves the cursor up five lines.

[Ctrl][PgDn] moves the cursor down 80% of the screen height.

[Ctrl][PgUp] moves the cursor up 80% of the screen height. (If the beginning of the document is showing or is less than 80% of the screen height above the top of the screen, [Ctrl][PgUp] will result in less cursor movement, or no movement at all if the zero line of the document is displayed on the screen.)

[Tab] advances the cursor across the screen, to the right, in ten character steps. [Shift][Tab] moves the cursor to the left in ten character steps. In an equation region, [Tab] moves the cursor within the region.

Two other motion commands are important but do not apply to the student version. These are [Ctrl]G and [Ctrl]E, which move 80 columns to the right and left, respectively.

Spend several minutes trying each one of the movements. Knowledge of these operations makes working within MathCAD much smoother. (Amateurs lean on the arrow keys.)

Range Variables.   Sometimes, it is desirable to perform an operation with a sequence of values. MathCAD can perform such a series of operations using either range variables in combination with some function or vectors.

A range variable takes on a series of values. The values are specified by providing the first, second, and last values of the sequence. The step size is the difference between the second and first values. If no second value is specified, the default step size is one. For example, if θ is to vary over the range from 0 to 2π, in steps of 0.1, the specification is typed [Alt]Q:0,0.1;2*[Alt]P. This would appear as

$$\theta := 0, 0.1 \ldots 2 \cdot \pi.$$

Note the ellipsis, $\ldots$, is achieved by typing a semicolon, **not** by typing a sequence of periods.

• • Enter the above specification for θ and then evaluate it. The results appear in tabular form.

A general method to generate a sequence starting at a and going to b taking N steps is

$$\text{step} := \frac{(b - a)}{N} \qquad x := a, a + \text{step} \ldots b.$$

A function of a range variable, func(range var), will, when processed, evaluate the function for each value of the range variable.

• • Evaluate $\sin(\theta)$; the above definition for θ must be specified first.

User-Defined Functions. A user-defined function is expressed as

$$name(variable\ list) = any\ legal\ combination\ of\ terms.$$

A simple example is associated with the free fall of an object near the surface of the earth. A function is desired to express the position of the object as a function of time. We write

$$x(t) = \frac{1}{2} a t^2.$$

Any parameter not in the variable list must be defined before the function is defined. In this case, a value would need to be assigned to a, the acceleration.

• • Reset the screen. Let $a = 9.81$. Define t as a range variable as described above (let t take on 10 or 15 values). Evaluate $x(t)$. Evaluate $x(2.645)$.

Subscripts. A variety of operations can be performed on variables with subscripts. By subscript, we mean active subscripts, not the literal subscripts that are part of a variable name. Avoid using both literal and active subscripts in one variable name.

A subscript is called using the left-hand square bracket, [. A subscript must be an integer. A subscript cannot have a value less than the value of ORIGIN.

• • Evaluate ORIGIN.

The default value is zero. You should not change this quantity without a clear understanding of the implications. Given the default value for ORIGIN, negative subscripts are out of range.

• • Define a range variable i, i:0;5. Then let $x[i:i^2$. Evaluate x_i and x_2.

Sum all the x-values. The summation operator is called by pressing [Shift]4 (that is, $).

• •   Press [Shift]4, and type $x[i$ at the place marker to the right of the summation sign. Press [Tab] to move the cursor to the place marker beneath the summation sign. Enter the index over which the summation is to take place, i. Evaluate the sum. (Ans. 55)

Vectors.   The variable x_i, above, is a vector; it holds a sequence of values. The subscript is a range variable with integral values. The vector values can be computed according to some function, or they can be entered by hand. Values are entered by hand, by typing, for example, $x[i$:value1, value2, ... , last value. The data appear in tabular form, once the first comma is entered.

• • Type $y[i$:0,1,2,4,7,8,10. Define i in a region previous to y and then sum the y_i.

To get some sense of what happens if values are not specified over the entire range of the subscript, try the following.

• • Define i:0;5. Let $z[i$:1,2,3. Sum the $z[i$. Did you get an error message? Cut the region defining the z_i. In its place type $z[5$:1. Evaluate z_i and the sum. Change the $z[5$ in the z-term to $z[6$ and sum.

The vector must be specified for the largest value of its subscript. Values of zero are assigned to all unspecified values of the variable.

Frequently, several different subscripts and ranges are required in one document. If a double subscript is required use parentheses, as in $x[(i,j)$ for $x_{i,j}$.

Disabling an Equation.   Sometimes it is useful to write an equation in a MathCAD document that can be seen but is not to be processed. Equations can be disabled by pressing [Esc], typing eq (equation), and pressing return. When an equation is disabled, a small box, □, resembling an unfilled place marker, ■, appears to the right of the equation. This Esc/eq process is a toggling operation; repeat it to enable a disabled equation. Find regions cannot be disabled.

1.3.2 Plot Regions

In MathCAD, plotting is done easily and, generally, quickly. We are going to want to plot many things, so become very familiar with the process.

To create a plot region, press the @ sign. An open rectangle, the plot space, with six place markers appears on the screen. The place markers at the extremes of the abscissa and ordinate specify the plot range. Those at the center of each axis specify the quantities to be plotted.

A quick way to create a plot region and enter the names of the variables to be plotted when plotting $y(t)$ vs. $x(t)$ is to type $y(t)$ @ $x(t)$. This saves some moving around. MathCAD will set limits automatically in all cases. Sometimes, it is convenient to set limits to control the display. If limits are set and one (or more) of the points is outside the defined region, lines will not be drawn to/toward the point(s) or from it (them). This feature has its uses.

There is control over the size of the plot space, the number of subdivisions within the region, whether the plot is linear or log, and the plot type, that is, the plot symbol and whether the points are connected with a line or not.

To select the different plot options, move the cursor inside the plot rectangle and press the letter f (format). On the command line, the following information is displayed (in the student version; the professional version gives the same information in menu form): logs=0,0 subdivs=1,1 size=6,15 type=l. To change any of the values, move the blinking underscore cursor to the point where changes are to be made, delete the undesired values, type in the desired ones, and press return. In the professional version, use the up or down arrow keys to highlight the quantity of interest, press return, enter the new values, press return, highlight and change other areas if desired, and finally press d (for done) and press return once more.

Size. Size can be changed to suit the character of a particular graph. The first number represents the vertical dimension (equivalent to that number of lines in the document), the second the horizontal (equivalent to that many spaces in the document). Because of the aspect ratio of the screen, equal values do not result in a square plot region. Roughly speaking, one down is equal to two or two and one-half across (for a PC), depending on the monitor. Note that square on the monitor does not necessarily

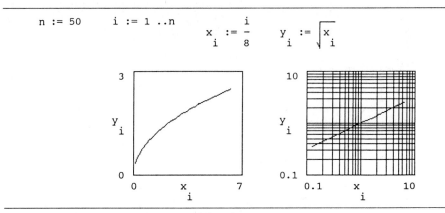

Figure 1.4 A linear and log-log plot of the square root function.

mean square on the printer. If the printed form is important, spend a few minutes and determine what ratio on the screen results in square on the printer. Other ratios can be determined from this.

• •   Reset the screen. Open a plot region. Change the size to 10,20. Change the size to 12,50. Press [Ctrl]V to see the region of the screen that is occupied. Remove the outline.

Linear/Log.   Zero in the log designation means the plot is linear; a positive integer means that the plot is a log plot (see Fig. 1.4). The number has nothing to do with the plot range and nothing to do with the number of log cycles. The number corresponds to the subdivisions. It is not a bad idea simply to use zero for linear and one for log plots. Of course, each axis is designated separately; a semilog plot would be 1,0 (or possibly, 0,1).

MathCAD does not select limits on a log graph to be powers of ten. Note the range of values and then replace them with the appropriate powers of ten. Be sure that the number of cycles you select corresponds to the range; otherwise, some points may be excluded.

Of the values to be plotted, values less than or equal to zero are not permissible with log plots. If a plot that had been linear is changed to a log plot, it may be necessary to delete the limits even though MathCAD specified them. Units cannot be used with log plots.

Subdivisions.   Subdivs, subdivisions, are best left at 1,1 when exploring. Once you have a graph that you want to refine, work with the subdivision possibilities. On a linear graph, choose the range and the number of subdivisions so that they are convenient multiples. For a log graph, of

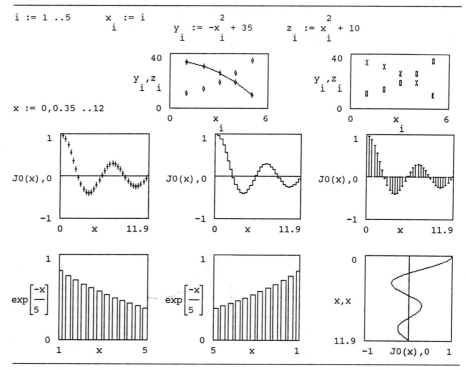

Figure 1.5 The plot types of the regions are, in order, Vv, xo, pl, sl, e, b, b, and l. In the exponential plot, the abscissa limits were reversed for a different presentation. Similarly, the last plot shows the function rotated from its previous orientation.

the choices 1,2, or 9, choose either one subdivision per cycle or nine. Two subdivisions are of little value.

Plot Types.   Graphs are not restricted to lines connecting points. Dots, Xs, pluses, rectangles, or diamonds, can be specified to appear at the plot points. They can be freestanding or connected by lines. In addition, there are some special cases: step, error bars, and bar charts. In a single plot region, different plot types may be used for different curves.

The plot types are l (line), d (dots), s (step), e (error bars), b (bar chart), x (xs), X (xs with lines), p (pluses), P (pluses with lines), o (rectangles), O (rectangles with lines), v (diamonds), and V (diamonds with lines)(see Fig. 1.5).

• •   Clear the screen. Enter specifications so that θ ranges from 0 to 2π in steps of 0.1. Enter θ in the abscissa place marker and $\sin(\theta)$ in the ordinate place marker. Press [F9].

Two curves can be plotted in the same plot region. Move the cursor to the ordinate name. Place the cursor so that it half-surrounds the) in $\sin(\theta)$. Enter a comma followed by the numeral 0. Press [F9]. The second curve is a plot of 0 vs. θ. Remove the 0 and plot $\cos\theta$ together with the sine function.

To gain experience using subscripted variables, let's repeat the sine plot using vectors. Both methods are useful; the subscripted method will be used more frequently. In the subscripted case, it is necessary to write the angle in terms of a range variable that takes on integral values. (See the discussion of vectors above.)

A specific procedure is outlined with possible variable names. The actual code is presented in the next paragraph. Specify the number of points to be plotted (N); assign a range variable (i) integral values that go from zero to the number N. If we wish to plot one cycle of the sine function, the angle should range from 0 to 2π. The quantity i/N will take on equally spaced values between 0 and 1 in N steps as i goes from 0 to N; multiply this quantity by 2π and the range is then scaled to 0 to 2π. Assign these values to the angle variable, ϕ_i. Finally, set the subscripted ordinate, y_i, equal to the sine of the subscripted angle variable, $\sin(\phi_i)$.

• •   Enter $N := 30$ (what equals symbol should be used?). Enter $i := 0\ldots N$ (how is the ellipsis entered?). Enter $\phi_i := i/N \cdot 2 \cdot \pi$ (which key is used for subscripts?). Enter $y_i := \sin(\phi_i)$. Type @ to create a plot region. Enter ϕ_i and y_i as the labels of the quantities to be plotted. Press [F9]. A zero line could be plotted in the same way as the previous example. Instead, use the format controls described above and change the number of subdivisions.

To plot two curves simultaneously that do not have a variable in common — for example, y vs. x and yy vs. xx (with appropriate subscripts or variable information) — on the same graph, enter x,xx and y,yy at the appropriate place markers. Three or more graphs can be plotted in the same region. How many curves can be successfully plotted may depend on the version of MathCAD that you have. At the present time the PC professional versions have the greatest flexibility.

Histogram.   In a histogram, data (from a data set) that fall within specified intervals are counted and presented in bar form. For example, assume that a number of measurements are performed with outcomes distributed over the range from zero to one. No outcome has a value, for example, of 0.212534796. We can't plot the number of events at a given value on the line. But if we divide the line between zero and one into a series of intervals, then we can count the outcomes that fall within those intervals. The histogram process determines the number of outcomes within a certain interval and displays the outcomes in bar graph form.

Three vectors are involved in creating a histogram: the data, the points defining the edges of the intervals, and the sums determined in the histogram process. A different index is needed for each of these vectors. For example, the data vector is x_i; the interval edge points are y_j; and the counts per interval, z_k. If there are to be M intervals, then j should range from $0, \ldots, M$ and k should range from $0, \ldots, M - 1$. The intervals could be determined by taking the difference between the maximum and minimum values of the data and dividing by the number of desired intervals. That always works; however, it may be desirable to refine the range and the number of intervals or, equivalently, the width of the intervals.

The histogram command is $z := \text{hist}(y, x)$; that is, counts per interval equals hist(intervals, data). In the histogram call, no indices are shown. When plotting the histogram, plot z_k vs. y_k. Let the plot type be b for bar. In the histogram plot, the bars are not centered about the interval value but begin at the lower edge and span about two-thirds of the region. A more attractive plot can be obtained by adding half the width of one interval to all the y values. See Fig. 1.6.

1.3.3 Text Regions

In MathCAD, you can place text in regions throughout the document. However, MathCAD is not a word processor. The text ability is principally useful in documenting code. It is well known that undocumented code written by one person is difficult for another to read. In addition, undocumented code returned to after an interim may not be easy reading for the author, either. You should take the time necessary to document your work as you go. In this text, much of the documentation will be in the text, where the file is discussed, rather than in the file itself.

A text region is created by typing a single quotation mark. Type normally. Use the [Backspace] (deletes the character to the left) and [Del] (deletes the character at the cursor location) keys to make changes. Exit the region using the arrow keys or page down, for example. You cannot exit a text region using the return key. In text regions, the cursor is a blinking underline.

• • Create a text region. A pair of quotation marks should appear on the screen. These quotation marks remain visible as long as the cursor remains within the text region; they disappear when the cursor is removed from the region.

Type the following sentence: "No matter where you go, there you are" (*Buckaroo Banzai*). Exit the region using the [↓]. [PgDn] would also move

```
Histogram of random data.  Values range from 0 to 10.  Divide the space into
ten intervals.

Generate the data.        n := 199      i := 0 ..n       x  := rnd(10)
                                                          i

Specify the intervals     m := 10       j := 0 ..m       y  := j
                                                          j

(If the data had been in the range from zero to one, y[j would be j/m.)

The interval index        k := 0 ..m - 1

Sort the data by size.     z := hist(y,x)
```

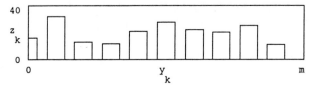

```
To center the bars in the intervals, add a half interval to each.
```

$$hlfint := \frac{y_1 - y_0}{2} \qquad y_j := y_j + hlfint$$

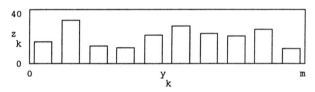

Figure 1.6 A histogram plot of randomly generated data.

the cursor out of the text region; however, the cursor moves five lines for each operation of the key, rather than one line for each operation of the [↓] key. Exit and enter by both methods.

Cursor Moves in Text Regions. Move the cursor to the end of the text region. Type the following sentence: "Poltergeists, not pair-production, make up the principal type of spontaneous material manifestation." Press [Home]. This operation moves the cursor to the beginning of the region.

[Ctrl]F and [Ctrl]B result in no cursor movement; they specify the direction of cursor movements that follow, forward or backward.

[Ctrl]W advances the cursor one word (W).

[Ctrl]L advances the cursor one line.

[Ctrl]S advances the cursor one sentence.

[Ctrl]P advances the cursor one paragraph.

Use all the operations; get used to them. They are convenient when editing text.

Incutting, Incopying, and Inpasting.   A section of text can be marked and cut or copied without cutting or copying the entire region. [Ctrl]X marks the beginning or ending of a subregion. [Ctrl][F2] is the incopy command; [Ctrl][F3] is the incut command; [Ctrl][F4] is the inpaste command. Incut or incopied text can be pasted only into an existing text region. One can type a single quote to create a text region and paste the incut text into that empty text region.

• •   Copy a phrase from one of the lines of text and paste it into a new region.

If a text has been edited and the lines are irregular, a rough wrap is obtained with [Ctrl]N (nudge).

If there is to be lots of text, break it up into a series of smaller text regions. Make the regions paragraph sized, no larger. Adjustments are slower with large regions.

1.3.4   *More Equation Features*

Brief descriptions of a number of more advanced features are presented with examples and some exercises.

Solve Blocks: Given Find and Given Minerr.   The given-find solve block can be used to solve simultaneous equations (see Fig. 1.7). The given-minerr procedure is used to minimize the errors in the solution of a set of equations and constraints.

The procedure is as follows:

Define all parameters. Supply guess values for the unknown(s); choose these as carefully as possible. Variables with active subscripts may not be used within a solve block. That is, a value such as x_i, where i is a range variable, may not be used; however, a specific value such as x_1 may be used within the block. No subscripted values can be used in the argument list of a find statement. (A series of values can be obtained using the block as a user-defined function. This is addressed below.)

The solve block begins with the word "given". This word is entered in equation mode, not text mode. The single word is a complete equation region.

```
Solution of a one-dimensional kinematic equation.

Parameters         g := 9.8     yo := 2        vo := 11     a := -g      y := 8

Guess values       t := 0.5
```

given
$$y \approx yo + vo \cdot t + \frac{1}{2} \cdot a \cdot t^2 \qquad t := find(t) \qquad t = 0.934$$

check
$$yo + vo \cdot t + \frac{1}{2} \cdot a \cdot t^2 = 8$$

```
Solution of a two dimensional kinematic problem.    rad := 1
```
$$deg := \frac{\pi}{180} \cdot rad$$

```
Parameters      vo := 15                    θ := 30·deg

                vox := vo·cos(θ)            voy := vo·sin(θ)      y := 0

Guess values    x := 10                     t := 1
```

given $x \approx vox \cdot t$
$$y \approx voy \cdot t + \frac{1}{2} \cdot a \cdot t^2 \qquad \begin{bmatrix} x \\ t \end{bmatrix} := find(x,t)$$

x = 19.883   t = 1.531

check
$$voy \cdot t + \frac{1}{2} \cdot a \cdot t^2 = 0 \qquad vox \cdot t = 19.883$$

Figure 1.7 Examples using the given-find solve block.

The equations and constraints to be solved are specified after the given statement, which begins the block, and before the find statement, which ends the block. Equations must use the relational equality $\approx$, [Alt]=. (In some versions of MathCAD, if a different equality is typed, deleted, and the proper relational equality entered, the solve block will not function. The entire statement must be reentered correctly.) The equations need not be in a vertical pattern. Expressions that are equivalent to $0 = 0$ may cause difficulty.

The solve block ends with a find (or minerr) statement.

If there is one unknown, the form is

$$x := find(x) \qquad or \qquad diffname := find(x)$$

If the equation is successfully solved, the find statement returns the solution value and assigns it to x, or as in the second case, the value can be assigned to a different name, for example, diffname, if desired. If a solution is not found, the unknown retains the guess value assigned to it. A minerr statement is of the same form as a find statement: $x := minerr(x)$.

If there is more than one unknown, the form is

$$\begin{bmatrix} var_1 \\ \vdots \\ var_n \end{bmatrix} := \text{Find}(var_1, var_2, \ldots, var_n)$$

The left-hand side is a column matrix, created by pressing [Alt]M. Enter the number of variables, n, and press return. Enter the names of the variables at the place markers; move from location to location using the [Tab] key.

Always check your results. Never accept any results blindly.

The procedure is the same whether one uses a find or minerr statement.

● ●   The following data are measurements from a free-fall experiment. Find the best value for g, using the given-minerr procedure. Let $k := 1 \ldots 4$. The data are: $s_k := 0.29, 1.8, 3.9, 12.7$; $t_k := 0.25, 0.6, 0.9, 1.6$. Enter the values for s and t; for example, type $s[k{:}0.29,1.8,\ldots$. Enter a guess value for g. Following the given statement, write the equation, $s \approx 1/2\, g\, t^2$ four times, with the subscripts on s and t taking on the k-values of $1, \ldots, 4$ (one equation with subscript 1, one equation with subscript 2, and so on). Complete the block with a minerr statement and evaluate g. To check, evaluate the $(1/2)\, g\, t_k^2$ terms and compare with the s_k.

It is possible to use a given-find solve block to determine a set of solutions instead of a single solution. However, subscripts cannot be used in the find statement of a solve block. Instead, the solve block is specified so as to be equivalent to a user-defined function. In particular, the find statement is written in functional form. Once this function is defined, the function can be evaluated for a series of arguments just as any other function.

The procedure to find a set of solutions using a given-find solve block is the same initially as it is to find a single value: (1) specify parameters and guess values for the unknowns; (2) provide the given statement; (3) specify the equations to be solved (so far the same); and (4) write the find statement, not with the unknowns on the left-hand side, but as a function — fcn($variable$) := find($unknown(s)$). (The variable must be a variable in the equations.) For example, you might complete a solve block with a statement of the form f(x) := find(y). After the solve block, if you specify a value(s) for x, evaluating f(x) returns the corresponding value(s) for y. A series of values can be obtained using subscripted variables.

The find statement is not restricted to one unknown. For example, with two unknowns, we write

$$f(b) = \text{find}(c, d),$$

where b is a parameter and c and d are the unknowns to be determined, all explicitly shown in the equations within the solve block. When a value(s) for b is specified, $f(b)$ returns the values of the unknowns. The individual values are addressed by the form

$$c = (\text{f}(b))_0 \qquad \text{and} \qquad d = (\text{f}(b))_1.$$

A series of values can be determined by defining parameter values b_i and evaluating $f(b_i)$. A solution must be sought for each parameter value so this may be a slow process.

Root Function.   The root function is designed to determine the roots, the zero points, of an expression. The more carefully you guide it with "good" guess values, the more quickly it will respond with the desired values.

The root function has two arguments:

$$root(expression, variable).$$

The first is the expression for which MathCAD attempts to find a zero. The second is the variable for which a value is sought, a value that, when substituted into the expression, makes it zero.

A good procedure before using the root function is to plot the expression against the variable. Observing the graph is an aid in selecting guess values. Specify a guess value for the variable before evaluating the root function. In many functions, there is more than one root. They can be found sequentially, using different properly selected guess values.

• •   Find the first two roots of $\sin(x)$. Express the results with π in the place marker. Repeat for the cosine. Express results with $\pi/2$ in the place marker. How far off can the guess be and still return the desired root? Describe how far in terms of the shape of the curve.

Derivative.   The derivative operation returns a numerical value for the slope of a curve; it does not return a function. The derivative operation is invoked by pressing the question-mark key. On pressing the key,

$$\frac{d}{d\blacksquare}\,\blacksquare$$

appears. At the place marker in the denominator, enter the variable name of the differentiation variable. At the expression place marker, enter the expression for which the derivative is to be taken. Specify a value for the differentiation variable and any other variables or parameters included in the expression before taking the derivative.

The acceleration could be defined as

$$a(t) := \frac{d}{dt}\left(\frac{d}{dt}x(t)\right),$$

where $x(t)$ gives the position as a function of time. However, derivatives of derivatives are quite slow and hence are not recommended.

• •   Take the derivative of the following functions:

$$y(x) = x^2,$$
$$y(x) = \exp(-x^2),$$
$$y(x) = \sin(x).$$

Plot each function with its derivative as x ranges from 0 to 2 in steps of 0.2.

Finding a Maximum or Minimum.   Use the root function in combination with the derivative operation to locate a maximum or minimum. When successful,

$$\text{root}\left(\frac{d}{dx}y(x),\ x\right)$$

returns the x-value for which $d/dx\, y(x)$ is zero. If you are sensible about guess values, the process works effectively. Plotting a function, especially if it is unfamiliar, is a useful aid in selecting guess values.

• •   For the expression $y(x) = x^2 \cdot \exp(-x^2)$, find the maximum. Try a few different guess values and see how robust the process is for this function.

• •   Find the first maximum and first minimum of $\sin(x)$, for $x > 0$. Express results with π in the place marker.

• •   Find the maximum of the function $y(t) = 10 \cdot t - 5 \cdot t^2$.

Integration.   The integration process can be considered a method for determining the area under a curve. You do not need to know the methods of calculus to perform these integrations. MathCAD will compute definite integrals numerically. Limits must be specified; a value, not a function, is returned.

The integration operation is called by pressing [Shift]7 (the ampersand, &):

$$\int_{\blacksquare}^{\blacksquare} \blacksquare \, d\blacksquare.$$

Four place markers are shown. The limits of integration must be specified; a variable name that has a defined value is fine. Infinity is *not* an acceptable limit.

Enter the desired expression for the integrand at the place marker after the integral sign. Enter the variable name, over which the integration is to be carried out, in the place marker after the *d*. All variables other than the integration variable must be defined before the integration.

● ●   Let $a := 5$. Perform the following integration:

$$xx := \int_0^2 a \cdot t \, dt.$$

● ●   Integrate $\sqrt{1 - x^2}$ as x goes from 0 to 1.
Integrate $\sin(x)$ as x goes from 0 to π.
Integrate $\sin^2(x)$ as x goes from 0 to π.

To find the average for some function, simply integrate the function over the region and divide by the size of the region:

$$f_{\text{av}} = \frac{\int_{\text{range}} f(x) \, dx}{\int_{\text{range}} dx}.$$

● ●   What is the average value of $\sin(x)$ and $\sin^2(x)$ as x goes from 0 to π?

If and Until.   The if statement is MathCAD's limited form of an if-then-else statement. The form is

$$x = if(condition, \, x - value \; if \; condition \; true,$$
$$x - value \; if \; condition \; false)$$

For example,

$$y(x) := if(\sin(x) > 0, \, \sin(x), \, -\sin(x))$$

will return $\sin(x)$ when $\sin(x)$ is greater than zero and will return minus $\sin(x)$ when $\sin(x)$ is less than zero. The net result is than the function $y(x)$ looks like the absolute value of $\sin(x)$.

It is possible to nest if statements:

$$x = if(condition\ I,\ value - condition\ I\ true,$$
$$if(condition\ II,\ value - condition\ I\ false\ II\ true,$$
$$value - condition\ I\ and\ II\ false))$$

For example, for

$$y(x) = if(x > 10,\ x,\ if(x < -10,\ x\ ,0)),$$

if $-10 < x < 10$ the function returns zero; otherwise it returns x.

The until statement permits a process to continue until an expression, not a condition, becomes negative:

$$y = until(expression,\ y - value)$$

For example, for

$$y(x) = until(\sin(x), \sin^2(x)),$$

if x takes on a series of increasing values, $y(x)$ takes on the value of $\sin^2(x)$. However, when x is large enough that $\sin(x)$ becomes negative, the process stops no matter how far x ranges. The values need not be in order; any x value rendering $\sin(x)$ negative stops the process.

Complex Numbers.   Complex numbers are of the form $a + bi$, where $i = \sqrt{-1}$. In MathCAD, when writing the imaginary part of a complex number, do not insert a multiplier between b and i; instead, write bi. If the value desired is just i, write $1i$.

If z is the complex number, $a + bi$:

Re(z) returns the real part of z, a.

Im(z) returns the imaginary part of z, b.

$|z|$ returns the magnitude of z, $\sqrt{a^2 + b^2}$.

$\bar{z}$ returns the complex conjugate of z, $a - bi$. The conjugate is called with a quotation mark, for example, $y"$.

arg(z) returns the angle in the complex plane. The angle is measured from the positive real axis to z. The value returned is in the range from $-\pi$ to π. (Similar to the atan function.)

• •   Enter the complex number $z = 3 + 4i$. Evaluate the real part of z, the imaginary part of z, the complex conjugate, the magnitude, and the argument.

Let $u = 1i$. Observe the 1 as the cursor is moved in and out of the region.

How would one enter the value, when the imaginary part of the complex number has magnitude $5.2 \cdot 10^4$?

Floor, Ceil, and Mod.   These three functions can all be used to modify the value of some quantity. The floor function returns the largest integer less than or equal to some argument. The function call is floor(arg), where the argument is real.

• • Determine the floor of π, e, -0.7, and 10.

The ceil function returns the smallest integer greater than or equal to some argument. The function call is ceil(arg), where the argument is real.

• • Determine the ceil of π, e, -0.7, and 10.

Is ceil(floor(π)) the same as floor(ceil(π))?

The mod function has two arguments. The function returns the remainder resulting from dividing the first argument by the second. The function call is mod($arg1, arg2$).

• • Evaluate mod(1,2), mod(5,3), mod(10,2), mod(11,2), and mod(-5,2).

Let i range from 0 to 15 in steps of 1. Let $s = 4$. What is mod(i, s)? What is floor(i/s)?

Heaviside Function, Φ.   The Heaviside function returns 1 if the argument is greater than or equal to zero. The function returns 0 if the argument is less than zero. The function call is $\Phi(x)$.

The Φ function can be used, for example, to express logic functions such as AND, OR, or XOR. In these logic expressions the arguments are restricted to two values, 0 and 1. For example,

$$\text{OR}(A, B) := \Phi(A + B - 0.5),$$

$$\text{AND}(A, B) := \Phi(A \cdot B - 0.5).$$

• • If A and B can each be zero or one, there are four possible combinations for the arguments for OR and AND. Evaluate the four outcomes for each.

In logic, a 1 is often referred to as true and a 0 as false. For what values of A and B is the expression OR(A, B) $-$ AND(A, B) true?

Bessel Functions.   These functions are not typically encountered in general physics. They belong to a class of functions known as orthogonal functions. You will have fun meeting them in more advanced courses. In this text, there is only one mention of them, in the optics section. You do not need to know the details of these functions to plot them or make use of them any more than you need to know the details of the integration process to determine the numerical result of an integration.

- • Plot $J0(x)$ and $J1(x)$ as x goes from 0 to 12.

1.4   General MathCAD Features

1.4.1   *Processing*

When operating in manual mode, MathCAD does not process anything unless requested. In automatic mode, MathCAD processes everything as the document is assembled. This can often be a nuisance as well as a waste of time. Let MathCAD calculate only when the document is ready, or at least ready to troubleshoot.

MathCAD makes two passes through a document when processing. On the first pass, it processes all global assignments; on the second pass, it processes all local assignments. The order is from left to right and from top to bottom. The topmost edge of the region, not the line on which the equality sign appears, determines its location.

Pressing [F9] is the quickest way to start MathCAD processing a document. The effect of pressing [F9] is to process the document up to and including the present screen. Regions below the bottom of the screen (with higher line numbers) are not processed. When additional regions become visible, pressing [F9] again results in further calculation.

However, pressing [F9] does not always have the desired effect. For example, sometimes after having made corrections to a document, when you press [F9] nothing happens. If this occurs, move the cursor to the next region and try again. Repeat if necessary, each time moving to successive regions. If processing still does not take place, press [Esc], type pro (process), and press return. This sequence always works, but it means recalculating the entire document.

If it is desired to halt a calculation because it is taking too long, or you see an error in the document that must be corrected, or for any other reason, press [Ctrl][Break]. On the command line appears the query, "interrupt calculation?" Type y and press return. The calculation is then halted and the comment "interrupted" descends from the region where processing had been taking place. If it is decided to continue and no changes have been made in the document, place the cursor in the interrupted region and press [F9]. Processing will continue from the point at which it had been interrupted. If the document has been changed, processing starts from the changed regions.

When working with a newly created document, processing may result in a huge string of error messages descending for almost every region in the document. Don't panic. And don't separate the regions. (If the error message is covered up by subsequent regions, press [Ctrl]R, and watch closely.) The number of error messages does not indicate that everything is wrong. If a region contains an error, all subsequent regions depending on the region containing the error will register error messages as well, even though nothing may be wrong with them. Usually, the problem is associated with something at the beginning that, when fixed, will then permit the processing to continue, at least to the next region containing an error.

It is not difficult to exceed the memory addressing capabilities of MathCAD 2.0 or 2.5. The operation [Alt][F2] will show the amount of memory used and the total amount available.

After many changes have been made in a document, the memory may become fragmented. Attempts to process a document may result in the error message "out of memory." If this happens, there are only two choices; quit MathCAD or save the file and then quit MathCAD.

When dealing with large documents, it is not a bad idea to exit MathCAD, reload, and then bring up the document. Although reset does clear the screen, it does not take care of the memory fragmentation problem.

1.4.2   Files

A number of files are on the disk that accompanies this text. These documents are basic to the text material and are often starting points for further exploration.

Loading a File from the Disk into the Computer.   To load a file into the computer from a floppy disk, insert the disk in the drive and press [F5]. To the query on the command line, enter the name of the file including the drive name and press return. For example, if the disk is in drive b and the filename is arith1.mcd, enter b:arith1 and press return. Typing the filename without the extension, .mcd, is proper. MathCAD looks for files with that extension. If the extension is anything else, then it must be entered as part of the filename.

If you are not sure of the filename, after pressing [F5] you can look at a list of all files ending in .mcd on a particular disk by typing the wild card character * when asked for the filename. If you need to specify a path

— that is, if you want to look at a disk other than the default — you would type, for example, b:*, and press return. A list of all files ending in .mcd on the disk in the b drive would then be listed. If the number is large, not all the filenames can be shown at once; use the [↑], [↓], [PgUp], and [PgDn] keys to move through the list. As you move, one file is always highlighted. When the file you wish to load is highlighted, press return. If the file being sought is not on the disk, press [Esc]. This will return control to MathCAD's main shell.

If a document is to be loaded into the computer while another document is still on the screen, the process is similar. Press [F5]. If the document on the screen has not been saved, the command line statement is, "changes not saved. OK to discard?" If it is OK to delete the resident document, type y and press return. This process dumps the present document and loads the file requested. If it is not OK to delete the present document, do not type y, press return; this halts the load request operation. Now save the document in the standard way as described below. Then press [F5] again.

Saving a File: Writing from Computer to Disk.   A document can be written to your disk so that it can be saved and retrieved later for computational purposes or for further development. If the document is new and has not been saved before, press [F6]. On the command line appear the words, "save as:" followed by a blinking cursor. This is a request for a filename. Enter the name without any extension (MathCAD will automatically add .mcd) and press return. For example, type myfile and press return. Math-CAD automatically stores the file under the name myfile.mcd. In general, try to use a descriptive name that suggests the contents. The filename is limited to eight characters.

If the file is to be saved on your personal floppy disk, then, depending on the specific computer arrangement, it may be necessary to precede the filename with a drive name. For example, b:myfile would store the file on a disk in the b drive. Without such a pathname, the file is stored on the disk in the default drive. This means that with a hard disk system, the document would normally be written to the hard disk, the standard default disk.

If a file already exists with the name you have selected, a message will appear saying, "file exists, OK to overwrite?" If you do not want to lose the file previously saved under that name, do not type y; simply press return. Then press [F6] again and choose a different name, one not already used. To respond in the affirmative, type y and press return. The effect is to erase the previously existing file with that name and store the present document under that filename.

"Thirtieth floor, sir. You're expected."

"Er, don't you want to search me?"

"No, sir."

"My ID cards."

"No need, sir."

"But I could be anybody."

"No you couldn't sir. This is Information Retrieval."

Brazil

1.4.3 Printing

Before attempting to print a document, be certain that the printer is turned on. Attempting to print with the printer power not on may cause the system to lock up. If this occurs, the system will have to be rebooted. If the document you were attempting to print was not previously saved, it is lost. It is not a bad habit to save a document before printing it.

Before printing, a printer driver must be selected. To select a driver, press [Esc], and type sel . Then, using the [↓] or [↑] keys, highlight the appropriate printer type, and press return. If this is to be the standard printer, the printer selection can be saved and the select process need not be repeated. To save this information, press [Esc], type configsave and press return. The MathCAD disk must not, of course, be write protected when you perform this operation.

Once a printer type has been selected, to print out the document, press [Ctrl]O ("oh", not zero). On the command line will appear the words "Print area" and the range over which the document extends. The cursor is at the command line. If you wish to print the entire document, simply press return twice. The document will start printing right away. If you wish to print only part of a document, change the line and/or column numbers to correspond to the range of interest. (This area can be determined in advance by moving the cursor to the beginning and the end of the area you wish to print and noting the line and column numbers of the cursor.) Then press return.

If something goes wrong, or if for any reason you want to stop the printing process, press [Ctrl][Break] or [Ctrl][Scroll Lock] (these are different names for the same thing) or [Ctrl]A.

1.5 On Processing Documents

Most of the documents on the disk cannot be observed in their entirety at one time on a computer screen. This means that you must move back and forth in the document while you are using it. Therefore, it is essential that you become familiar with the cursor movements so that you can jump about in the document without leaning on the arrow keys.

Frequently, the document is divided into sections. The key parts of a section may fit on a single screen. Be sensitive to which variables or parameters need to be changed and what plot regions or values you want to be able to see. Position the document so that as little movement as possible is needed. Delete lines between regions if that will help; but don't make things so crowded that you have difficulty discriminating between subscripts from one line and superscripts from another.

With each document a number of questions are asked; pursue them. Work out the details. Then step back and ask, "What does this mean?" Your efforts will be rewarded with insight. There is no victory without a struggle.

"Easy, it's an innocent question."

"No question from you is innocent, Mr. Gittes."

Chinatown

Each document is devoted to a particular problem. The techniques are general and you should have no qualms about cannibalizing the documents and restructuring them in any way that will be useful to you.

"I ate his liver with some fava beans and a nice Chianti."

Silence of the Lambs

CHAPTER
2

Curves and
Curve Fitting

Visualization is often a key to insight and intuitive understanding. The graphical presentation of either analytic expressions or data — one form of visualization — is essential for taking in quickly large-, medium-, and small-scale features. Such information is crucial in both theoretical and experimental physics. Tables of numbers do not convey their information as readily as does a graphical presentation of data. Even appreciating a curve as basic as that of the sine would be difficult without any graphical representation.

In this chapter, we demonstrate how to plot families of curves, examine some frequently used approximations, consider very briefly the use of interpolation, and finally examine some methods for curve fitting, that is, finding analytic expressions that correspond to data.

2.1 Families of Curves

In this section, you are encouraged to build yourself a library of graphical representations of common mathematical functions. It is one thing to look at a book and see what someone else has plotted; it is rather different to control the process and create your own graphs.

As you go through each case, calculate a sufficient number of examples so that you have a clear understanding of the general form associated with the functions and of the role of the parameters. When data are examined, such knowledge is very useful.

2.1.1 *Plotting Families of Curves*

We wish to plot sets of curves. To do this we need to examine some features of user-defined functions. In particular, it is important to determine the order in which values are presented when there are two arguments associated with a function.

•• First let's do an exercise. Plot each of the following three functions in three different plot regions (you will have nine plot regions in all). Let the first set of regions be linear, log cycles (0,0), the second set semi-log, log cycles (1,0), and the third set log-log, log cycles (1,1). The three functions to be plotted are (1) $y = 0.9\,x + 1$, (2) $y = 1.2\,e^{-2x}$, and (3) $y = 3\,x^2$.

Plot the functions either as a function, $y(x)$, or as a vector, y_i. Explore the effect of changing coefficients, powers, and sign. Notice especially which functional forms are straight lines among the various plot types. Many functions are not straight lines in any of these representations.

Frequently, it is useful to define a function in MathCAD. The procedure is straightforward: *function-name(variable list) := function* in terms of variables and other parameters previously defined. Quantities in the variable list need not be assigned values before the function is defined. When values are substituted in the variable list and the defined function is evaluated, the results are the same as when those values are substituted into the function itself.

For example, the roots of a quadratic could be written as

$$x(a, b, c) := \frac{-b \pm \sqrt{b^2 - 4\,a\,c}}{2\,a}.$$

Entering values for a, b, and c in x returns the root values. An interesting function that appears in optics is

$$\mathrm{sinc}(x) := \frac{\sin(x)}{x}.$$

The advantage is clear if the function is used more than once. The function need not be rewritten, only the function name with variable list. Note that when a function is defined, identical names on the left- and right-hand sides refer to the same quantities, regardless of the order.

When the function is evaluated, the order in the list, not the name is crucial.

• •   Explore the effect of the order of variables in the list. Reset MathCAD. Let $n := 1, m := 2, i := 0 \ldots n, j := 0 \ldots m, y(i,j) := (m+1) \cdot i + j,$ and $z(j,i) := (m+1) \cdot i + j$. Predict first the sequence of numbers (six in each case; why?) and then evaluate, one by one, $y(i,j), y(j,i), z(j,i), z(i,j)$. Note, in comparing the two sets of y and z values, that the values are not just in a different order; they are different.

Frequently, it is less confusing if functions are defined with variable names not used in the problem. For example, let $w(a,b) := (m+1) \cdot a + b$ and $w'(a,b) := (m+1) \cdot b + a$. This is less confusing when you evaluate the combinations such as $w(j,i)$. How do these functions compare with the expressions for y and z that were just evaluated?

Change the values of n and m to 2 and 4. Plot $y(i, j)$ vs. j and $y(i, j)$ vs. i in two different plot regions. Explain the plots. Reduce the upper limit of the abscissa from 4 to 3.9 in the y vs. j plot. What effect does this have?

The function y contains points that lie along three straight lines with different intercepts. When y vs. j is plotted, we see the three lines, but they are all connected. If you reduce the plot range (from 4 to 3.9), the points at 4 cannot be plotted and MathCAD's plot line is broken, starting again with the next value. In this way, we see the curves independently. Include values slightly beyond the region that is to be plotted, and then plot only the restricted region. The curves will then not show the interconnecting lines.

In describing families of curves, typically, a function (for example, f) is expressed in terms of an independent variable (for example, x) and a set of parameters (for example, $a, b, \ldots$). To plot a family of curves, define a function in terms of the independent variable and one of the parameters. A sequence of curves can then be plotted in one plot region by specifying a sequence of values for the parameter. This can be repeated for each parameter simply by changing the parameter name in the definition.

Begin the family of curves with the case of a straight line.

$$y = mx + b \quad \text{can be written} \quad y(m, x) := m \cdot x + b.$$

On plotting $y(m, x)$ vs. x, each curve shows the general functional form; stepping through a sequence of four or five values of m shows the m-dependence for a given b. Another function, $y1(b, x)$, can be defined and plotted showing the b-dependence for a given m.

In each case below, indicate on the printout from your work the parameter values associated with each curve.

• • Load CURV1, a family of straight-line curves (see Fig. 2.1).

Check the programming steps. Process, then change the document so that for a slope of one ($m = 1$), b takes on the values -4, -2, 0, 2, 4.

• • Plot the family of parabolas, $y(a, x) = ax^2$. Let a_i take on the values 0.4, 1, 4. Include positive and negative values for x. Repeat for a_i equal to -0.4, -1, and -4. Could both sets of values of a be plotted simultaneously?

• • Plot the family of parabolas, $y(a, x) = a \cdot (x - x_o)^2 + y_o$. Consider the cases $x_o = 1$, $y_o = 0$, and $x_o = 0$, $y_o = 1$; let a take on the positive values from the previous example.

• • Plot the family of sine curves, $y(\theta, \phi) = \sin(\theta + \phi)$. Let θ range from 0 to 4π. Let ϕ take on values of 0, $\pi/6$, and $\pi/3$ radians. Consider the same values for $\sin(\theta - \phi)$. Examine the cosine similarly.

• • Plot the single curves, x, y, and z vs. α where $x(\alpha) = \sinh(\alpha)$, $y(\alpha) = \cosh(\alpha)$, and $z(\alpha) = \tanh(\alpha)$. Let α take on the values $\alpha = 0, 0.05, \ldots, 2$.

• • Define $r(t) = \sqrt{t^2 - 1}$. Plot $r(t)$ vs. t for $t = 1, 1.05, \ldots, 3$. Plot $x(\alpha)$, $y(\alpha)$ vs. α; set the limits to be the same as for the plot of $r(t)$ vs. t.

```
Family of curves          case - straight line

b := 0              y(m,x) := m·x + b

The i index varies the slope.   The j index controls the position.

i := 0 ..4      m  := -1 + 0.5·i      j := 0 ..14      x  := -7 + j
                 i                                      j

Restrict the x range of the plot to avoid connecting the lines.

xl := 6                    xl
                     ⌈  x ⌉
                   y ⎮m , ⎮
                     ⎣ i  j⎦
                          -xl
                        -xl    x    xl
                                j
```

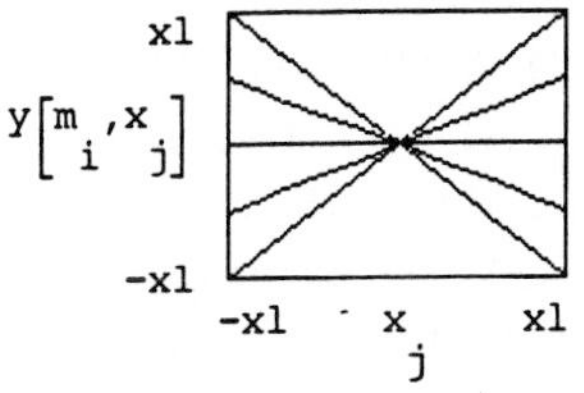

Figure 2.1 CURV1, a family of straight-line curves.

• • Plot the family of exponentials $y(x, a) = y_o \exp(-ax)$. Let x range from 0 to 4. Let a take on values 1, 2, and 4. Repeat with $\exp(+ax)$.

• • Plot the family of exponentials rising to a fixed level, $y(x, a) = y_o \cdot (1 - \exp(-ax))$. Use the same ranges for a and x as in the previous example.

• • Examine the curves $\exp(-x^2/a^2)$ and $\exp(-(x - x_o)^2/a^2)$.

• • Plot the three curves $\ln(x)$, x, and $\exp(x)$ vs. x in the same plot region. Avoid the singularity at $x = 0$.

2.2 Approximations

For most problems in the real world, exact answers are not possible. Simplifications and approximations are frequently useful and necessary. The criterion is always whether the approximation maintains the accuracy necessary for the problem at hand. Consequently, the practitioner must know the range of validity of the approximations and the range of errors introduced. Misunderstood approximations may lead to incorrect results and false conclusions.

Some commonly encountered approximations are (either for small angle or small x)

$$\sin(\theta) \approx \theta, \tag{2.1}$$

$$\tan(\theta) \approx \theta, \tag{2.2}$$

$$\cos(\theta) \approx 1, \tag{2.3}$$

$$\exp(x) \approx 1 + x, \tag{2.4}$$

$$(1 + x)^n \approx 1 + nx. \tag{2.5}$$

Typically, the value of making an approximation is that the mathematical form of the approximation is simpler than the function, permitting analytic solutions to problems (for some limited range) that might otherwise be far more complicated or even intractable. The motion of the simple pendulum, for example, can be solved easily in the small-angle approximation. The general analytic solution is far more complicated. Numerical techniques also permit a kind of general solution that is readily accessible. The latter will be discussed subsequently.

The approximations shown above all represent the first term of a series expansion. For the case of the sine function, the expansion is

$$\sin(\theta) \simeq \theta - \frac{\theta^3}{3!} + \frac{\theta^5}{5!} + \cdots .$$ (2.6)

• • Load CURV2. Compare of the sine curve with approximations (see Figs. 2.2 and 2.3).

Three different functions are defined which include the first one, two, or three terms of the sine series. The first two approximating functions are plotted together with the sine curve as a function of angle. Observe the quality of fit and the sign of the deviation as the angle increases.

To help see the differences, $d1$ and $d2$ and percent differences $pd1$ and $pd2$ are plotted. To find the angle that corresponds to a given percent error (we consider the $f1$ approximation), it is necessary to solve the transcendental equation

$$\frac{x - \sin(x)}{\sin(x)} \cdot 100 = \text{percent error}.$$

MathCAD's given-find solve block handles this equation readily. Finally, a given-find solve block is used to determine a set of solutions for percent error given a series of values for the angle. (See the section in Chapter 1 on solve-blocks.) In the case at hand, we have fcn(*percent error*) = find(*angle*); specify the percent error, evaluate fcn(*percent error*) and return the corresponding angle. A series of values can be obtained using subscripted variables. See the end of CURV2 for an example.

• • Using the first given-find (for the function, $pd1(x)$), determine the angles at which the error is 2%, 5%, and 10%. Use the functional form of the solve block at the end of the document to generate the sequence of percent errors vs. angle for the second approximation.

• • Plot $f3(x)$ and $\sin(x)$ as x goes from 0 to π, just to see what improvement occurs.

• • For the approximation $\tan(\theta) \approx \theta$ (the expansion starts $\theta + \theta^3/3 + \cdots$), what angles correspond to percent errors 1, 2, 5, and 10? How does the approximation compare with the $\sin(\theta) \approx \theta$ approximation in terms of accuracy?

• • For the approximation $\cos(\theta) \approx 1$, what angles correspond to percent errors 1, 2, 5, and 10?

Approximations to the sine curve.

$rad \equiv 1$

$deg \equiv \dfrac{\pi}{180} \cdot rad$

Examine the range from 0 to π. $x := 0, .1 .. \pi$

Three approximations.

$f1(x) := x$

$f2(x) := x - \dfrac{x^3}{3!}$ $f3(x) := x - \dfrac{x^3}{3!} + \dfrac{x^5}{5!}$

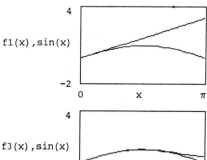

$f1(x), \sin(x)$

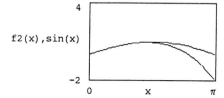

$f2(x), \sin(x)$

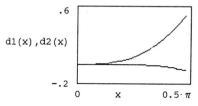

$f3(x), \sin(x)$

Differences and % differences between the function and the approximations.

$d1(x) := f1(x) - \sin(x)$ $d2(x) := f2(x) - \sin(x)$

$pd1(x) := \dfrac{d1(x)}{\sin(x)} \cdot 100$ $pd2(x) := \dfrac{d2(x)}{\sin(x)} \cdot 100$

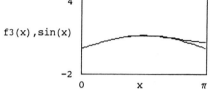

$d1(x), d2(x)$

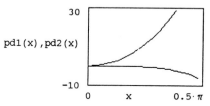

$pd1(x), pd2(x)$

To find where a given value of error occurs, use the given find and solve.

$x := 1$

Given $|pd1(x)| \approx 1$ $x := Find(x)$ $x = 13.986 \cdot deg$

Given $|pd2(x)| \approx 1$ $x := Find(x)$ $x = 57.742 \cdot deg$

Figure 2.2 CURV2, comparison of the sine curve with approximations. (See the next figure for the rest of the document.)

```
To compare, find the percent error for approximation 1 at 57.7 deg.

pd1(x) = 19.173          Remember at this point approx 2 was off only by 1%.
                         We can use x since the last value assigned to it is
                         the desired one.
```

```
A series of values is generated by writing the find statement as a function.
```

```
The equation in the solve block relates y, the angle, with some percent error,
pe.  Provide a percent error in the argument for f and the find statement
returns the angle.   A sequence of value for percent error are provided in
pe[i.  Ang[i are the returned angles that correspond to those percent errors.
```

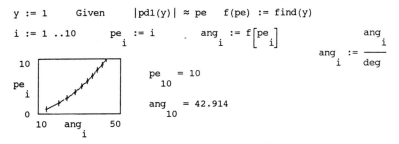

Figure 2.3 items:

```
y := 1      Given     |pd1(y)| ≈ pe    f(pe) := find(y)

i := 1 ..10       pe  := i       ang  := f⎡pe ⎤                        ang
                    i               i    ⎣  i⎦                            i
                                                             ang  :=  ─────
   10                                                           i      deg
 ┌──────────────────┐         pe   = 10
 │              ⟋    │           10
pe│          ⟋       │
 i│        ⟋         │         ang   = 42.914
   │    ⟋             │           10
 0 └──────────────────┘
   10    ang      50
            i
```

Figure 2.3 CURV2 *continued.*

● ●   Find the 1, 2, 5, and 10 percent error points for

$$e^x \approx 1 + x,$$
$$(1+x)^n \approx 1 + nx,$$
$$ln(1 \pm x) \approx \pm x.$$

● ●   Is $\cos(x) + \sin(x)$ a reasonable approximation for e^x, where $0 < x < 1$? Test quantitatively.

2.3   Interpolation

Data are frequently taken at intervals suitable to the measurement at hand. If values are to be obtained from these data, for points intermediate to the sampled data, some kind of interpolation process is required. Math-CAD provides both linear and cubic-spline interpolation. Linear interpolation is, as the name suggests, equivalent to drawing straight lines between adjacent data points and extracting values from those lines. Cubic-spline interpolation, on the other hand, fits a series of smooth curves from one data point to the next. In each case, the first and second derivatives of the fitted curves are continuous.

Besides the obvious application of simply finding an intermediate value, it is frequently useful to determine the area under a curve. This process typically requires integration. However, MathCAD's integration feature requires a functional form, not discrete data values, to integrate. The interpolation process can be used to create a function that can be integrated.

For example, if you had data that were values of velocity at discrete time intervals, it might be of interest to know the corresponding distance travelled. To determine the distance, you would like to integrate the velocity over time to find the area under the curve, which on a $v - t$ curve represents the distance travelled.

The function, linterp, for linear interpolation requires three arguments. They are (1) x-axis data vector with values in ascending order (such as time), (2) y-axis data vector (such as velocity), and (3) an x-value for which the interpolation is to be carried out (for example, a particular time). Linterp returns the y-value (for example, velocity) corresponding to the specified x-value (in this example, time) based on a linear interpolation between adjacent points.

The cubic-spline interpolation process requires two steps: first the creation of a spline vector, and second the interpolation process itself. There are three choices for the creation of a spline vector. The differences between these different spline choices become more evident near the endpoints.

The lspline function generates a spline curve that approaches linearity near the endpoints; pspline approaches a parabola; and cspline approaches a cubic. The functions are lspline(x, y), pspline(x, y), and cspline(x, y), where x and y represent the abscissa and ordinate vectors. The interp function is almost identical to the linterp function already described, except there are four arguments. The first argument is a vector generated by one of the spline functions. The next three arguments are the same as those used in linterp.

• • Load CURV3, examination of linear interpolation (see Fig. 2.4).

The first line of Fig. 2.4 shows three data points. The second shows a call to the linterp function and the interpolated values of two points. The three arguments of the linterp function are the x vector (values in ascending order), the y vector (notice the vectors are called by name and that there are no subscripts), and the variable v, which represents the value of x where the interpolation is to be made.

In the third line, calls are made to each of the spline functions; normally only one would be called. The interp function is called by $z1$, $z2$, and $z3$; each is based on a different spline function. Note that there are four arguments. The first is the spline vector of choice; the rest are as for the linterp case. The plot shows the original data as two straight-line segments. The three curves represent the different spline functions. Between $x = 0$ and $x = 1$ and between $x = 1$ and $x = 2$, the linear interpolation function results in the least curvature. In the range from 0 to 1, the cubic spline has the greatest curvature, but in general it need not. Clearly,

---

Interpolation. An example demonstrating linear, parabolic and cubic splines.

The data

$x_0 := 0$ $x_1 := 1$ $x_2 := 2$ $y_0 := 0$ $y_1 := 1$ $y_2 := 4$ $i := 0 ..2$

Interpolating using the linear interpolation function, linterp.

$z(v) := linterp(x,y,v)$ $z(.7) = 0.7$ $z(1.5) = 2.5$

Create the spline vectors. Normally, only one would be selected.

$s1 := lspline(x,y)$ $s2 := pspline(x,y)$ $s3 := cspline(x,y)$

$s1 = \begin{bmatrix} 0 \\ 3 \\ 0 \end{bmatrix}$ $s2 = \begin{bmatrix} 2 \\ 2 \\ 2 \end{bmatrix}$ $s3 = \begin{bmatrix} 4 \\ 2 \\ 0 \end{bmatrix}$ $v := 0,.05 ..2$

Interpolate and plot.

$z1(v) := interp(s1,x,y,v)$ $z2(v) := interp(s2,x,y,v)$

$z3(v) := interp(s3,x,y,v)$

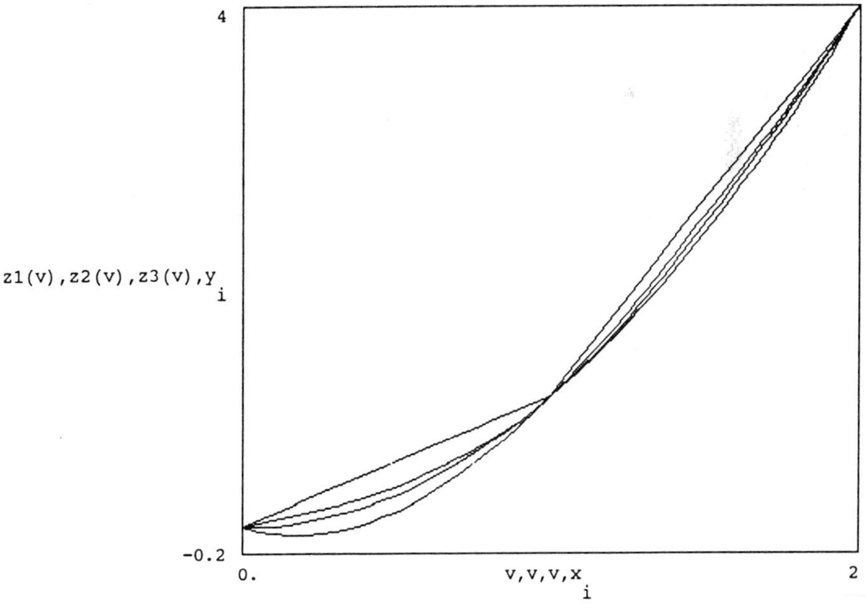

In the 0-1 region. From left to right the curves are y, z1, z2, z3.
In the 1-2 region. From left to right the curves are y, z3, z1, z2.

Figure 2.4 CURV3, examination of linear interpolation.

the interpolation processes can give rather different results depending on the method used. Extrapolation beyond the range where the data exist is more error prone than interpolation between points in the data set.

• •   Load CURV4, the area under a curve (see Figs. 2.5 and 2.6).

The v_j represent any arbitrary data set taken at the corresponding t_j. The area under the curve can be determined by taking the appropriate sums or by integrating.

The area is determined first, in effect by counting the number of rectangles beneath the curve. This can be done by a variety of methods. We

---

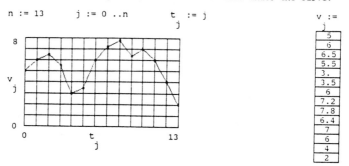

Given some arbitrary data, determine the area under the curve.

n := 13   j := 0 ..n t_j := j v_j :=

5
6
6.5
5.5
3.
3.5
6
7.2
7.8
6.4
7
6
4
2

The area can be approximated by imagining that the area is constructed from thirteen rectangles of unit width and a height determined by f. Since the width is unity in this case, the area is the sum of the f's for each rectangle.

A sum over all v's would be in error, an overestimate, as it counts both end points and results in one extra rectangle.

$$\sum_j v_j = 75.9 \qquad \text{A sum over all f's, an overestimate.}$$

If the v's are evaluated at the left hand side of the rectangles the area is suma; if the v's are evaluated at the right hand side the area is sumb.

k := 0 ..n - 1

$$\text{suma} := \sum_k v_k \qquad \text{sumb} := \sum_k v_{k+1} \qquad \begin{array}{l} \text{suma} = 73.9 \\ \text{sumb} = 70.9 \end{array}$$

If the average of v values on either side of the interval is used the result is sumc.

$$\text{sumc} := \sum_k \frac{v_k + v_{k+1}}{2} \qquad \text{sumc} = 72.4$$

Figure 2.5 CURV4, the area under a curve. (See the next figure for the rest of the document.)

```
Integration in Mathcad requires a function, not a series of points.  A
function based on the points can be created using an interpolation process.
```

```
A linear interpolation between points is achieved by calling the linterp
function.  The arguments are the abscissa vector, the ordinate vector, and
the position variable specifying where the interpolation is to be evaluated.
```

$$vv(x) := linterp(t,v,x) \qquad a := 0 \qquad b := 13 \qquad vint := \int_a^b vv(t') \, dt'$$

$$vint = 72.4$$

```
A smoother interpolation is possible using a cubic spline approximation,
instead of a linear one.  Two calls are necessary.  The first creates a
spline vector based on the abscissa and ordinate vectors.  The second
performs the interpolation; the arguments are the spline, abscissa, and
ordinate vectors, and the position variable specifying where the
interpolation is to be evaluated.
```

$$sp := cspline(t,v) \qquad vv'(x) := interp(sp,t,v,x)$$

$$vint' := \int_a^b vv'(t') \, dt' \qquad vint' = 72.63 \qquad \frac{vint'}{vint} = 1.003$$

```
A comparison of results.
```

$$suma = 73.9 \qquad sumb = 70.9 \qquad sumc = 72.4$$

$$vint = 72.4 \qquad vint' = 72.63$$

```
One would expect sumc and vint to be the same, since in this case determining
the average v value is equivalent to a linear interpolation.  If one
assumes the data are representative of a smooth curve which does not make
rapid changes relative to the spacing of the v values, then the cubic spline
approximation is the "best" of the values from the set described.
```

Figure 2.6 CURV4 *continued.*

could say that for each interval of t, we take the value of v at the beginning of the interval as the representative value. If the curve increases as much as it decreases, on the average, the overestimates and underestimates will tend to cancel. Alternatively, we could take the values at the end of the intervals. Of course, we do not want simply to sum the v values as there is one more value than there are intervals.

Three sums yield different values, as Figs. 2.5 and 2.6 show. Notice the use of the index to switch from the beginning of the interval to the end of the interval. (For $k := 0 \ldots n - 1$, the sum for the value at the beginning of the interval is over v_k, whereas the sum over the final values is v_{k+1}.) Finally, an average value is used, resulting in an intermediate value. This provides the most reliable value of these various sums.

To integrate the curve, we must have a functional representation of the data. The interpolation processes can provide that representation. It is desirable to define an interpolation function in terms of the variable to be integrated — for example, $vv(x)$ — and then to integrate that function. Integration using the linterp process (not lspline) is a straight-line interpolation and gives the same result as did the averaging sum.

Integrating with one of the spline functions is similar. First define the spline vector, then the interpolation function. Finally, integrate the interpolation function. A cubic-spline interpolation provides a smoother fit to the data than does the linterp process and yields a modestly different result. If we assume that the function does not make very rapid variations compared to what is seen, then the cubic-spline interpolation is the more reliable value.

• •   The area determined by the suma process can be visualized as follows. In the plot region of v vs. t, change the ordinate to read v_j, v_j. Then change the plot type to ls and process.

• • A cubic spline was used in the integration. How much different would the results be if the linear or parabolic splines were used instead?

• • Since the integrals using the two different splines do not yield identical results, they cannot be the same at all points. Plot $vv(t_j)$ and $vv'(t_j)$ vs. t_j in the same plot region. Are the differences obvious?

Define a new index, m, that ranges from 0 to 39. Define a new abscissa vector $h_m = m/3$. Now plot $vv(h_m)$ and $vv'(h_m)$ vs. h_m. Differences should be more apparent. Locate a range when the integration based on the linear interpolation would be greater than that of the cubic spline interpolation. Change the limits of integration, a and b, to enclose this region. Check the results from the two integrations. Repeat for a different range where the integration of the cubic spline should be greater.

• •   Extrapolation is dangerous. Let $m = 0, \ldots, 20$ and $h_m = m$. Plot $vv(h_m)$ and $vv'(h_m)$ vs. h_m. Be sure that MathCAD is able to set all the plot limits (if you are using an existing plot region, delete any limits that may have been entered). Is this extrapolation what you expected? Would it be significantly different if you used pspline or lspline?

2.4   Curve Fitting

A fundamental goal of physics is to describe nature in elegant mathematical form. A concise, elegant theory is more likely to be correct than a complex, messy one (Occam's razor). In the ongoing evolution of understanding, theoreticians attempt to formulate mathematical descriptions of nature; experimentalists attempt to measure the properties of nature. The ideas of each group, of course, influence and direct the other.

Physics and the other natural sciences must be founded on experiment. Observations of the night sky led to many theories of the extraterrestrial world. Plato's theory of homocentric spheres and Ptolemy's theory of epicycles are two such theories. Copernicus put forth still another theory. Still mired in spheres, it was complex and incorrect, yet it contained a key idea of placing the sun near the center of the universe. Kepler seized the idea and, after Herculean labors, solved the age-old problem of the planets, giving for the first time the correct description of planetary motion. Newton's laws of motion and his law of universal gravitation provided a powerful and commensurate mathematical description.

Good data are good data, but in and of themselves are uninterpretable. A theory provides a mechanism for the interpretation of data, yet the theory can be correct only insofar as it agrees with experiment. The cosmic theories of Plato and Ptolemy, and even the sphere-laden theory of Copernicus, were abandoned because they did not correspond sufficiently to the data. Kepler's and Newton's laws do correspond more precisely. In addition, their laws have an overarching simplicity and elegance that suits. Yet the ultimate truth is elusive. For all practical purposes, what is "true" in science is what the community of experts believes to be true.

Do not confuse the last statement with the idea that personal opinion reigns supreme and that we have only pure subjectivity as a guide to the truth. Once the experiments have been performed, the data analyzed, and the theories expounded, judgments are not freely made. The real world is out there whether we approve or not. The universe was evolving long before people appeared on the scene; the universe will continue should we cease to exist. The judgments that we make are about the adequacy of our models to correspond to nature.

> "— but we do at least know that the universe has some shape and order and that you know, trees do not turn into people or goddesses, and there are very good reasons why they don't, and you can't just believe absolutely anything."
>
> *My Dinner with André*

One element in this process of comparing data with a theoretical model is curve fitting. This helps to address the question of how well the suggested mathematical description fits the data. In the following sections, we describe how to perform a linear least squares fit, a polynomial fit, and a general fit.

2.4.1   *Linear Least Squares Fit*

MathCAD includes two built-in functions, slope(x, y) and intercept(x, y), where x and y are subscripted vectors. These functions return the slope and y-intercept of the linear least squares regression line for the data represented by x and y. Note that the index i must start from 0. If the index had started at 1, MathCAD would include the numbers in the zeroth element in the fit. If the zeroth element or any other element is not specified, it is treated as a zero and will skew the results.

• •   Load CURV5, glider on an air track — linear least squares example (see Fig. 2.7).

Velocities of a glider on an air track tilted at 10° are calculated in CURV5 (Fig. 2.7). Random errors are calculated and added to the velocities. Slopes and intercepts are calculated for both cases.

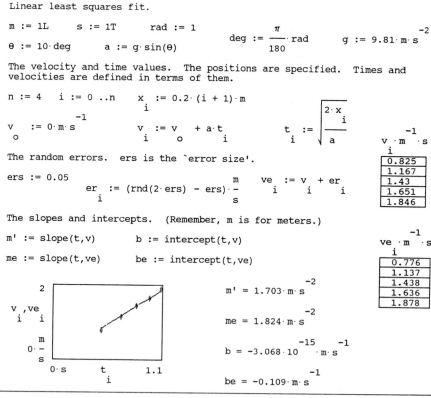

Figure 2.7   CURV5, glider on an air track — linear least squares example.

The error size is represented by *ers*. The errors are randomly distributed over the range $\pm ers$. It is important to include errors. It is one thing to fit to perfect data and another to do the calculation with values closer to what an experimentalist might encounter. Even small errors can have a greater effect on the outcome than one might expect. By examining both cases, one begins to get a sense of the limitations of the curve-fitting process. The errors are uniformly distributed; one could calculate errors with a Gaussian distribution but the extra calculation would gain us little at this stage.

The v_i are the precise values; the ve_i include errors; m' and b are the slope and intercept, respectively, for the ideal data. Why do we use m' and not m?

Move the place marker to the er_i region and press [F9] to recalculate with a new set of random values. Repeat several times. Observe the data in the plot region and the numerical values of slope and intercept.

Change the error size (ers) and repeat. In this example, how do error size and uncertainty in slope compare? How small must the random errors be to determine the slope consistently within 1%?

Examine when $v_o \neq 0$.

The linear least squares functions of slope and intercept can be extended to exponentials and one-term polynomials. To apply these functions for the case

$$y = Be^{mx}, \tag{2.7}$$

take the natural log of both sides:

$$\ln(y) = \ln(B) + mx. \tag{2.8}$$

The m-value could be determined by evaluating the equation for two sets of coordinates (x_1, y_1) and (x_2, y_2), subtracting one equation from the other, and solving for m:

$$\ln(y_1) = \ln(B) + m\,x_1, \tag{2.9}$$

$$\ln(y_2) = \ln(B) + m\,x_2, \tag{2.10}$$

$$m = \frac{\ln(y_2) - \ln(y_1)}{x_2 - x_1}. \tag{2.11}$$

However, by identifying $Y = \ln(y)$ and $b = \ln(B)$, equation 2.8 can be rewritten as

$$Y = mx + b.$$

Performing a least squares fit to this equation, using the slope and intercept functions, yields values for m and b based on all the data points rather than just two as in the previous example. Because $B = \exp(b)$, the fit process provides the two parameters, m and B, of the original equation.

A similar method can be applied to expressions of the form

$$x = Bt^m. \tag{2.12}$$

Taking the natural log of both sides yields

$$\ln(x) = \ln(B) + m\ln(t). \tag{2.13}$$

The m-value could be determined by evaluating the equation for two sets of coordinates, (x_1, y_1) and (x_2, y_2), subtracting one equation from the other, and solving for m. The B term drops out as in the previous case, and we have

$$m = \frac{\ln(x_2) - \ln(x_1)}{\ln(t_2) - \ln(t_1)}. \tag{2.14}$$

If we identify $X = \ln(x)$, $b = \ln(B)$, and $T = \ln(t)$, the log form of the equation becomes

$$X = mT + b.$$

Performing a least squares fit to this equation yields values for m and b. Because $B = \exp(b)$, the fit process provides the two parameters, m and B, of the original equation.

• •   Load CURV6, free-fall data with errors (see Fig. 2.8).

The data used here could be from a photograph of a falling ball bearing illuminated with a stroboscope. The expected form of the data is

$$x = \frac{1}{2}at^2.$$

This example is like the case just described, $x = Bt^m$, where $B = a/2$ and $m = 2$. We assume the time data are accurate. Errors are added to the x-data — $\pm ers$.

We take natural logarithms of the xe and t data and fit these data with a linear least squares fit. After $\exp(be)$ is computed, the values are compared with the ideal values.

Place the cursor in the er_i region and process several times. Note that even when the plot seems to fit well, the value for ac may differ significantly from the ideal value. The error is unrealistically large. Reduce it and process several times. How small does the error need to be to determine the value ac with 5% of its ideal value? What is the difference between a linear plot of X_i vs. T_i and a log-log plot of xe_i vs. t_i? (We refer to this problem in the next section.)

```
Linear least squares fit.      Log-log case.

Generate the data.

g := 9.81      a := g

n := 5          i := 0 ..n      t  := 0.05·(i + 3)              1   2
                                 i                        x  := - ·a·t
Add in an error signal.                                   i   2    i

ers := .05      er  := rnd(2·ers) - ers    xe  := x  + er
                  i                          i     i    i

X  := ln⌈xe ⌉   T  := ln⌈t ⌉
 i      ⌊  i⌋    i      ⌊ i⌋

Extract the parameters.

me := slope(T,X)        be := intercept(T,X)        be := exp(be)    ac := 2·be

          me           me = 2.553                 Ideal me value 2
z  := be·t
 i         i           ac = 18.15                 Ideal ac value 9.81
```

Display of the data, data plus errors, and fit. Show the log values, which, if they represent a polynomial, should lie along a straight line.

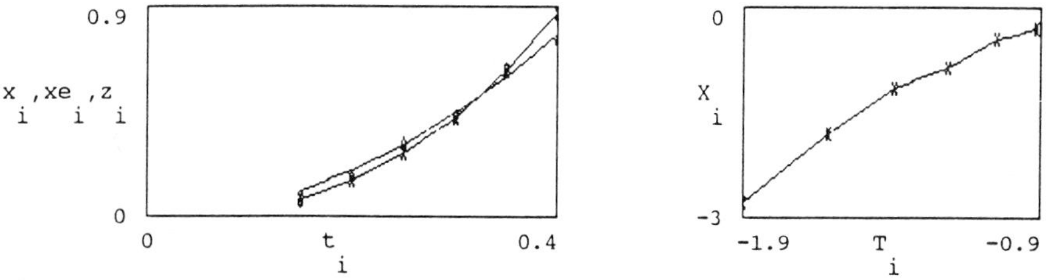

Figure 2.8 CURV6, free-fall data with errors.

2.4.2 *Polynomial Curve Fitting*

Here, we want to extract the coefficients of a polynomial,

$$y = a_0 + a_1 x + a_2 x^2 + \cdots, \tag{2.15}$$

given data sets (x_i, y_i). Matrix methods are helpful in determining the coefficients. An understanding of matrices is *not* necessary to use the results.

Let Y be a vector containing all the y-values; let A be a vector containing all the coefficients; and let X be a matrix whose zeroth column is all 1s (x_i^0), whose first column is the x_i s, whose second column is the x_i^2s, etc. Thus we can write the entire set of conditions (one for each data set)

$$Y \simeq XA \tag{2.16}$$

where $\simeq$ implies a least squares fit. (See Section 2.4.3 on general fit for a discussion of the least squares process.) If we interchange rows and columns, the form would be AX; these two forms are equivalent. We wish to solve for the coefficients.

The solution is given by

$$(X^T X)^{-1}(X^T Y) = A. \tag{2.17}$$

Thus we have the coefficients in terms of the x and y values. In case matrix operations are unfamiliar to you, these operations will be demonstrated for a small data set.

• •   Load CURV7, quadratic fit (see Figs. 2.9 and 2.10).

```
Demonstration of matrix operations for quadratic fit.

Four data points.
```

$n := 3$ $i := 0 \, .. n$ $x_i := i$ $y_i := 3 + 2 \cdot x_i + 1 \cdot x_i^2$

x_i	y_i
0	3
1	6
2	11
3	18

```
The equation for y in terms of x, is in effect, four
equations.  One for each i.  The coefficients remain
the same.
```

The equations can be written in matrix form as $Y := X \cdot A$ ▫ where

$$X := \begin{bmatrix} 1 & 0 & 0 \\ 1 & 1 & 1 \\ 1 & 2 & 4 \\ 1 & 3 & 9 \end{bmatrix}$$ ```The columns represent x to the zeroth, first, and second power.```

$$A := \begin{bmatrix} 3 \\ 2 \\ 1 \end{bmatrix}$$ ```The coefficients for the zeroth, first, and second powers of x.```

$Y := X \cdot A$ $$Y = \begin{bmatrix} 3 \\ 6 \\ 11 \\ 18 \end{bmatrix}$$ ```The same as the y values.```

```
So far, all that has been demonstrated is that the equations can be written
in matrix form.

Now let's construct the X matrix more automatically and form the various
products on the way to determining the coefficients.  (If there were lots
of data points, entering them by hand would be a waste of time.)

We wish to use the same names, so set them to zero before we start.
```

$X := 0$ $Y := 0$ $A := 0$

```
The zeroth column of the X matrix is one, x to the zeroth power.  Y is
similarly defined.
```

$X_{i,0} := 1$ $Y_i := y_i$

```
The first column is x to the first power; the x's are already specified.
```

$X^{\langle 1 \rangle} := x$ ```The column name is called using [Alt]6.```

Figure 2.9 CURV7, quadratic fit. (See the next figure for the rest of the document.)

```
The second column is x squared, that is, each element of x squared.
Calculate them using the vectorize operation.  This is a method to do element
by element calculations rapidly.  The operation is called with [Alt]-.
```

$$X^{\langle 2\rangle} := \overline{\begin{bmatrix} x^2 \end{bmatrix}}$$

$$X^{\langle 0\rangle} = \begin{bmatrix} 1 \\ 1 \\ 1 \\ 1 \end{bmatrix} \qquad X^{\langle 1\rangle} = \begin{bmatrix} 0 \\ 1 \\ 2 \\ 3 \end{bmatrix} \qquad X^{\langle 2\rangle} = \begin{bmatrix} 0 \\ 1 \\ 4 \\ 9 \end{bmatrix} \qquad X = \begin{bmatrix} 1 & 0 & 0 \\ 1 & 1 & 1 \\ 1 & 2 & 4 \\ 1 & 3 & 9 \end{bmatrix}$$

```
The transpose is called by [Alt]1.
The transpose interchanges rows
and columns.
```

$$X^{\mathsf{T}} = \begin{bmatrix} 1 & 1 & 1 & 1 \\ 0 & 1 & 2 & 3 \\ 0 & 1 & 4 & 9 \end{bmatrix}$$

$$A := (X^{\mathsf{T}} \cdot X)^{-1} \cdot (X^{\mathsf{T}} \cdot Y)$$

$$A = \begin{bmatrix} 3 \\ 2 \\ 1 \end{bmatrix}$$

```
The data are exact, so one would expect a `perfect' fit.
```

Figure 2.10 CURV7 *continued.*

A quadratic equation, a polynomial, is expressed in CURV7 (Fig. 2.9). A small number of x values are defined. The corresponding y-values are calculated. The x and y values are shown in tabular form. The matrix X has three columns; the values of each column are x to the zeroth, first, and second powers. The values were entered by hand. [Alt]M is the command to create a matrix. The number of rows and columns are requested. Enter the values and press return. The matrix appears filled with place markers; move the cursor to each marker and enter the desired value. The coefficients a_i are listed in the vector A. We perform the matrix multiplication $X \cdot A$ just as if the matrices were numbers. We evaluate Y and see that the values are identical with the y_i.

To construct our matrix more directly from the data, we set

$$X_{i,0} := 1$$

because we want the zeroth column to be all ones. We set the next column equal to the x_i and the next to x_i^2 (type [Alt]^ for a superscript):

$$X^{\langle 1\rangle} = x, \qquad X^{\langle 2\rangle} = x^2.$$

A vectorize operation is used in the document to speed the term by term squaring (the vectorize operation is called by typing [Alt]–).

We examine each column of X and the assembled matrix. We also look at the transpose, which is matrix X with the rows and columns interchanged. Finally, we calculate the values for A in terms of X, its transpose, and Y. The values for A are the ones we seek. Simple examples are useful to make the various steps concrete.

Show that $X^T X$ is a square matrix. Is $(X^T X)^{-1}$ also square? What is the form $X^T Y$?

• •   Load CURV8, free-fall data with errors — as in CURV6/Fig. 2.8. (See Fig. 2.11.)

This procedure is no different from that used in CURV7 (see Figs. 2.9 and 2.10). Without all the details, it can be expressed with only a few statements.

Note that this x and t information is the same as that of CURV6. Errors are defined in the same way but are, of course, random. Does the polyfit procedure yield better, the same, or worse values for ac, on average, than the methods used in CURV6? Provide a semiquantitative answer. How does this document provide information about the value of the exponent of t? Is there a t term in the original expression?

Remove the term linear in t and fit with just a constant and a squared term. Delete $T^{\langle 1 \rangle}$. Rename $T^{\langle 2 \rangle}$ as $T^{\langle 1 \rangle}$ so there are only two columns instead of three. Again examine quality of fit with the same error sizes.

• •   Load CURV9, example with a noninteger exponent (see Fig. 2.12).

Data are created as the square root of x; errors are added. In the first plot region, the exact data and the data with errors are plotted as plot type ℓv. A fit is attempted using the form

$$y = a_0 + a_1 x + a_2 x^{0.5}.$$

Quadratic fit   Free fall data with errors.

$g := 9.81$ $a := g$

$n := 5$ $i := 0 \ ..n$ $t_i := 0.05 \cdot (i + 3)$ $x_i := \frac{1}{2} \cdot a \cdot t_i^2$

$ers := .05$ $er_i := rnd(2 \cdot ers) - ers$ $xe_i := x_i + er_i$

$T_{i,0} := 1$ $T^{\langle 1 \rangle} := t$ $T^{\langle 2 \rangle} := \overrightarrow{\left[t^2 \right]}$

$b := (T^T \cdot T)^{-1} \cdot (T^T \cdot xe)$ $b = \begin{bmatrix} -0.244 \\ 1.668 \\ 2.133 \end{bmatrix}$ $ac := 2 \cdot b_2$

$ac = 4.267$

Figure 2.11 CURV8, free-fall data with errors — as in CURV6.

Fit with a **square root term.**

The data $n := 10$ $i := 0 ..10$ $x_i := \sqrt{i}$

The error $ers := 1$ $er_i := rnd(2 \cdot ers) - ers$

The signal $y_i := x_i + er_i$

The expression to be fitted

$Y := a_0 + a_1 \cdot x + a_2 \cdot x^{0.5}$ □

In matrix form $Y := X \cdot a$ □

$X_{1,0} := 1$ $X^{<1>} := x$ $X^{<2>} := \left[\overrightarrow{\sqrt{x}}\right]$

$a := (X^T \cdot X)^{-1} \cdot (X^T \cdot y)$

$a = \begin{bmatrix} -0.624 \\ 0.893 \\ 0.118 \end{bmatrix}$

$fit0_i := a_0$ $fit1_i := a_1 \cdot x_i$ $fit2_i := a_1 \cdot x_i^{0.5}$

The contribution of each term.

The fitted expression $Y := X \cdot a$

A comparison of fit and data

A comparison of the data with the square root term.

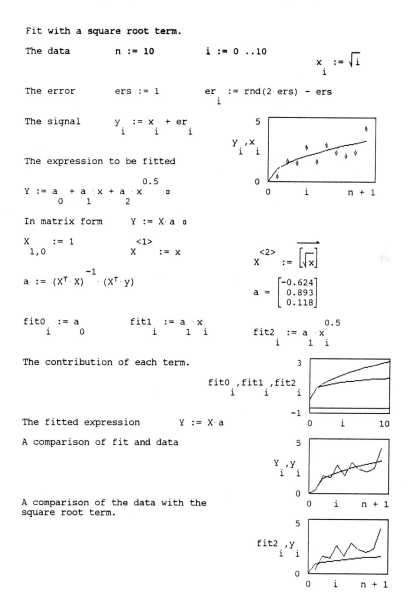

Figure 2.12 CURV9, example with a noninteger exponent.

To examine the fit, we plot each term separately. We plot the data and the fit. We plot the data and the square root term only.

Process several times to get a sense of the fit. Change error size and repeat.

Remove the x term and try to fit the data with the form $a_0 + a_1 x^{0.5}$. How does this change the quality of the fit? Change error size and repeat.

Try fitting the data with $a_0 + a_1 \ln(x)$. Compare the quality of fit with the previous case. Change error size and repeat.

• •   The following data are from an accelerator experiment performed by John Davis. Fit energy (as abscissa) and change in energy per thickness (ordinate) with a quartic, $f(x) = a_0 + a_1 x + a_2 x^2 + a_3 x^3 + a_4 x^4$. In matrix form our equations would appear as

$$dEdx = EA.$$

Plot the data and fitted curve vs. energy as energy increases from 0 to 3 in steps of 0.1:

$$E_i = 0.4, 0.6, 0.8, 1.2, 1.3, 1.4, 1.7, \text{ and } 2.3,$$
$$dEdx = 54.4, 61.2, 63.9, 63, 62.9, 62.4, 60, \text{ and } 54.8.$$

2.4.3   General Fit

In a least squares fit process, the mean squared error is minimized. You can express the mean squared error for any functional form and ask MathCAD to minimize the quantity.

Values for x and $y(x)$ are known. The analytic expression, $f(x)$, represents the data. Under ideal circumstances, the measured values $y(x)$ and the computed values $f(x)$ would be the same. The function f includes some parameters. The fit process determines the values for the parameters that best fit the data.

The general procedure would be as follows:

Define the functional form for the expression to be fitted, for example, $f(a, b, x) = \cos(ax) \cdot e^{-bx}$.

The functional form, f, defines the function that will be used to fit the data. The parameters, whose values are sought, must be included in the list of arguments together with the variables.

Define the total mean squared error. Take the difference between the data and the function evaluated at the corresponding point (use the same subscripts), for example,

$$sse(a, b) = \sum_i \Big[(y_i - f(a, b, x_i)) \Big]^2. \tag{2.18}$$

Use a given-minerr (not a given-find) solve block to determine the parameters that make the sum of the squared errors as close to zero as possible, for example,

$$\text{given}\quad sse(a, b) \approx 0 \quad 1 \approx 1 \quad \begin{bmatrix} a \\ b \end{bmatrix} = \text{minerr}(a, b).$$

The $1 \approx 1$ is required because there must be as many equations in the solve block as there are quantities for which values are sought.

This process can be very slow to converge. Choose the guess values carefully.

Consider a simple example from optics. A beam of light directed toward a smooth surface of water is bent as it passes from the air to the water. Ancient data from Ptolemy give the incident and refracted angles of a beam of light as it passes from air to water. The beam angles are measured from the perpendicular to the surface, the normal. Centuries later a relation, a functional form, between angles and indices of refraction was determined. Snell's law states that

$$\sin(\theta_{\text{inc}}) = n_{\text{ir}} \sin(\theta_{\text{refr}}),$$

where n_{ir} is the index that is sought.

The goal is to fit the data according to Snell's law and determine the best value for the index of refraction, which is a measure of the change in speed of the beam as it passes from air to water. In other words, given the data, what is the value of n_{ir} that makes the mean squared error between the data and the functional form the least?

• • Load CURV10, general fit to determine the index of refraction (see Fig. 2.13).

The data consist of the incident and refracted angles. The values used for the incident angle, θi_i, are every 10 degrees from 10 to 80. The values for the corresponding refracted angles, θr_i, are shown in tabular form.

Snell's law, which you are not expected to know at this point, can be written $\sin(\theta i) = n_{\text{ir}}\sin(\theta r)$.

The procedure is as follows. First, express the functional form and include the unknown in the argument list. For this problem, solve Snell's law for θi; include in the argument list the parameter for which a value is sought. This appears as

$$\theta i'(n_{\mathrm{ir}}, \theta r) := \mathrm{asin}(n_{\mathrm{ir}}\sin(\theta r)).$$

The prime is used to distinguish the calculated value from the original data (θ' is equivalent to f, above).

Define the sum of the squares of all the errors:

$$sse(n_{\mathrm{ir}}) := \sum_i \left[\theta i_i - \theta i'\big[n_{\mathrm{ir}}, \theta r_i\big]\right]^2.$$

In a given-minerr solve block, request MathCAD find the value for n_{ir} that makes sse a minimum. Note that sse is set equal to zero, but in general a value of zero is unattainable. In effect, a minimum is determined for sse.

Make comparisons of the fitted values and the data.

What is the average deviation of θi and $\theta i'$? (MathCAD summation is [Shift]4.)

Observe the original data. Plot θr vs. θi. To get some sense of the quality of the fit, plot $n_{\mathrm{ir}} \cdot \sin(\theta r)$ vs. $\sin(\theta i)$. Finally, look at the angles

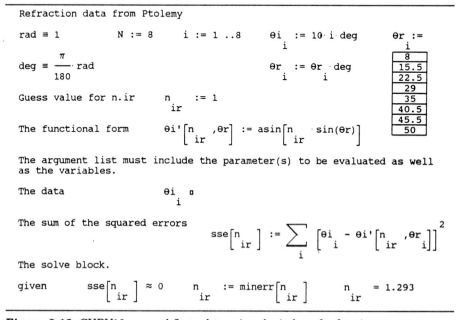

Figure 2.13 CURV10, general fit to determine the index of refraction.

themselves. Plot $\theta i'(n_{ir}, \theta r_i), \theta i_i$ vs. θi_i. Set plot types to be vl for this last plot.

• •   It is suggested that the data $y_i = 0, 3.4, 8, 11.9, 10.2, 5.5, 3.6, 1.1,$ $0, 0.2, 0.1,$ and $v_i = i$ can be fitted with a curve of the form

$$f = av^2 e^{-bv^2}.$$

Plot y_i, f vs. v_i. Try different values for the parameters a and b in the function f until a close fit is obtained. Use the best values for a and b from this process as guess values for a and b in a general fit. Perform a general fit. Plot the final results with the data as a function of v_i. How close were your values of a and b?

• •   In the file TORNADOS are listed the approximate number of tornados that occurred in the United States during the first six months of each year from 1950 to 1990. (April through June is tornado season.) What curve best represents the data? Based on your results, how many tornados will there be in the year 2001?

"I believe I deserve an explanation, Dave."

2001

CHAPTER

3

Units and Dimensional Analysis

In this chapter you learn how to include units in your calculations, to simplify the fundamental units resulting from a calculation, to determine units of unfamiliar variables in equations, and to consider other systems of units. The choice of units is to some extent arbitrary. The fundamental units of length, mass, time, and current, for example, are the choices of the SI system of units (used almost exclusively in general physics texts today). Among some professional physicists, however, the cgs system of units is still popular. The English system is, of course, in use. Still other systems of units are possible.

Dimensional analysis is then considered. This is a very useful technique that permits you to find dimensionless groups and to learn dependences between variables. Several examples, including Kepler's harmonic law and a nuclear explosion, are considered.

3.1 Units

Every term in an equation must have the same physical units. We cannot equate kilograms and meters or add an acceleration and a velocity and expect anything meaningful. This gives us a very powerful handle. We can look at an equation that we have never seen before and know if it is a possible equation. If every term has the same units, the equation could be true; if the terms do not agree, then we know the equation cannot be correct.

Remember the following points when dealing with units:

1. If there are derivatives or integrals, the ds and integral signs have no units and should be ignored. When thinking units, act as if these symbols did not exist.

2. Similarly, in dimensional analysis, differentials and integrals do not appear. A ratio such as A/T might be dA/dT. This is a limitation of the method.

3. Don't be put off by an equation that you don't know. Units are units. Several equations and quantities in this section may be unfamiliar to you. These procedures provide a means of brushing up against unfamiliar ideas and learning something about them.

In this section, we wish to concentrate on the units, not the numerical values associated with physical quantities. In particular, we will see how to set up a system of units, how to use units in calculations, how to convert units generated by MathCAD in a calculation to a particular form, what happens if we try to combine units improperly. Then we look at some further examples with units.

The general procedure for handling units using MathCAD is straightforward:

1. Define the base units.

2. Define any other needed units in terms of the base units or in terms of any other unit already defined (that is, to the left or above in the document).

3. Assign appropriate units to all variables by attaching them as multipliers or divisors of the numerical values.

4. Perform any MathCAD operations in the standard way. (Some Math-CAD operations must use unitless values; these include logarithms and exponentials.)

Results are given with units attached. Initially, the units are given in terms of the fundamental quantities mass, length, time, and charge. However, these can easily be converted into the particular units that were defined in steps 1 and 2 above. The numerical value of the quantity is automatically adjusted to correspond to the units selected.

There are four basic units in MathCAD. (It would be preferable if there were a fifth so that temperature was automatically included; if you think so too, write Mathsoft a letter.) When defining units, if the global equality, the tilde ˜, which appears as ($\equiv$), is used, then no matter where they are placed, MathCAD notices them first. The global equalities are also read left to right, top to bottom, so that units defined in terms of other units still need to be in the proper order. Examples of definitions of base units are

$$m \equiv 1L \qquad kg \equiv 1M \qquad s \equiv 1T \qquad and \qquad C \equiv 1Q.$$

Notice that there is no multiplication between the numerical value 1 and the fundamental unit name in the definition process.

• •   Reset MathCAD. Perform the operations in this section as they are discussed in the text. Start by defining the units as described.

Let m represent length in meters, kg mass in kilograms, s time in seconds, and C electric charge in coulombs. In general, define only those units that are needed for a particular application. For example, when treating problems in mechanics, charge is generally not needed.

The name associated with a unit can be changed. For example, one might prefer to change the unit for charge, the coulomb, to the unit of current, the ampere. Or it could be changed to temperature. The name is changed by pressing [Esc], typing dimension, and pressing return. Move to charge, for example, delete the name (backspace over it), and type temperature, current, force, or whatever is desired.

Note that it is not possible to use one variable name to represent two quantities. If m is used for length (meter), then it *cannot* be used later for some generic mass. Similarly, if N is used for the unit of force, the Newton, then it cannot be used later, for example, for normal force or for the upper limit of some index. Instead, use a more complex name, m1, or Nf, or use a literal subscript, m_a, of the $x.x$ variety. For literal subscripts, avoid using numbers for the subscripts; it is difficult to distinguish them from active subscripts.

Once the base units are defined, then any other units to be used are expressed in terms of units already defined. Two commonly used quantities are force (think of it as mass times acceleration; acceleration has units of velocity per time) and energy (think of it as force times distance).

The units of the Newton and the Joule, the SI units of force and energy, respectively, are

$$N := kg \cdot \frac{m}{s^2} \qquad J := N \cdot m.$$

Similarly, cm and gm would be defined cm := 0.01 · m and gm := 0.001 · kg, respectively. To write force equals ten Newtons, use the units as a multiplier. Then evaluate F:

$$F := 10 \cdot N \qquad F = 10 \cdot \text{mass} \cdot \text{length} \cdot \text{time}^{-2}.$$

The result is expressed in terms of the fundamental units, not in terms of the units that were defined. (The numerical value is, however, associated with the base units that were defined.) In general, this is not the desired form for the units. It is necessary to convert the fundamental units into those that have been defined. This process is easy.

Evaluate F (the various different forms which follow are all different expressions of the same F and would occupy only one line in MathCAD):

$$F = 10 \cdot \text{mass} \cdot \text{length} \cdot \text{time}^{-2}.$$

Move the cursor to the place marker, the ■ near the end of the region. The cursor must be in the region for the place marker to be visible. Type kg at the place marker and process. When a quantity is entered at the place marker, press [F9] to process; otherwise, it may appear as if no change occurs:

$$F = 10 \cdot \text{length} \cdot \text{time}^{-2} \cdot kg.$$

At the place marker, include the unit for length, meter; multiply kg · m:

$$F = 10 \cdot \text{time}^{-2} \cdot kg \cdot m.$$

At the place marker, include the units for time; multiply kg · m · s⁻²:

$$F = 10 \cdot kg \cdot m \cdot s^{-2}.$$

Finally, at the place marker, remove all the units and replace them with gm · cm · s⁻²:

$$F = 1 \cdot 10^6 \cdot gm \cdot cm \cdot s^{-2}.$$

The numerical value automatically adjusts to correspond to the chosen units. If we try once again, this time typing $m^2 \cdot s^2$ at the place marker (try it), the units are not simplified. But no harm is done. What is wrong can be seen and changed.

Often in problems or in performing measurements, all quantities are not specified in one system of units. For example, some values may be in meters, some in centimeters, and some in microns. However, as long as you follow the procedure of defining and assigning, MathCAD will take care of all the details. For example, you have

$$x1 := \text{m} \qquad x2 := \text{cm} \qquad x := x1 + x2 \qquad x = 1.01 \cdot \text{length},$$

or, by using the place marker, you have

$$x = 1.01 \cdot \text{m} \qquad \text{or} \qquad x = 101 \cdot \text{cm}.$$

If we try to add quantities of mixed units, MathCAD responds with an error message and will not process any improperly defined quantities, such as the quantity "mess" below:

$$y := \text{m} \qquad t := \text{s} \qquad \text{mess} := y + t.$$

The error message "incompatible units" is displayed. If mess is evaluated after this region (try it), the resulting error message is "undefined". If mess had been defined previously in a legitimate statement in the document, the first assignment would still apply.

If an incompatible-unit problem appears and it is not immediately obvious what the problem is, evaluate the quantities that are incompatible side by side. A comparison will show how the units differ.

No matter how complex or how simple the terms, units must be the same if the quantities are to be added or equated.

•• If units are mixed in a problem, as long as the proper assignments are made, MathCAD will make the necessary conversions. For example, the velocity of a falling object is to be determined using the equation $v = \sqrt{2\,g\,h}$ with g the acceleration due to gravity and h the height through which the object falls. Calculate the final velocity of a mass that is permitted to fall through a distance of 125 cm, near the surface of the earth. To keep the arithmetic trivial, let $g = 10 \cdot \text{m} \cdot \text{s}^{-2}$. Perform the calculation in your head or on a scrap of paper. Then perform the calculation using MathCAD and incuding units. Express the results in m/s and in cm/s, using the place marker.

•• Assume that the equation below relating force, F, electric charges, $q1$ and $q2$, and the distance, r, is valid. The equation is a statement of Coulomb's law and is similar to Newton's law of gravitation. The constant ϵ_o is analogous to the constant G in the law of gravitation:

$$F = \frac{1}{4\,\pi\epsilon_o} \cdot \frac{q1 \cdot q2}{r^2}.$$

Let $q1$ and $q2$ each have a magnitude of one microcoulomb. Let r be one meter. Let the force be 0.009 Newtons. Simplify the units for ϵ_o:

$$\epsilon_o = \frac{q1 \cdot q2}{4 \cdot \pi \cdot r^2 \cdot F}.$$

Use the Newton to simplify the units. Try the Joule. The volt is equal to 1 Joule per coulomb; incorporate it. The farad is expressed in coulombs per volt; incorporate it. (At this point the only units should be farad and meter.)

Don't be afraid to question and test. The computer makes that process easier. These examples help to reinforce the idea that we have another means of thinking about concepts and equations.

• • Verify that the equation $F = mv^2/r$ is dimensionally correct. F is force, m is mass, v is velocity and r is a radius.

• • If the following equation, which relates kinetic energy and absolute temperature, is correct, what are the units of k?

$$\frac{1}{2}mv^2 = \frac{3}{2}kT.$$

A given-find solve block is one way to determine the units.

• • In cgs units, the unit of force — the counterpart to the SI unit, the Newton — is the dyne, where dyne $:= \text{gm} \cdot \text{cm} \cdot \text{s}^{-2}$. The cgs energy unit — the counterpart to the SI unit, the Joule — is the erg, where erg $:= \text{gm} \cdot \text{cm}^2 \cdot \text{s}^{-2}$.

Define the Newton, Joule, dyne, and erg. Evaluate each and make use of the place markers. In the place marker of the Newton, put dyne; in the place marker of the dyne, put Newton. In the place marker of the Joule, put erg; in the place marker of the erg, put Joule.

It is very simple to go back and forth between any desired sets of units.

• • The units that we choose are to some degree arbitrary. The SI system uses meter, kilogram, and second; the cgs system uses centimeter, gram, and second; the English system uses foot, something, and second. The most familiar "something" unit is the pound, but in common usage the pound is a unit of force, not mass. Of course, there is the slug as a unit of mass and there is also a pound mass. Would it make any difference if the basic units were length, force, and time, instead of length, mass, and time? If the concern is simply a matter of analytic expression, then the answer is no. In terms of experimentally determining the values, however, it does make a difference.

Let us select for the mechanical units distance, time, and force. Change the unit name associated with M to force. Define X $:=$ 1L, T $:=$ 1T, F $:=$ 1M. Evaluate them so that the names are displayed.

Now define v as X/T, a as X/T^2, m as F/a, ρ as m/X^3, E as $F \cdot X$, and $mmtm$ (momentum) as $m \cdot v$. The expressions for energy and momentum are useful reminders that quantities should be considered in more than one light.

• • Express the same quantities in a system in which the base units are mass, momentum, and time.

3.2   Dimensional Analysis

We have already seen how useful units are, but now we want to use units in a different way. We want to use them to find how one variable depends on another for some particular set of circumstances. For example, in free fall, starting from rest near the earth's surface, velocity depends on the square root of the distance fallen. This is a very useful relationship. We would like to be able to determine relationships like this using only dimensional analysis.

The basic approach that we are going to use is this: we form a dimensionless group from the essential variables of a problem (dimensionless means that all units of length cancel, all units of time cancel, and so on). We set that dimensionless group equal to a constant of unknown magnitude. Despite the uncertainty with regard to the constant, this grouping tells us the dependence of one variable on another and permits us to see how changing one variable will affect others.

Before we consider specific examples, we provide a more detailed statement of the general procedure:

1. Determine the variables associated with the problem. This is likely to be the most difficult part; it cannot be done blindly. There are no simple rules that tell us when all the necessary variables have been selected. Some physical sense of the nature of the problem is required.

2. Count the number of variables in step 1. Call that number $N(V)$.

3. Count the number of fundamental units associated with the variables in steps 1 and 2. Call that number $N(U)$.

4. Subtract $N(U)$ from $N(V)$. This is the number of independent groups that can be formed. Typically in the problems we deal with, there will be only one group; but that is not always so.

5. Show the variables in each (for us, generally only one) group in a row, one multiplying the other. Set the exponent of the first variable to 1,

the exponent of the second to a, the exponent of the third to b, and so on.

6. Form a similar group, but replace all the variables with their fundmental units. Keep the same exponents.

7. Write $N(U)$ equations, one for each fundamental unit, in which we add all the exponents for the particular fundamental unit and set that sum equal to zero.

8. Solve the equations of the previous step and determine the values of the exponents a, b, ...

9. Form the group with the now known exponents and set it equal to a constant.

10. Examine the interdependence of the variables.

Consider again the free-fall problem, mentioned just before the procedure list. For this case, we will enumerate the steps of the procedure as we go through it.

1. First decide which variables are to be included. The problem explicity mentions velocity and distance fallen. Use h for height and v for velocity. Gravitational forces act on the mass, making it fall with acceleration g. Let's try out this set — h, v, and g — and see what dependence results.

2. The number of variables is three: h, v, and g.

3. The number of fundamental units is two: length and time.

4. There is one group: $3 - 2 = 1$.

5. Group variables in a row, multiplying each other, with exponents 1, a, b, ... Disable the equation:

$$h \cdot v^a \cdot g^b \; \square.$$

6. Form a similar group but replace the variables with the associated fundamental units. Keep exponents the same. Disable the equation. Velocity has units of L/T and acceleration has units of L/T^2:

$$L \cdot \left[\frac{L}{T}\right]^a \cdot \left[\frac{L}{T^2}\right]^b \; \square.$$

7. See step 8.

8. Here we combine two steps into one. Use a given-find structure to determine the values for the exponents.

Enter guess values for the variables: $a := 1$, $b := 1$.
Given
$1 + a + b \approx 0$. The L-equation; exponents sum to zero.
$-a - 2 \cdot b \approx 0$. The T-equation; exponents sum to zero.
$$\begin{bmatrix} a \\ b \end{bmatrix} := \text{Find}(a, b) \qquad a = -2 \qquad b = 1.$$

9. Form the group; show original form; replace a and b with values from step 8:
   $$h \cdot v^a \cdot g^b \,\square \qquad \text{or} \qquad h \cdot v^{-2} \cdot g := C1\,\square.$$

10. Recall that earlier it was suggested that velocity was proportional to the square root of the distance fallen. Could we extract that information from this group? (Constants $C1$, $C2$, and $C3$ are not equal, but they are all constant. At the moment, that is all that is important about them.) Rewrite the equation as follows:
    $$v^2 := C2 \cdot h \cdot g\,\square.$$
    Since g is a constant near the surface of the earth, we have
    $$v := C3 \cdot \sqrt{h}\,\square.$$
    The suggested relationship is revealed through dimensional analysis.

• •   Plot linear and log-log plots of v vs. h and, for comparison, h vs. h with h as the independent variable.

• •   If the distance h were fixed and measurements were made at different locations, how would the final velocity depend on g?

3.2.1 Kepler's Harmonic Law

As another example of dimensional analysis, consider the case of what we now refer to as Kepler's third law, or his harmonic law. It relates the period — the time for a satellite to complete one full orbit — with its mean distance from the attracting body about which it orbits. Kepler discovered this law starting with raw data. It required an enormous intellectual struggle on his part to find the law. By assuming that we know the units of G, the universal gravitational constant, we can find his third law — at least the proportionalities — just using dimensional analysis. And it will be vastly easier to do.

First we must decide which variables should be included in the group. Clearly, we want the period, T, and the mean distance, r. Gravity is the attracting force, so include G. The massive central body is essential to all the orbiting bodies, so include M. Group the variables, permitting the first to have an exponent of 1. Disable the group:

$$T \cdot r^A \cdot G^B \cdot M^C \; \square.$$

Then write the same group in terms of fundamental units and disable. This will be our guide for the individual equations for mass, length, and time that must be solved:

$$T \cdot L^A \cdot \left[\frac{L^3}{M \cdot T^2}\right]^B \cdot M^C \; \square.$$

Now determine the exponents, A, B, and C. We solve for them using a given-find procedure. Equations are written for each of the fundamental units, M, L, and T. The equations express the fact that the sum of the exponents for the associated variable must add to zero since the quantity we seek is dimensionless.

$$A := 1 \qquad B := 1 \qquad C := 1 \qquad \text{Guess values for the variables.}$$

$$\text{Given}$$
$$-B + C \approx 0,$$
$$A + 3 \cdot B \approx 0,$$
$$1 - 2 \cdot B \approx 0.$$

These are the equations for M, L, and T, respectively,

$$\begin{pmatrix} A \\ B \\ C \end{pmatrix} := \text{Find}(A, B, C) \qquad A = -1.5 \qquad B = 0.5 \qquad C = 0.5.$$

Since the values of A, B, and C are all multiples of $1/2$, double all exponents. This operation includes the term with exponent 1. We justify this operation by noting that a dimensionless quantity squared is still dimensionless. Write the group:

$$T^2 \cdot r^{-3} \cdot G \cdot M := \text{const} \; \square.$$

Solving for the T term, we get

$$T^2 := \frac{\text{const}}{G \cdot M} \cdot r^3 \; \square.$$

We do not know what the constant is: dimensional analysis will not provide us with that information. But it does provide us with proportionalities, which is very useful information. For example, for a satellite at a given r, if the central mass were to increase by a factor of 4, the period would decrease by a factor of 2.

We've obtained quite a bit of useful information from this exercise. We now know that the square of the period is proportional to the cube of the mean distance. One thing this knowledge will permit us to do is to scale our solar system or any similar system. A procedure would be to measure the periods of the various orbiting bodies and then to calculate all the distances relative to one of them. We can perform some calculations because the constant cancels out when we take the ratio of similar proportionalities.

For example, if we write separate equations, one for the case of the earth and one for Jupiter, we obtain

$$T_e^2 = C \cdot r_e^3 \qquad T_j^2 = C \cdot r_j^3.$$

Note that the constant is the same in each of the above equations; also note that G and M are now included in the constant. When we take the ratio of the two equations, the constant disappears:

$$\frac{T_e^2}{T_j^2} = \frac{r_e^3}{r_j^3}. \tag{3.1}$$

If we assume that we know the periods of the planets and the distance from the earth to the sun, we can calculate the distances to all the other planets. To do this, we simply use the previous equation relating the period and mean distance for the earth and one of the other planets. Knowing the periods $T1$ and $T2$ and a distance $r1$, we can calculate an unknown mean distance $r2$.

The period of Jupiter's orbit is 11.85 earth years. If we specify the distance from the earth to the sun as one astronomical unit, 1 AU, we calculate the distance from Jupiter to the sun in AU:

$$r_j = r_e \cdot \left(\frac{T_j}{T_e}\right)^{2/3}. \tag{3.2}$$

This yields the value $r_j = 5.198$ AU. Knowing the periods of all the planets, we can calculate the distances to all the planets and, in effect, construct a scale model of the solar system with true relative distances. Of course, if we knew the distance from the earth to the sun from some independent measurement, then we would be able to calculate the actual distances.

One additional useful way to appreciate the relationship of Kepler's third law is to plot a universal curve (see Fig. 3.1). In this case, since r and T are both raised to powers, a log-log curve is appropriate. If the plot is in terms of earth years and astronomical units, r_e and T_e are both 1. This expression then becomes

$$r(T) = T^{2/3}. \tag{3.3}$$

One year implies 1 AU, eight years implies 4 AU, and so on. For Jupiter, $r(11.85) = 5.198$.

```
Planetary orbits          Kepler's Harmonic Law

In compact form               A  B  C
                            T·r ·G ·M  ≈
                                                       A │ ⎡  L³  ⎤ B│ C
                                                    T·L │ ⎢ ──── ⎥  │·M  ≈
                                                        │ ⎣ M·T² ⎦  │

A := 1     B := 1          C := 1

given       -B + C ≈ 0     A + 3·B ≈ 0     1 - 2·B ≈ 0

⎡A⎤                        A = -1.5        B = 0.5           C = 0.5
⎢B⎥  := find(A,B,C)
⎣C⎦

Squaring all terms and solving for the T term        2     const  3
                                                    T  := ─────·r
                                                           G·M

or              2         T := 0.2,1 ..100
                ─
                3
     r(T) := T
```

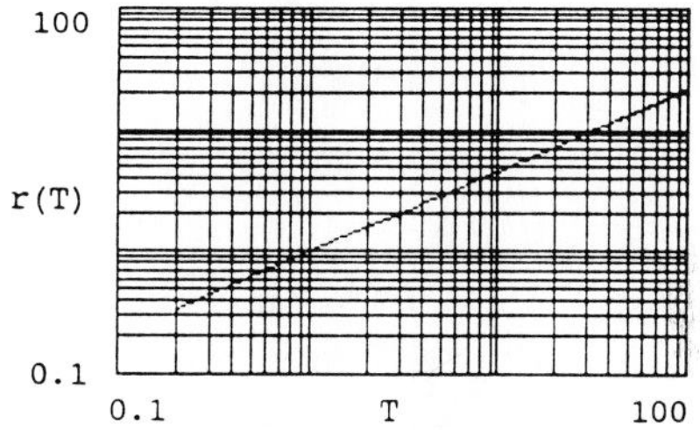

Figure 3.1 DA1, Kepler's third law. Period is in earth years and distance is in astronomical units.

• • Define the function $T(r)$ and plot $T(r)$ vs. r on a log-log plot.

• • Load document PLANET. The periods of the planets and the distances from the sun in AU are included. Using the periods, calculate the distances to the planets and compare with the table. Do you think the distances to the planets are measured or calculated?

3.2.2 Nuclear Explosion

A sequence of 16 photographs were made at intervals of 1/8 ms (millisecond) following the explosion of a nuclear weapon in the atmosphere. Measurements were made from the photographs to determine the size of the shock wave in each case; r_i give the radii in meters. The first photograph corresponds to a time of 1/8 ms after detonation.

• • Load DA2, nuclear explosion (see Fig. 3.2).

Determine whether the relationship between r and t is linear, exponential, or a power law. We plot data in four different ways and determine in which case the plot is closest to a straight line. (In general, curves need not be straight in any of these representations.)

```
Dimensional Analysis - Nuclear Explosion              r :=
                                                       i
A sequence of 16 photographs were made an intervals of 1/8 ms    ┌────┐
following the explosion of a nuclear device in the atmosphere.   │ 31 │
Measurements were made from the photographs to determine the     │ 35 │
size of the shock wave in each case.  r[i gives the radius in     │ 45 │
meters.  The first photograph corresponds to a time 1/8 ms        │ 50 │
after the detonation.                                             │ 52 │
                                                                  │ 55 │
                                                                  │ 61 │
n ≡ 16    i ≡ 1 ..n                          -3                   │ 67 │
                              t   := 0.125·10  ·i                 │ 70 │
                               i                                  │ 74 │
                                                                  │ 74 │
                                                                  │ 77 │
                                                                  │ 78 │
                                                                  │ 82 │
                                                                  │ 84 │
                                                                  │ 87 │
                                                                  └────┘
```

Figure 3.2 DA2, nuclear explosion.

Determine the slope on a log-log plot by selecting two representative points (for example, #3 and #15); refer, if necessary, to the section on linear least squares fit. Determine the inverse of the slope as well. Repeat with several other points; see how much variation appears in the slope.

Finally, determine the slope and its inverse based on the entire data set rather than on just one pair of points. Use the slope function. See the section on linear least squares fit.

•• Given the value just determined for the slope, what approximate proportionality is expected between r and t? Express the relationship in such a form that the exponents of both r and t are integers.

•• Find a dimensionless group that will relate the size of a shock wave from a nuclear detonation as a function of time. Compare this with the results from the previous example.

Four variables are needed. They are the density of air before being affected by the explosion, the radius of the shock wave, the corresponding time, and the energy from the explosion. The four variables and three fundamental units (mass, length, and time) imply one group.

Determine the exponents for the variables. Form the group. Solve for r in terms of the other variables. In this particular case the constant is approximately 1.

Compare the results from the dimensional analysis with those of the exploratory study of the explosion using the time and size measurements.

Using the experimental values for r and t, and taking the density of air to be $1.2 \, \mathrm{kg \, m^{-3}}$, calculate the energy released in the explosion.

Determine the energy E_i in terms of the r_i and t_i. Then determine the average energy value.

If the energy were available in a controlled manner, how many 100-watt bulbs could be powered for a year? A watt is equal to 1 Joule per second.

If we use the mass-energy relationship, $E = m c^2$, where c is the speed of light, what mass is converted to energy in the process?

Calculate the radius vs. time for the energy associated with the mass conversion of 5 gm of matter.

Double the energy and calculate new radii for the same time intervals. Compare these results with the previous calculations. Plot the two together. Would they be easy to distinguish?

If the lengths, r_i, from the DA2 file were in error by 10%, by how much would the apparent energy of the blast change?

• • Find the dimensionless group associated with the variables radius, velocity, and acceleration. (Recall this result when dealing with circular motion.)

• • Find the dimensionless group associated with the variables energy, mass, acceleration, and height. (Recall this result when dealing with gravitational potential energy.)

• • Investigate, using dimensional analysis, the variation of pressure with depth in water. How many additional variables are needed? What are they? Find the dimensionless group. (Recall this result in the chapter on statics in the section "tower of bricks.")

3.3 Estimation Problems

Estimation problems are just what they sound like, estimations; precision is not the objective. These problems are useful because they encourage the solver to apply basic concepts freely. Often a simple model is required, which you must envision.

You can approximate shapes with spheres or cylinders or other basic geometrical shapes. You should know a few values for density and, of course, the relationship between density, mass, and volume.

• • How many buckets of water does it take to fill a swimming pool?

● ●   How many drops of rain fall on an area the size of a football field when an inch of rain falls?

● ●   Estimate the surface area of your body. (You'll need this later when we discuss wind chill.)

● ●   What is the mass of all living human beings?

● ●   What is the volume of the earth's oceans?

● ●   How many atoms are there in the earth's atmosphere?

● ●   How many meters of electrical wire are in a typical household?

● ●   How many kilometers of sidewalk are there in Manhattan?

● ●   For a one-way commute of 10 miles, over a normal lifetime of work how far would a person drive? Express the result as a multiple of the circumference of the earth?

● ●   For the same commute, how many days of the worker's life are spent commuting?

● ●   For the same commute, how many gallons of gas are consumed?

● ●   For the same commute, how many pounds of tire rubber are worn away?

"Mistake? We don't make mistakes."

"Bloody typical. They've gone back to metric without telling us."

Brazil

CHAPTER

4

Vectors

Many physical quantities can be represented as vectors. Displacement, velocity, acceleration, force, momentum, and electric and magnetic fields are examples of vector quantities. Each has a direction and a magnitude. Temperature, on the other hand, has magnitude but no direction and is not a vector; temperature is a scalar. (Is energy a vector or a scalar?) Other quantites such as the moment of inertia, if treated in complete generality, cannot be described as vectors in the sense referred to above but require a more complex description in terms of tensors. All the physical quantities considered in this text will be treated as scalars or vectors.

4.1 Vector Sums and Components

Vector quantities can be represented with an arrow. The length of the arrow is proportional to the magnitude of the vector. Any vector can be represented as the sum of other vectors. Draw an arrow to represent a displacement vector — a displacement, for example, associated with walking across a field. Now think of this displacement as one side of a triangle, any triangle. From the original point, you could reach the final point by an unlimited number of pairs of vectors. The other two sides lead from the starting point to the final point; the vector sum of the two displacements is the same as the original displacement — same magnitude, same direction.

One especially important case is that of a right triangle, with the legs of the triangle parallel to the x- and y-axes of the coordinate system. The legs, in this case, are referred to as the x and y components of the vector. The hypotenuse is the resultant.

It is important to be able to describe a vector in terms of magnitude and direction — polar coordinates (r, θ) — or in component form — Cartesian coordinates (x, y). The relationships that express one form in terms of the other are

$$(r, \theta) \leftarrow (x, y) \qquad r = \sqrt{x^2 + y^2} \qquad \theta = \tan^{-1}\left(\frac{y}{x}\right). \qquad (4.1)$$

The arctangent operation returns angles within a restricted range. MathCAD's angle function, which returns a value between 0 and 2π, is often preferable:

$$\theta = \text{angle}(x, y). \qquad (4.2)$$

The reverse description giving the x and y components in terms of the magnitude and angle is

$$(x, y) \leftarrow (r, \theta) \qquad x = r\cos(\theta) \qquad y = r\sin(\theta). \qquad (4.3)$$

• • Load VEC1, the addition of two vectors, V1 and V2, in polar coordinates; paralellogram rule (see Fig. 4.1).

Just above the four plot regions are LV1, LV2 and AV1,AV2, the magnitudes and angles associated with vectors V1 and V2. Displayed in the four plot regions are the two vectors individually, the vector sum in two forms (order exchanged), and (in the last frame) the two sums plus the resultant, exemplifying the parallelogram law of vector addition. Enter values for (LV1, AV1, LV2, AV2). Angles are in degrees; change the numerical value; leave the ·deg. Try data sets (2,30,3,60), (2,210,3,60), (−2,30,3,60), (1,0,1,90), (1,0,1,180), (1,0,1,170). (Sums of vectors with equal magnitude are of interest, for example, in optics when we consider phasors.)

A convenient method for conversion between radians and degrees is to define them as shown near the beginning of the document. Define radians equal to 1 (the radian is not a true unit). Define degrees in terms of radians. Any angle returned by MathCAD will be in radians. Entering

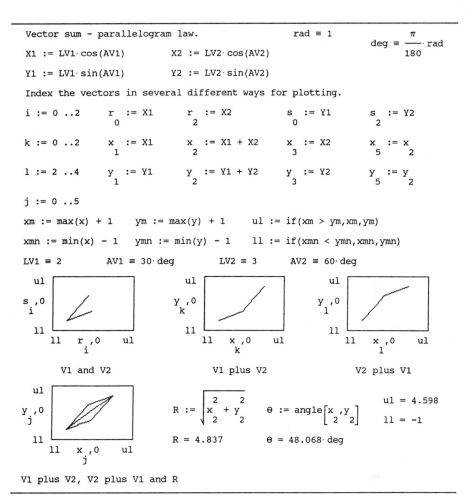

```
Vector sum - parallelogram law.                    rad ≡ 1
                                                              π
                                                   deg ≡ ───· rad
X1 := LV1· cos(AV1)        X2 := LV2· cos(AV2)                 180

Y1 := LV1· sin(AV1)        Y2 := LV2· sin(AV2)

Index the vectors in several different ways for plotting.

i := 0 ..2    r  := X1     r  := X2        s  := Y1      s  := Y2
               0            2               0             2

k := 0 ..2    x  := X1     x  := X1 + X2   x  := X2      x  := x
               1            2               3             5    2

l := 2 ..4    y  := Y1     y  := Y1 + Y2   y  := Y2      y  := y
               1            2               3             5    2

j := 0 ..5

xm := max(x) + 1    ym := max(y) + 1    ul := if(xm > ym,xm,ym)

xmn := min(x) - 1   ymn := min(y) - 1   ll := if(xmn < ymn,xmn,ymn)

LV1 ≡ 2        AV1 ≡ 30· deg       LV2 ≡ 3     AV2 ≡ 60· deg
```

V1 and V2

V1 plus V2

V2 plus V1

V1 plus V2, V2 plus V1 and R

$$R := \sqrt{x_2{}^2 + y_2{}^2}$$

$$\theta := \text{angle}[x_2, y_2]$$

$$R = 4.837 \qquad \theta = 48.068 \cdot \text{deg}$$

ul = 4.598

ll = -1

Figure 4.1 VEC1, the addition of two vectors, V1 and V2, in polar coordinates, paralellogram rule.

deg at the place marker will automatically convert the angle to degrees (press [F9]).

Four different indices, i, j, k, and l, are used here. This permits different plotting combinations. Note that each plot region uses a different index. Although MathCAD will automatically set plot limits, on some occasions it is preferable to specify them. Applied to vectors, max and min return the maximum and minimum values. The if statement lets us choose between cases.

• • Load VEC2, practice with components (see Fig. 4.2).

In the first graph, are shown the coordinate axes, a vector randomly choosen, and a marker which shows whether the angle given is measured clockwise or counterclockwise from the x-axis. The length of the vector is l.v; the angle shown is ang.

Enter values for the x and y components, using sine and cosine functions with angles less than 90 deg. Do not use lengths greater than lim(it). To see what you specified, process (F9), and look at the second graph. In the second graph will be the x and y components specified plus the original vector.

To repeat, press esc, type pro, press return, and press F9 with cursor in x.comp or y.comp region. You can then enter new values and press [F9] to check as often as needed. Remember trig function arguments are in radians unless otherwise specified.

$$l_v = 1.4 \qquad x_{comp} := 0 \qquad\qquad y_{comp} := 0$$

$$ang = 340 \cdot deg \qquad lim = 1.54 \qquad x_7 := x_{comp} \quad y_8 := y_{comp} \qquad x_8 := x_7$$

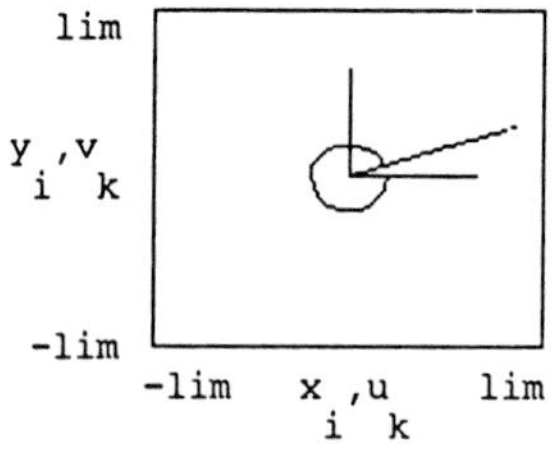

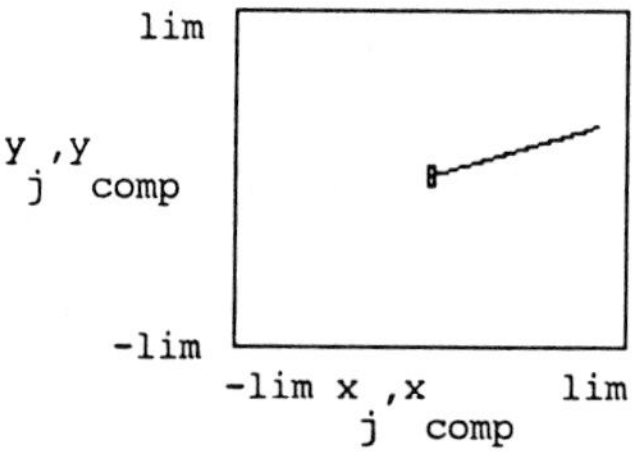

$$rad \equiv 1$$
$$deg \equiv \frac{\pi}{180} \cdot rad$$

The x,y from 0 to 4 draw the coordinate axes.
The 5th value is for the randomly directed vector.

$$i \equiv 0\,..5 \qquad x_i \equiv 0 \qquad y_i \equiv 0 \qquad \theta_{rng} \equiv 2 \cdot \pi \qquad \begin{bmatrix} \theta_v \\ z \\ l_v \end{bmatrix} \equiv \begin{bmatrix} floor(rnd(320)) \cdot deg \\ rnd(1) \\ .1 \cdot floor(rnd(7)) + 1 \end{bmatrix}$$

$$x_1 \equiv 1 \qquad y_3 \equiv 1 \qquad lim \equiv 1.1 \cdot l_v$$

$$\theta_v \equiv \theta_v + 20 \cdot deg$$

$$x_5 \equiv l_v \cdot \cos[\theta_v] \qquad y_5 \equiv l_v \cdot \sin[\theta_v] \qquad l_v = 1.4$$

$$ang \equiv if[z > .5, \theta_v, \theta_{rng} - \theta_v] \qquad \theta_v = 20 \cdot deg$$

$$x_6 \equiv 0 \qquad y_6 \equiv 0 \qquad y_7 \equiv 0 \qquad j \equiv 4\,..8 \qquad ang = 340 \cdot deg$$

$$N \equiv 23 \qquad k \equiv 0\,..N$$
$$\Phi_k \equiv if\left[z > .5, \theta_v \cdot \frac{k}{N}, -\left[\left[2 \cdot \pi - \theta_v\right] \cdot \frac{k}{N}\right]\right]$$

$$u_k \equiv .3 \cdot \cos[\Phi_k] \qquad v_k \equiv .3 \cdot \sin[\Phi_k]$$

Figure 4.2 VEC2, practice with components.

A vector is selected at random and plotted in the first plot region. The length of the vector is l_v; the angle, ang, may be measured either clockwise or counterclockwise from the x-axis. At x_{comp} and y_{comp}, enter values for the x and y components using the sine and cosine functions and angles less than 90°; for example, $2 \cdot \sin(15 \cdot \deg)$. In the second plot region, the original vector and the vector based on the values supplied for x_{comp} and y_{comp} are plotted. If correctly specified, the two vectors will appear as one. Try several examples. To make sure that the program repeats properly, press [Esc], type pro, press return, move the cursor to either the x or y component region, and press [F9]. Then you can enter new component values repeatedly while the original vector remains unchanged.

A number of statements are required to specify the length, angle, and direction of rotation of the vector, and to draw the axes, the vector, and the arc indicating the direction of rotation. To avoid clutter at the beginning of the document, the code is presented at the end. These statements all require the use of the global equality so that these statements will process before the segment that is shown at the beginning.

• •   Load VEC3, sum from two to six vectors (see Fig. 4.3).

Specify the number of vectors to be summed, N. (Move the cursor to the region; delete the existing value, and enter the new value.) Enter the magnitude and direction of the vectors r_i and θ_i (in degrees). When entering values for r and θ, it is necessary to insert a comma between successive numbers.

In the left-hand plot region shown in Fig. 4.4, the individual vectors are displayed. In the right-hand plot region, the vectors are summed in the order entered. The origin is marked with an open rectangle; the resultant ends in a diamond. The magnitude and direction of the resultant are shown. (We have limited ourselves here to six vectors simply for reasons of space in the document and to keep the graph from getting too muddled. There is no inherent reason for the number.)

Possible data sets include:

$$N = 4; r = 1, 1, 1, 1; \theta = 0, 90, 180, 270;$$

$$N = 5; r = 1, 1, 1, 1, 1; \theta = 0, 90, 180, 270, 360;$$

$$N = 3; r = 1, 1, 1; \theta = 0, 120, 240;$$

$$N = 4; r = 1, 1, 1, 1; \theta = 0, 120, 240, 360.$$

The number of vectors to be summed. $N := 4$ $i := 1 .. N$ $rad \equiv 1$

Enter the values for r and θ.

$$r_i := \qquad \theta_i := \qquad deg \equiv \frac{\pi}{180} \cdot rad$$

r_i	θ_i
2	40
3	10
2.5	-60
3.5	165

$\theta_i := \theta_i \cdot deg$

The vector components.

$$x_i := r_i \cdot \cos\left[\theta_i\right] \qquad y_i := r_i \cdot \sin\left[\theta_i\right]$$

Select a reference point. In this case the origin is selected.

$$X_0 := 0 \qquad Y_0 := 0 \qquad\qquad j := 1 .. 2 \cdot N$$

Vx and Vy represent a means to draw all the vectors starting from the reference point.

$$Vx_{2 \cdot i-1} := X_0 \qquad Vx_{2 \cdot i} := X_0 + x_i \qquad Vy_{2 \cdot i-1} := Y_0 \qquad Vy_{2 \cdot i} := Y_0 + y_i$$

X and Y represent a running sum of the vectors entered in r and θ above.

$$X_i := X_{i-1} + x_i \qquad Y_i := Y_{i-1} + y_i \qquad n := 0 .. N$$

To give the two graphs the same scale, calculate limits.

$xm := max(X) \qquad ym := max(Y) \qquad vxm := max(Vx) \qquad vym := max(Vy)$

$ul := if(xm > ym, xm, ym) \qquad\qquad ul' := if(vxm > vym, vxm, vym)$

$ul := if(ul > ul', ul, ul') + 1 \qquad ul = 6.737$

$xmn := min(X) \qquad ymn := min(Y) \qquad vxmn := min(Vx) \qquad vymn := min(Vy)$

$ll := if(xmn > ymn, ymn, xmn) \qquad\qquad ll' := if(vxmn > vymn, vymn, vxmn)$

$ll := if(ll < ll', ll, ll') - 1 \qquad ll = -4.381$

The resultant $k := 0 .. 1$ $RX_0 := X_0$ $RY_0 := Y_0$ $RX_1 := X_N$ $RY_1 := Y_N$

$$\delta RX := X_N - X_0 \qquad \delta RY := Y_N - Y_0$$

$$R := \sqrt{\delta RX^2 + \delta RY^2} \qquad \theta_R := angle(\delta RX, \delta RY)$$

Figure 4.3 VEC3, sum from two to six vectors. (See the next figure for the rest of the document.)

• •   Load VEC4, find the third vector so that the sum is zero (see Fig. 4.5).

Two vectors are selected at random. Specify the third vector, r_3, θ_3, which when added to the others yields a resultant of zero magnitude.

It is necessary to jump back and forth from the r_3, θ_3, regions to the end, where the plots are shown. ([Ctrl]home and [Ctrl]end are useful here.) The two random vectors are shown in the first plot; all three vectors are shown in the second plot. In the final plot, the three vectors are added and the resultant is shown (ending in a diamond). Observe the third vector,

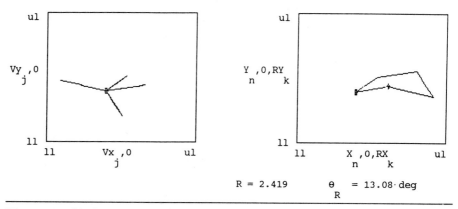

Figure 4.4 VEC3 *continued.*

which you control. Decide what length and angle are needed and jump to the beginning. Then change values and jump back to see how successful you were with your choices. Each time press [F9] to see the results.

To start over with a new set of random vectors, place the cursor in the r_1 region and press [F9].

4.2   Vector Products

Vectors can not only be added and subtracted, but also they can be multiplied. There are two types of products; one results in a scalar and one in a vector. They are known as scalar and vector products, or dot and cross products. The products can be specified in terms of polar or Cartesian coordinates.

The dot (scalar) product is defined as

$$\mathbf{a} \cdot \mathbf{b} = a \cdot b \cdot \cos(\theta_{ab})$$

or

$$\mathbf{a} \cdot \mathbf{b} = a_x \cdot b_x + a_y \cdot b_y + a_z \cdot b_z = \sum_i a_i \cdot b_i.$$

The cross (vector) product is defined as

$$\mathbf{a} \times \mathbf{b} = a \cdot b \cdot \sin(\theta_{ab})$$

or

$$\mathbf{a} \times \mathbf{b} = \begin{pmatrix} i & j & k \\ a_x & a_y & a_z \\ b_x & b_y & b_z \end{pmatrix}.$$

Two vectors are selected at random; magnitudes between 0 and 2, any angle. The user should enter a magnitude and direction (degrees as measured counterclockwise from the x-axis for vector C.

$rad \equiv 1$

$$deg \equiv \frac{\pi}{180} \cdot rad$$

$r_1 \equiv rnd(1.5) + 0.5 \qquad \theta_1 \equiv rnd(2 \cdot \pi) \qquad r_2 \equiv rnd(1.5) + 0.5 \qquad \theta_2 \equiv rnd(2 \cdot \pi)$

Enter magnitude and direction of a third vector r[3 that will result in the vector sum being zero.

$\theta_3 := 295 \cdot deg \qquad r_3 := 1.7$

$r_1 = 0.502 \qquad \theta_1 = 69.592 \cdot deg \qquad r_2 = 1.377 \qquad \theta_2 = 126.106 \cdot deg$

$X_0 := 0 \qquad Y_0 := 0 \qquad N := 3 \qquad i := 1 \ .. N \qquad j := 1 \ ..2 \cdot (N - 1)$

$x_i := r_i \cdot \cos\left[\theta_{\cdot i}\right] \qquad y_i := r_i \cdot \sin\left[\theta_i\right] \qquad Vx_{2 \cdot i-1} := X_0 \qquad Vy_{2 \cdot i-1} := Y_0$

$Vx_{2 \cdot i} := x_i + X_0 \qquad Vy_{2 \cdot i} := y_i + Y_0 \qquad k := 1 \ ..2 \cdot N$

$\begin{bmatrix} X_i \\ Y_i \end{bmatrix} := \begin{bmatrix} X_{i-1} + x_i \\ Y_{i-1} + y_i \end{bmatrix} \qquad Rx_0 := X_0 \qquad Rx_1 := X_N \qquad Ry_0 := Y_0 \qquad Ry_i := Y_N$

$m := 0 \ ..1 \qquad n := 0 \ ..N \qquad \delta Rx := Rx_1 - Rx_0$

$\delta Ry := Ry_1 - Ry_0 \qquad \qquad \theta_R := angle(\delta Rx, \delta Ry)$

$$R := \sqrt{\delta Rx^2 + \delta Ry^2}$$

$\qquad \qquad R = 0.092 \qquad \theta_R = 27.521 \cdot deg$

$mVx := max(Vx) \qquad mX := max(X) \qquad mVy := max(Vy) \qquad mY := max(Y)$

$ul1 := if(mVx > mVy, mVx, mVy) \qquad ul2 := if(mX > mY, mX, mY)$

$ul := if(ul1 > ul2, ul1, ul2) + 0.2 \qquad ul = 1.783$

$nVx := min(Vx) \qquad nX := min(X) \qquad nVy := min(Vy) \qquad nY := min(Y)$

$ll1 := if(nVx < nVy, nVx, nVy) \qquad ll2 := if(nX < nY, nX, nY)$

$ll := if(ll1 < ll2, ll1, ll2) - 0.2 \qquad ll = -1.741$

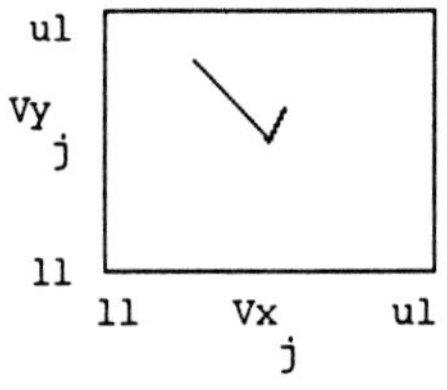

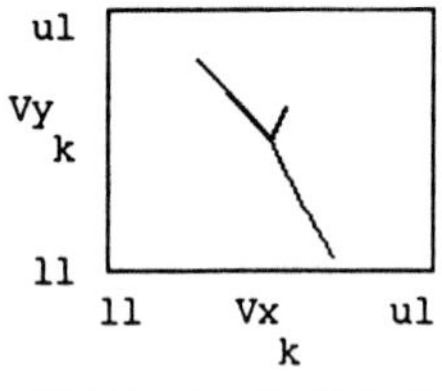

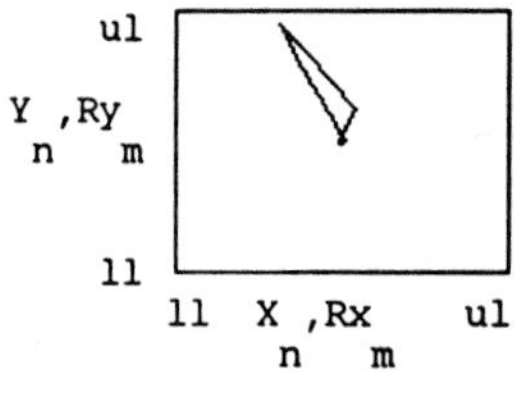

Figure 4.5 VEC4, find the third vector so that the sum is zero.

The angle θ_{ab} is the angle between the vectors **a** and **b**. $a \cdot \cos(\theta_{ab})$ is the component of vector a in the direction of b. Similarly, $b \cdot \cos(\theta_{ab})$ is the component of vector b in the direction of a. Thus the dot product $a \cdot b \cdot \cos(\theta_{ab})$ can be visualized either as the component of a in the direction of b times b or the component of b in the direction of a times a. The dot product is a maximum when the two vectors are aligned. The cross product, on the other hand, is zero when the vectors are aligned and is maximum when the vectors are at right angles.

The cross product as expressed in Cartesian coordinates includes the unit vectors, i, j, k, which have length 1 (unity; hence they are unit vectors) in the x, y, and z directions, respectively. The direction of the resultant is determined by constructing the resultant from the components. The expression for the cross product in polar coordinates does not include the direction. An additional piece of information, such as the right-hand rule, is needed to determine the direction.

It is important to note that the dot product is commutative

$$\mathbf{a} \cdot \mathbf{b} = \mathbf{b} \cdot \mathbf{a}$$

but the cross product is not. Changing the order changes the sign.

$$\mathbf{a} \times \mathbf{b} = -\mathbf{b} \times \mathbf{a}$$

Dot products occur in the formula for work, $F \cdot dl$; in Gauss's law, $E \cdot dA$ and $B \cdot dA$; in Faraday's law, $E \cdot dl$; and in Ampere's law, $B \cdot dl$. Cross products occur in the formulas for torque, $r \times F$; angular momentum, $r \times p$; force on a charged particle in a magnetic field, $q(v \times B)$; and in the Poynting vector (which describes energy flow in terms of electric and magnetic fields, $(1/\mu_o) \cdot (E \times B)$. These examples are noted to indicate that the dot and cross products are used frequently and are necessary tools in our description of the physical world.

• •   Load VEC5, vector products (see Figs. 4.6 and 4.7).

The dot and cross products are defined in two ways. In this example, the two vectors lie in the $x - y$ plane. Note how the vectors are specified in each case (polar and Cartesian). Observe that both forms of the dot and cross products are equal. Regarding the cross product, only the z-component is nonzero. Why is this?

Try several different values of a, b, θ_{ab}, and ϕ. Make sure you understand the role of each. If the r_i and s_i have values (1,1,1); (2,2,2), what is the value of the cross product? If the values are (1,0,0); (0,1,0), what is the cross product?

Now turn to the cross product in determinant form. We define the unit vectors, ii, jj, kk (we use double letters to avoid confusion with

```
Vector products                                    rad := 1
                                                                    π
                                                        deg := ——— · rad
Polar coordinate forms                                             180

dot product       dp(a,b,c) := |a|·|b|·cos(c)      a scalar

cross product     cp(a,b,c) := |a|·|b|·sin(c)      a vector

Rectangular coordinate forms

dot product       dp'(a,b) := a·b        Components of a and b must be
                                         specified.
cross product     cp'(a,b) := a × b      The cross is alt 8.

Example      a := 2        b := 3        θab := 30·deg   The angle between a & b.

Same vectors in component form.   Vectors in x-y plane.

i := 0 ..2   φ := 45·deg     φ is any arbitrary angle between vector a and
                             the x-axis.
     r :=             s :=
      i                i
   ┌─────────┐     ┌──────────────────┐
   │ a·cos(φ)│     │ b·cos(θab + φ)   │
   │ a·sin(φ)│     │ b·sin(θab + φ)   │
   │    0    │     │        0         │
   └─────────┘     └──────────────────┘

Plot the two vectors.      x := r     x := s      y := r      y := s
                            0    0     2    0      0    1      2    1

Plot limits.      ul := if(max(x) > max(y),max(x),max(y)) + 0.4

                  ll := if(min(x) < min(y),min(x),min(y)) - 0.4
```

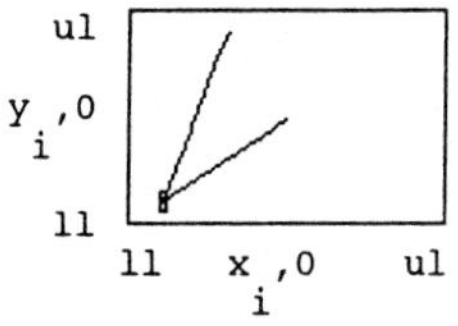

```
Evaluate the products       dp(a,b,θab) = 5.196        dp'(r,s) = 5.196

                            cp(a,b,θab) = 3                       ┌ 0 ┐
                                                      cp'(r,s) =  │ 0 │
                                                                  └ 3 ┘

It is not necessary to define functions.   r·s = 5.196   r × s = ┌ 0 ┐
                                                                 │ 0 │
                                                                 └ 3 ┘
```

To appreciate the operations implied by these products, note that for the dot product

```
r·s □    is equivalent to     r ·s  + r ·s  + r ·s  = 5.196
                               0  0    1  1    2  2
```

where 0, 1, 2 refer to the x, y, and z components of the vectors.

Figure 4.6 VEC5, vector products. (See the next figure for the rest of the document.)

The individual terms of the cross product can also be expressed.

The cross product requires unit vectors and scalars within the determinant which Mathcad does not allow.

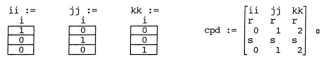

Write separate determinants for each component and multiply the magnitudes by the corresponding unit vector.

$$cpdx := \begin{bmatrix} 1 & 0 & 0 \\ r & r & r \\ 0 & 1 & 2 \\ s & s & s \\ 0 & 1 & 2 \end{bmatrix} \qquad cpdy := \begin{bmatrix} 0 & 1 & 0 \\ r & r & r \\ 0 & 1 & 2 \\ s & s & s \\ 0 & 1 & 2 \end{bmatrix} \qquad cpdz := \begin{bmatrix} 0 & 0 & 1 \\ r & r & r \\ 0 & 1 & 2 \\ s & s & s \\ 0 & 1 & 2 \end{bmatrix}$$

$|cpdx| = 0$ $\qquad\qquad |cpdy| = 0$ $\qquad\qquad |cpdz| = 3$

$cpd := |cpdx| \cdot ii - |cpdy| \cdot jj + |cpdz| \cdot kk$

$$cpd = \begin{bmatrix} 0 \\ 0 \\ 3 \end{bmatrix}$$

The place markers provide one means of showing the equivalence of the different definitions.

$r \cdot s = 5.196$ $\qquad\qquad r \cdot s = 6 \cdot \cos(30 \cdot deg)$ $\qquad r \cdot s = 1 \cdot a \cdot b \cdot \cos(\theta ab)$

The two forms are equivalent.

As there is only one component in the cross product, the place marker can easily be used in this case as well. The two product forms are equivalent.

$$r \times s = \begin{bmatrix} 0 \\ 0 \\ 3 \end{bmatrix} \qquad r \times s = \begin{bmatrix} 0 \\ 0 \\ 6 \end{bmatrix} \cdot \sin(30 \cdot deg) \qquad r \times s = \begin{bmatrix} 0 \\ 0 \\ 1 \end{bmatrix} \cdot a \cdot b \cdot \sin(\theta ab)$$

Figure 4.7 VEC5 *continued.*

subscripts). We write each component of the cross product as a separate determinant. Whatever the values of r and s that were specified, they are computed and the final vector is assembled by summing the components.

Finally, the place markers permit displaying the results in a variety of forms.

Form the cross products of $ii \times jj$, $jj \times kk$, $kk \times ii$, and $ii \times kk$. What are the corresponding dot products?

Evaluate an example of $a \times b$ and $b \times a$ to verify that the cross product is not commutative.

• • Load VEC6, rotating vectors (see Fig. 4.8).

Rotating vectors are often used to describe varying electric and magnetic fields. We observe in Fig. 4.8 a single rotation angle θ. Note that if

Vector rotation using complex notation.

The following equation includes 0, 1, π, e, i, plus and minus. That they all combine in such a simple way is marvellous.

$$e^{-i \cdot \pi} + 1 = 0$$

To appreciate this, we must explore the relation exp(i θ). It can be written

$$e^{i \cdot \theta} := \cos(\theta) + i \cdot \sin(\theta) \ \square$$

Observe that this is consistent by plotting the real and imaginary parts of the exponential.

θ := 0,.1 ..2·π

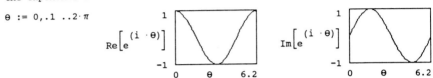

The trigonometric functions cosine and sine should remind you of x and y components. By varying θ, a vector with x-component cos(θ) and y- component sin(θ) is rotated.   In the complex plane, complex numbers are plotted with the real part along the x-axis and the imaginary part along the y-axis. The effect of multiplying a vector by exp(i θ) is to rotate it through the angle θ.

Let A be any complex number; A can be represented as a vector in the complex plane by plotting the real part of A along the x-axis and the imaginary part along the y-axis. Multiplying A by exp(i φ) rotates the vector A through the angle φ. NOTE: the complex part of A, or any complex number in Mathcad, must be of the form numberi, with no multiplier between. If the number is one, it must be written explicity, 1i; the number one will disappear when the cursor leaves the region.

rad := 1

$$\deg := \frac{\pi}{180} \cdot rad$$

n := 0 ..1 A_1 := 3 + i

φ := 30·deg

$$lm := \left| A_1 \right| + 0.5$$

$$B := A \cdot e^{i \cdot \phi}$$

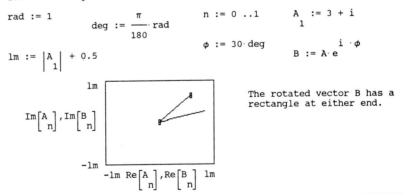

The rotated vector B has a rectangle at either end.

Figure 4.8 VEC6, rotating vectors.

θ were a function of time — ωt, for example — the rotation angle would be continuously increasing. We observe the projection and an example of the wave form generated as the projection varies in time.

As an alternate means of presentation, plots of the real and imaginary parts of $\exp(i\,\theta)$ show that they are indeed $\cos(\theta)$ and $\sin(\theta)$.

Below, we plot two vectors **A** and **B**. We specify the x and y components of **A** by writing them as real and imaginary parts of **A**. **B** is defined as **A** times a complex exponential term.

Let $\phi = 0$. Change the magnitudes and signs of the real and imaginary parts of **A**. Observe the plot region. Can you place **A** in any quadrant? Along any axis?

Let **A** $= 1$. Let ϕ equal $45°, 90°, 135°, 180°$. Observe vector **B** relative to **A**. Let **A** $= 1 + i$. Let ϕ equal $45°$, $90°$, and $135°$.

Multiplying by a complex exponential term is equivalent to a rotation.

> "Don't fight it son. Confess quickly. If you hold out too long you could jeopardize your credit rating."
>
> *Brazil*

CHAPTER

5

Kinematics

The kinematic equations describe how things move. They do not indicate the sources of those motions; they do not answer questions as to why things move as they do. The relationships that we discuss do not have complex forms; yet the equations do not, in general, correspond to the typical intuition. They require experience. A number of exercises are included in this chapter to help you gain that experience and build intuition.

5.1 One-Dimensional Motion

The fundamental kinematic definitions are

$$a = \frac{dv}{dt} \qquad \text{and} \qquad v = \frac{dx}{dt}, \tag{5.1}$$

where a is acceleration, v is velocity, x is displacement, and t is time.

For the case of constant acceleration, the variables are position, velocity, and time; the parameters are initial position, initial velocity, and acceleration. The basic kinematic equations each include two of the variables.

For one-dimensional motion, these equations are

$$x = x_o + v_o t + \frac{1}{2}at^2 \qquad (x, t) \qquad (5.2)$$

$$v = v_o + at \qquad (v, t) \qquad (5.3)$$

$$v^2 - v_o^2 = 2a(x - x_o) \qquad (v, x). \qquad (5.4)$$

Only two of these equations are independent. That is, any two contain all the information that is available.

The variables x and v represent the position and velocity at time t. The values for initial position and velocity are taken at $t = 0$. If the initial velocity and the acceleration were both zero, $x = x_o$ for all time. If the acceleration were zero and v_o were nonzero, the position would increase linearly with time. The units for x, v, a, and t are L, L/T, L/T^2, and T.

• •   Obtain the (x, t) equation by substituting for v from the (v, t) equation into the (v, x) equation.

• •   Load KIN1, an examination of the (x, t) equation (see Fig. 5.1).

In this document, we examine the (x, t) kinematic equation $x = x_o + v_o t + (1/2)at^2$ and consider the role of the individual terms. Each term in the equation is defined separately $(x1(t), x2(t), x3(t))$, so each term can be examined independently from the others. Two sums are defined which incorporate two of the three terms from the (x, t) equation.

Process the document. Examine the graphs; look at each curve and explain to yourself why the curve has the form that it does. Look at the specific values for the parameters and connect those values to the plots. The plot types remain consistent from one plot to the next.

Try various values for x_o, v_o, and a. Before processing, predict the shapes of the curves. Try cases until your predictions are consistently correct. Change both the magnitude and sign of the parameters. Initially, change one value at a time. Try these values (they are in the form (x_o, v_o, a)): (3,0,0), (9,0,0), (–9,0,0), (0,1,0), (0,5,0), (2,5,0), (2,–5,0), (–2,–5,0), (0,0,1), (0,0,5), (0,0,–5), (0,5,5), (0,5,–5), (0,–5,–5).

Let $x_o = 0$. In the second plot region, interchange the order of $x3$ and $x23$, so that the order is $x2$, $x23$, $x3$, 0. Change the plot type to e. The bars then show the difference between $x2$ and $x23$ and between $x3$ and 0. Why are the corresponding bar lengths equal? What does each represent?

Exercises with kinematic curves.

Enter values for the parameters x.o, v.o, and a. There are values already
assigned to these names. Run the document with these values, then try values
of your own choosing.

$$x_o := 10 \qquad v_o := 9 \qquad a := -8 \qquad t := 0,.3 ..3$$

The kinematic equation treated
here is

$$x := x_o + v_o \cdot t + \frac{1}{2} a \cdot t^2$$

The terms of this equation are written separately so that the effect of each
term can be observed independently from the others.

$$x1(t) := x_o \qquad x2(t) := v_o \cdot t \qquad x3(t) := \frac{1}{2} a \cdot t^2$$

Two sums are also specified.

$$x12(t) := x1(t) + x2(t)$$
$$x23(t) := x2(t) + x3(t)$$

The complete function is x(t).

$$x(t) := x1(t) + x2(t) + x3(t)$$

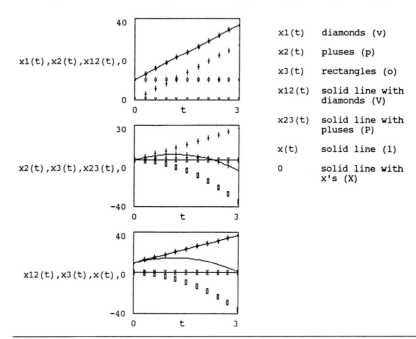

x1(t)   diamonds (v)

x2(t)   pluses (p)

x3(t)   rectangles (o)

x12(t)   solid line with
         diamonds (V)

x23(t)   solid line with
         pluses (P)

x(t)   solid line (l)

0   solid line with
         x's (X)

Figure 5.1 KIN1, an examination of the (x, t) equation.

•• Load KIN2, pick values and match the given curve (see Fig. 5.2).

Now see how well you can estimate the parameters associated with
a kinematic curve just by observing it. We consider the case of constant
acceleration.

Two kinematic curves are plotted.   You have control over the parameters in
(h); the function f is fixed.   Select values for x.0, v.0, and a so that
the two curves coincide.   Use integer values for each.

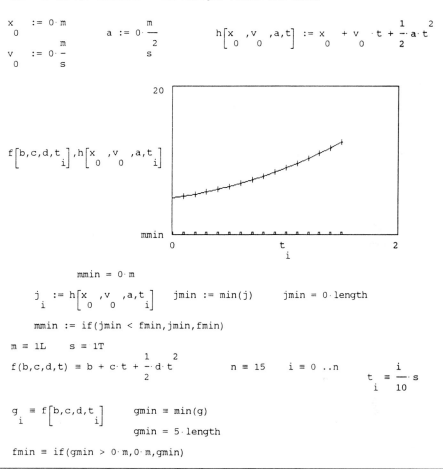

Figure 5.2 KIN2, pick values and match the given curve.

Two kinematic curves (x, t) are plotted. The functions f and h represent position as a function of time. For the function h, the stored values for x_o, v_o, and a are zero; the associated curve is the set of open rectangles. Examine the nonzero kinematic curve f; predict values for x_o, v_o, and a; enter the values and process. All values are integers. The challenge is to cause the two curves to overlap with the smallest number of tries. (1 to 2 tries are excellent, 3 very good, 4 good, 5 fair, 6 go back to KIN1 for more practice.) Repeat for KIN3 and KIN4.

KIN5 is a similar document except that the unknowns x_o, v_o, and a are determined by a random process. In KIN2, KIN3, and KIN4, the values are stored at the end of the document under the names b, c, and d rather than being determined randomly. Set values for your colleagues and let them try them. Have them set values for you. KIN5, of course, selects values at random, but because they are random, you have no control over the sequence in which the cases occur. To repeat KIN5 with new values, press [Esc], type pro, and press return. Place the cursor in the x_o region and press [F9]. Then enter integer values for x_o, v_o, and a; press [F9]. Change x_o, v_o, and a, each time pressing [F9], until a perfect overlap is achieved.

MathCAD's given-find procedure can be used in a general way to solve the kinematic equations. In these equations, there are three variables, position, velocity, and time (x, v, and t) and three parameters, initial position, initial velocity, and acceleration (x_o, v_o, and a). Only two of the three kinematic equations are independent; consequently, two unknowns can be determined.

> "OK. We're a hundred an' six miles from Chicago,
> we've got a full tank of gas, half a pack of cigarettes,
> it's dark, and we're wearing sunglasses."
>
> *The Blues Brothers*

• • Load KIN6, solve the (x, t) and (v, t) kinematic equations with a given-find procedure (see Fig. 5.3).

Once the kinematic equations are understood, it is helpful to be able to solve them in a general way, a way that is adaptable to a wide variety of cases. Using MathCAD's given-find procedure, we can set up the equations and solve for them in a variety of circumstances. Be aware, though, that equation solvers tend to be a bit fussy; unfortunately, you cannot write down any equations in just any form whatever and expect a solution.

A general procedure is to specify values for all six quantities (variables and parameters). Enter the kinematic equations in a solve block and solve for the unknowns.

You cannot simply enter six values at random. Four of the values must correspond to some possible physical situation. Two values are guess values; it is prudent to choose these values as close to the true values as possible.

```
General kinematic equation solving routine.

Given x, x.o, v.o and a, solve for v and t.

x := 0                 v := -10              t := 2

x    := 10             v    := 9             a := -8
 o                      o
```

$$x'(t) := x_o + v_o \cdot t + \frac{1}{2} a \cdot t^2 \qquad v'(t) := v_o + a \cdot t \qquad t' := 0, .1 .. 3$$

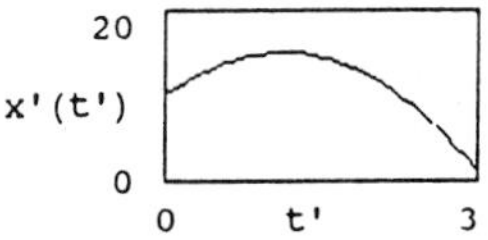

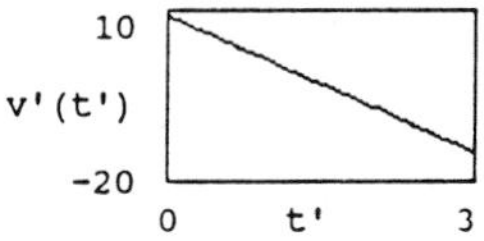

```
Be sure that the guess value for t is greater than the t where x is a max.
```

$$tt := 1 \qquad tmx := root\left[\frac{d}{dtt} x'(tt), tt\right] \qquad tmx = 1.125$$

```
t = 2

given
```

$$x \approx x_o + v_o \cdot t + \frac{1}{2} a \cdot t^2 \qquad v \approx v_o + a \cdot t \qquad t > 0$$

$$\begin{bmatrix} v \\ t \end{bmatrix} := find(v,t) \qquad\qquad\qquad x_o = 10$$

```
check                                                    v    = 9
                                                          o
```

$$x_o + v_o \cdot t + \frac{1}{2} a \cdot t^2 = 0 \qquad\qquad x = 0 \qquad\qquad a = -8$$

$$v_o + a \cdot t = -15.524 \qquad\qquad v = -15.524 \qquad t = 3.066$$

Figure 5.3 KIN6, solve the (x, t) and (v, t) kinematic equations with a given-find procedure.

Plotting x vs. t or v vs. t may help to indicate appropriate guess values. Keep the names of the functions plotted and the names of the variables distinct; note the use of primes.

If a curve has a maximum or a minimum, the guess value should be on that side of the maximum or minimum which will incline the solver toward the desired solution.

In this example, v and t are the unknowns. For the given parameter values and for $x = 0$ at some time t greater than zero, what are the values of velocity and time?

From the plot of x' vs. t', a value of t when $x = 0$ can be estimated. Using the root function with derivative to determine the time corresponding to the peak of the x vs. t curve, we have a reasonable sense of a guess value for t. A guess value of t less than tmx would not result in the desired solution. Try it.

It is always a good idea to check the results. Check by substituting the new values into the kinematic equations and comparing. Under certain circumstances the solver may return values as if they were solutions, yet the values are not the correct ones. Always think for yourself. Always ask if the results make sense.

Tabular Differences.   In experiments in which position is determined at regular intervals, using, for example, spark tape or photogates, velocity and acceleration information is readily extracted from the position data.

From a series of positions x_i and corresponding times t_i, average velocities over the time interval between measurements are given by

$$v_j = \frac{x_{j+1} - x_j}{t_{j+1} - t_j}. \tag{5.5}$$

Accelerations are similarly defined:

$$a_j = \frac{v_{j+1} - v_j}{t_{j+1} - t_j}. \tag{5.6}$$

If the time intervals between successive measurements are constant,

$$v_j = \frac{x_{j+1} - x_j}{\delta t} \tag{5.7}$$

and

$$a_k = \frac{v_{k+1} - v_k}{\delta t} = \frac{x_{k+2} - 2x_{k+1} + x_k}{\delta t^2}. \tag{5.8}$$

If the range of position measurements is for $i = 1, \ldots, n$, the range of j would be $1, \ldots, n-1$, and the range of k would be $1, \ldots, n-2$.

• •   Verify Eq. 5.8 for acceleration in terms of the various x-values.

• •   Load KIN7, tabular differences (see Fig. 5.4).

This example is performed using "perfect" data. Values for x_i are calculated from the (x, t) kinematic equation. Values for velocity and acceleration are determined according to the method of tabular differences.

Notice that the x-values are calculated using a familiar kinematic equation and the graphs look familiar. You should remind yourself that the velocity and acceleration values are determined in a completely different way than they were before.

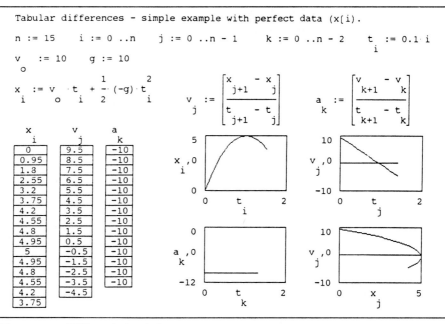

Figure 5.4 KIN7, tabular differences.

The x vs. t, v vs. t, and a vs. t graphs should all look familiar. Interpret the x vs. v curve.

Display the velocity and acceleration values in tabular form.

Calculate the acceleration values directly from the x information.

Calculate

$$J_l = \frac{a_{l+1} - a_l}{t_{l+1} - t_l} \tag{5.9}$$

where $l = 0, \ldots, n - 3$. If the initial data were not perfect, what would one expect here?

5.2   Simple Numerical Methods

If, instead of knowing the position information and extracting velocity and acceleration information, you knew the accelerations, then the velocity and position information could be constructed.

Consider first a familiar kinematic example:

$$a = \frac{dv}{dt}.$$

If, as in the previous section, we consider finite intervals, then the acceleration is written

$$a = \frac{\delta v}{\delta t} \quad \text{and} \quad \delta v = a\delta t.$$

Substituting $\delta v = v_{i+1} - v_i$ into the previous relation, we obtain

$$v_{i+1} = v_i + a\delta t. \tag{5.10}$$

This is a difference equation for velocity. Similarly,

$$v = \frac{dx}{dt} \quad \text{becomes} \quad v = \frac{\delta x}{\delta t},$$

which yields

$$x_{i+1} = x_i + v_?\delta t.$$

A problem now presents itself (the subscript ? is intentional) which did not occur when we were thinking about constant acceleration. (Remember that constant acceleration is a special case, one that is easily handled analytically. Many interesting applications do not have constant acceleration.) For constant acceleration the difference equation for velocity is correct. However, the difference equation for position, x, cannot have a constant velocity if acceleration is constant.

The question is, if the velocity is not constant then what velocity value should be used? A number of algorithms can be used here. We consider only some basic algorithms. Our goal is not great numerical precision; our goal is to learn how to approach problems numerically and gain insight.

Recall the integration that we performed in Section 2.3. We considered several sums. One took the value at the beginning of the interval, one took the value at the end, and one took the average. The algorithms that we consider make similar choices. The Euler method assumes that the rate of change (for example, of x) is constant over the interval (for example, δt) and evaluates the rate of change (for example, velocity) at the beginning of the interval. The Euler algorithm then indicates

$$x_{i+1} = x_i + v_i\delta t. \tag{5.11}$$

If the acceleration were not constant, the velocity equation would be similar:

$$v_{i+1} = v_i + a_i\delta t. \tag{5.12}$$

If the points are uniformly spaced in time, the values are given by

$$t_{i+1} = t_i + \delta t$$

or more simply

$$t_i = t_0 + i\delta t.$$

For motion that is periodic, such as an orbit, a modification of this algorithm by Cromer is useful. The equations for t and v are the same. The equation for x becomes

$$x_{i+1} = x_i + v_{i+1}\delta t. \tag{5.13}$$

The velocity is taken at the end of the interval rather than at the beginning. This variation by Cromer is sometimes referred to as the last point approximation (as opposed to first point, in the case of the Euler algorithm). We will refer to this variation as the Euler-Cromer algorithm.

When greater precision is required, we list without explanation a third algorithm, the velocity form of the Verlet algorithm:

$$v_{i+1} = v_i + \frac{1}{2}(a_{i+1} + a_i)\delta t, \tag{5.14}$$

$$x_{i+1} = x_i + v_i\delta t + \frac{1}{2}a_i\delta t^2. \tag{5.15}$$

The velocity equation uses for acceleration an average between the accelerations at the beginning and end of the interval. The position equation should have a familiar ring to it.

The size of the accumulated error is related to the size of the time interval. Smaller time intervals, in general, increase accuracy but take longer to compute. There is no general-purpose ideal choice of time interval.

• •   For the case of constant acceleration, does the difference equation for velocity, as expressed in the Euler algorithm, yield results identical to the true values? Calculate a sequence of half a dozen terms and compare the two sets of values. Explain the result.

• •   Load KIN8, Euler algorithm for the case of free fall (see Fig. 5.5).

This is a direct application of the Euler algorithm to the case of constant-acceleration free fall. Values for time, velocity, and position are calculated using the Euler algorithm (t, v, x) and using familiar analytic expressions (T, V, X). The results are presented in tabular form. How do the velocities compare? What did you determine in the previous exercise?

Euler algorithm - case of constant acceleration - free fall.

$n := 20$ $i := 0 .. n$ $t_0 := 0$ $v_0 := 0$ $x_0 := 0$

$g := 10$ $a := g$ $\delta t := .02$ $j := 0 .. 10$

The difference equations. In this case the equations are not coupled.
The v equation, for example, does not depend on x. They can be calculated
independently from each other as long as the v equation is before the x.

$t_{i+1} := t_i + \delta t$ $v_{i+1} := v_i + a \cdot \delta t$ $x_{i+1} := x_i + v_i \cdot \delta t$

Similar quantities determined by direct application of analytic expressions.

$T_i := i \cdot \delta t$ $V_i := v_0 + a \cdot T_i$ $X_i := x_0 + v_0 \cdot T_i + \frac{1}{2} \cdot a \cdot T_i^2$

t_j	v_j	x_j	V_j	X_j
0	0	0	0	0
0.02	0.2	0	0.2	0.002
0.04	0.4	0.004	0.4	0.008
0.06	0.6	0.012	0.6	0.018
0.08	0.8	0.024	0.8	0.032
0.1	1	0.04	1	0.05
0.12	1.2	0.06	1.2	0.072
0.14	1.4	0.084	1.4	0.098
0.16	1.6	0.112	1.6	0.128
0.18	1.8	0.144	1.8	0.162
0.2	2	0.18	2	0.2

Figure 5.5 KIN8, Euler algorithm for the case of free fall.

Plot the positions as determined according to the two methods. Plot
the difference between the two position calculations as a function of the
number of iterations. Plot the percent difference between the two methods
vs. i.

Determine the dependence of the difference between X_i and x_i as
a function of the magnitude of δt. For a specific value of t (not i) determine the dependence. As δt changes, it may be necessary to change
n in order to compare identical values of t. After each determination
(four or five points) type the values into two tables so the results can
be plotted when the sequence is completed. (For example, create two
tables $Xdiff_k$ and $tint_k$. After each calculation, enter new values in
these tables.) Plot $Xdiff$ vs. $tint$. If a functional form, $Xdiff \propto \delta t^n$,
is assumed, what value of n is expressed by these values? The exponent
characterizes a numerical method as nth order. What order is the Euler
method?

What would happen if the velocity were taken at the end of the interval instead of at the beginning? Try it.

For constant positive acceleration, does taking the velocity at the beginning of the interval overestimate or underestimate the value for position? By taking the value at the end of the interval, is the velocity overestimated or underestimated? In this particular case, how would you modify the difference equation for x to give exact results?

Although we approached the algorithms with a particular kinematic example, the algorithms are general. Any equation of the form

$$\frac{dy}{dx} = f(x) \tag{5.16}$$

can be written in difference form:

$$y_{i+1} = y_i + f(x_i)\delta x. \tag{5.17}$$

5.3   Two-Dimensional Motion

The equations for describing two-dimensional motion are essentially the same as those for one dimension, except that there is a set of equations for each dimension:

$$x = x_o + v_{ox}t + \frac{1}{2}a_x t^2, \qquad y = y_o + v_{oy}t + \frac{1}{2}a_y t^2, \tag{5.18}$$

$$v_x = v_{ox} + a_x t, \qquad v_y = v_{oy} + a_y t, \tag{5.19}$$

$$v_x^2 - v_{ox}^2 = 2a_x(x - x_o), \qquad v_y^2 - v_{oy}^2 = 2a_y(y - y_o). \tag{5.20}$$

Velocity and direction must refer to the direction being described.

The two sets of equations appear to be independent, sharing only the variable time. Whether they are independent or not depends on how a_x and a_y are specified. For motion near the surface of the earth and with no air resistance, $a_x = 0$ and $a_y = -g$ (if plus is up). Under these circumstances, the x and y motions are decoupled and each set can be treated independently of the other. However, if the acceleration is velocity dependent, as it is when air resistance is included, the motions are not completely independent. In this case, both a_x and a_y would depend on both v_x and v_y, adding considerable complexity to the problem.

Projectile Motion.   Special cases of projectile motion are worked out in most general physics texts. The range, R — the horizontal distance that the projectile travels when initial and final heights are the same — the maximum height that the projectile reaches, and the general trajectory equation are three particularly useful forms. The initial velocity is v_o; the initial angle with respect to the horizontal is θ_o:

$$R = \frac{v_o^2 \sin(2\theta_o)}{g} \qquad \text{(range)}, \qquad (5.21)$$

$$h = \frac{v_o^2 \sin^2(\theta_o)}{2g} \qquad \text{(maximum height)}, \qquad (5.22)$$

$$y - y_o = \tan(\theta_o)(x - x_o) - \frac{g}{2v_o^2 \cos^2(\theta_o)}(x - x_o)^2 \qquad \text{(trajectory)}. \quad (5.23)$$

The following three MathCAD documents aid in visualizing two-dimensional motion. Each presents the information in a different way.

• •   Load KIN9, visualizing projectile motion (1) (see Fig. 5.6).

```
Imagine a projectile fired horizontally in a uniform gravitational field.
The projectile will move in the plus x-direction; it will fall in the negative
y-direction.   The distance fallen in time t is  1/2 * a * t^2.

First, draw vectors which represent the distance fallen.  Space them
uniformly along the x-axis.   The code for y[(3*i+1) represents the distance
fallen.
```

$n := 4$ $\qquad i := 0 \; ..n$ $\qquad g := 9.8$ $\qquad a := -g$

$x_0 := 0$ $\qquad y_0 := 0$ $\qquad \delta t := .5$ $\qquad t_i := i \cdot \delta t$

$v_{ox} := 2$ $\qquad \delta x := v_{ox} \cdot \delta t$ $\qquad j := 0 \; .. (3 \cdot n + 2)$ $\qquad v_{oy} := 0$

$x_{3 \cdot i} := x_0 + i \cdot \delta x$ $\qquad x_{3 \cdot i+1} := x_{3 \cdot i}$ $\qquad x_{3 \cdot i+2} := x_{3 \cdot i}$

$y_{3 \cdot i} := 0$ $\qquad y_{3 \cdot i+1} := v_{oy} \cdot t_i + \frac{1}{2} \cdot a \cdot t_i^2$ $\qquad y_{3 \cdot i+2} := 0$

$xm := \max(x) \cdot 1.1$ $\qquad ym := \max(y) + (\max(y) - \min(y)) \cdot .1$

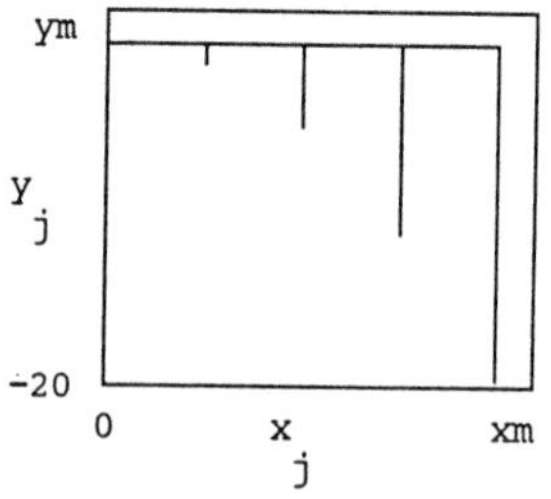

```
Note the rapidly increasing          If we connect the tips of these
distance which is proportional        y-displacements, the trajectory is
to time squared.                      roughly indicated.
```

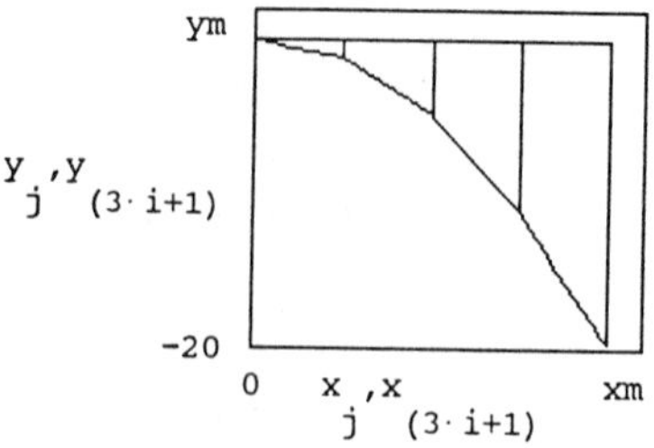

Figure 5.6 KIN9, visualizing projectile motion (1).

A particle is fired horizontally in a uniform gravitational field. There is no air resistance; $a_x = 0$. In equal increments of time, the projectile moves equal distances in the x direction. The net y displacement is proportional to t^2. We see the net movement in the x and y directions. We see a rudimentary trajectory. We note that $(1/2)a_y t^2$ is the distance the particle deviates from the path taken were $a_y = 0$.

Try different values for v_{oy} and v_{ox}. (Let v_{ox} be positive.)

Let $n = 8$ and $\delta t = 0.25$. The motion is shown at more points; the trajectory is smoothed. Let v_{oy} take on the values 8, 15, and –8. Observe and interpret the changes.

Compare plots of y vs. t and y vs. x. How are they similar? How are they different?

• •   How would the Euler method perform in evaluating the x motion? How would the Euler-Cromer method perform in the same case? Try them. Did you predict correctly?

• •   Load KIN10, visualizing projectile motion (2) (see Fig. 5.7).

The basic motion here is the same as in KIN9: projectile motion with no air resistance. In this case, the total displacement vector is displayed rather than the x and y components of the motion. An additional plot of y velocity vs. x velocity is included. The vectors are displayed at equal time intervals.

How are these graphs related to those in KIN9? Why are the shapes of the vx vs. vy graph and the x vs. y graph so different?

Examine the curves for $\theta_o = \pm 15°$.

• •   Load KIN11, visualizing projectile motion (3) (see Fig. 5.8).

The trajectory of a projectile is shown. At a series of points along the trajectory, the x and y components of the velocity vector are shown.

In this document and in the previous two, more complicated paths are drawn than are typical of the plots that have been discussed. A vector is created that contains not only the points on the trajectory (this example) but at each point takes a series of other steps. For example, in this case a step is taken in the x direction proportional in size to the x velocity, the same step is taken back to the trajectory point, a step is taken in the y direction proportional to the y velocity, the same step is taken back to the trajectory point, a step is taken to the next trajectory point, and the velocity vector process is repeated.

Velocity and displacement vectors associated with free fall.

$n := 4$ $i := 0 ..n$ $g := 9.8$ $\delta t := 0.1$ $a := -g$ $\delta vy := a \cdot \delta t$

$v_o := 5$ $\theta := 0 \cdot \dfrac{\pi}{180}$ $v_{ox} := v_o \cdot \cos(\theta)$ $v_{oy} := v_o \cdot \sin(\theta)$

$vx_{3 \cdot i} := 0$ $vx_{3 \cdot i+1} := v_{ox}$ $vx_{3 \cdot i+2} := 0$

$vy_{3 \cdot i} := 0$ $vy_{3 \cdot i+1} := v_{oy} + i \cdot \delta vy$ $vy_{3 \cdot i+2} := 0$

$j := 0 ..3 \cdot n + 2$ $vxm := v_{ox} \cdot 1.1$ $xtra := (\max(vy) - \min(vy)) \cdot .1$

$vymx := \max(vy) + xtra$ $vymn := \min(vy) - xtra$

$x_{3 \cdot i} := 0$ $x_{3 \cdot i+1} := v_{ox} \cdot i \cdot \delta t$ $x_{3 \cdot i+2} := 0$

$y_{3 \cdot i} := 0$ $y_{3 \cdot i+1} := v_{oy} \cdot i \cdot \delta t + \dfrac{1}{2} a \cdot (i \cdot \delta t)^2$ $y_{3 \cdot i+2} := 0$

The velocity vectors at equal time intervals.

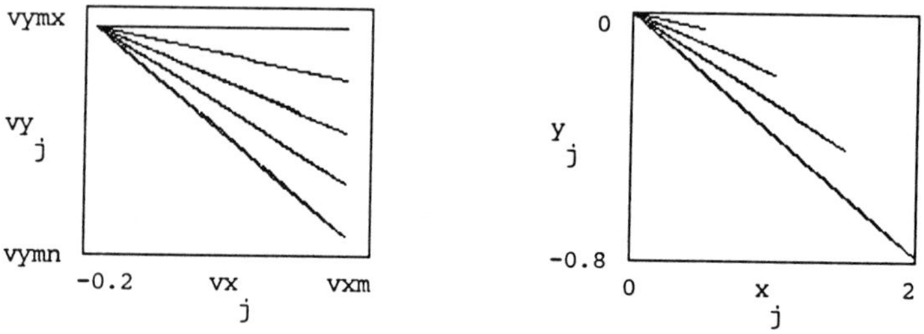

Figure 5.7 KIN10, visualizing projectile motion (2).

Try a number of different values for the initial velocity and angle as well as for a_x and a_y. For example, let $a_y = -g$ and $a_x = \pm g/2$.

• • Load KIN12, properties of a projectile (see Figs. 5.9 and 5.10).

Properties of projectile motion, range, maximum height, and time of flight can be determined in a variety of ways. It is always useful to be able to approach a problem in more than one way.

If the initial and final heights are the same, the range equation can be used. The equations for *xmax* and *ymax* are the range and height equations. Time is specified in terms of the range and the *x* velocity.

Whether or not the initial and final heights are equal, the kinematic equations may be solved directly to determine range, time of flight, and height.

The trajectory equation can be used in a variety of ways. Here we use it to determine the maximum height and in combination with the root-derivative process to find the x and y coordinates of the maximum of the trajectory.

If, instead of v_o, θ_o being known, the known values include v_o and $xmax$, find θ_o. Or given θ_o and $xmax$, find v_o. Will there be more

---

Examine the x and y components of the velocity along a two dimensional trajectory of a massive object near the surface of the earth. Try different input parameters. For best display, you may wish to change the limits on the graph.

$N := 10 \qquad i := 0 \; ..N$

Input parameters $\qquad v_o := 5 \qquad \theta := 60 \qquad \theta := \theta \cdot \dfrac{\pi}{180} \qquad dt := 0.1$

$\qquad g := 9.8 \qquad\qquad\qquad t_i := i \cdot dt$

position $\qquad\qquad x_o := 0 \qquad y_o := 0$

velocity $\qquad\qquad v_{ox} := v_o \cdot \cos(\theta) \qquad v_{oy} := v_o \cdot \sin(\theta)$

acceleration $\qquad\quad a_x := 0 \qquad\qquad a_y := -g$

$vx_i := v_{ox} + a_x \cdot t_i \qquad\qquad x_i := x_o + v_{ox} \cdot t_i + \dfrac{1}{2} a_x \cdot t_i^2$

$vy_i := v_{oy} + a_y \cdot t_i \qquad\qquad y_i := y_o + v_{oy} \cdot t_i + \dfrac{1}{2} a_y \cdot t_i^2$

$rf := 10 \qquad j := 1 \; ..5 \cdot N$

$vvx_{5 \cdot i} := x_i \qquad vvy_{5 \cdot i} := y_i \qquad vvx_{5 \cdot i+2} := vvx_{5 \cdot i} \qquad vvy_{5 \cdot i+1} := vvy_{5 \cdot i}$

$vvx_{5 \cdot i+3} := vvx_{5 \cdot i} \qquad vvy_{5 \cdot i+2} := vvy_{5 \cdot i} \qquad vvx_{5 \cdot i+4} := vvx_{5 \cdot i}$

$vvx_{5 \cdot i+1} := x_i + \dfrac{vx_i}{rf} \qquad vvy_{5 \cdot i+3} := y_i + \dfrac{vy_i}{rf} \qquad vvy_{5 \cdot i+4} := vvy_{5 \cdot i}$

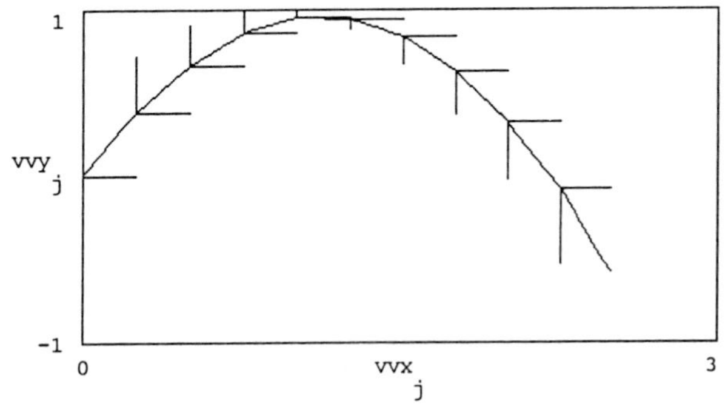

Figure 5.8 KIN11, visualizing projectile motion (3).

than one solution? Plot your results. Given $xmax$ and $ymax$, what are v_o and θ_o?

Projectile properties. The time, range, and maximum height are determined in more than one way.

parameters $v_o := 15$ $\theta_o := 30 \cdot deg$ $g := 9.81$ $rad \equiv 1$

$a := -g$

$deg \equiv \dfrac{\pi}{180} \cdot rad$

$v_{ox} := v_o \cdot \cos\left[\theta_o\right]$ $v_{oy} := v_o \cdot \sin\left[\theta_o\right]$

$v_{ox} = 12.99$ $v_{oy} = 7.5$

If the initial and final heights are the same, the range and height equations can be used. First, the range.

$$x_{max}\left[\theta_o\right] := \left[\dfrac{\left[v_o\right]^2 \cdot \sin\left[2 \cdot \theta_o\right]}{g}\right]$$ $x_{max}\left[\theta_o\right] = 19.863$

The time to cover the range.

$$t := \dfrac{x_{max}\left[\theta_o\right]}{v_{ox}}$$ $t = 1.529$

The maximum height.

$$y_{max} := \dfrac{v_o^2 \cdot \sin\left[\theta_o\right]^2}{2 \cdot g}$$ $y_{max} = 2.867$

The time and range using a given-find solve block. This method is more general and is not restricted to final and inital heights being equal.

$x := 2$ $y := 0$ $t := 2$

Given $x \approx v_{ox} \cdot t$ $y \approx v_{oy} \cdot t + \dfrac{1}{2} \cdot a \cdot t^2$ $y \approx 0$

$\begin{bmatrix} x \\ t \end{bmatrix} := Find(x,t)$ $x = 19.863$ $t = 1.529$

If the initial and final heights are the same, the maximum height can be determined from the trajectory equation, knowing that maximum height occurs at x/2.

$$y\left[x,\theta_o\right] := x \cdot \tan\left[\theta_o\right] - \dfrac{g}{2 \cdot v_o^2 \cdot \cos\left[\theta_o\right]^2} \cdot x^2$$ $y\left[\dfrac{x}{2},\theta_o\right] = 2.867$

Figure 5.9 KIN12, properties of a projectile. (See the next figure for the rest of the document.)

Or by using the (y,t) equation and evaluating at t/2.

$$y'(t) := v_{oy} \cdot t + \frac{1}{2} \cdot a \cdot t^2 \qquad\qquad y'\left[\frac{t}{2}\right] = 2.867$$

In general the maximum height can be determined by finding the x-value of the maximum using the root-derivative procedure and then find y-value.

$$x\_ymx := root\left[\frac{d}{dx} y\left[x, \theta_o\right], x\right] \qquad x\_ymx = 9.931 \qquad y\left[x\_ymx, \theta_o\right] = 2.867$$

If the final y-value is zero, the time could be determined using the root function.

$$t := 1.5 \qquad trng := root\left[v_{oy} \cdot t + \frac{1}{2} \cdot a \cdot t^2, t\right] \qquad trng = 1.529$$

A plot of the trajectory is always useful.

$$step := \frac{x_{max}\left[\theta_o\right]}{20} \qquad x := 0, step \, .. \, x_{max}\left[\theta_o\right]$$

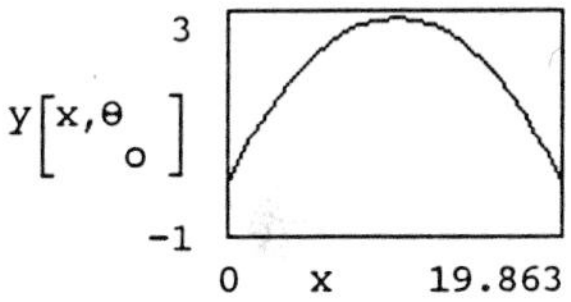

Figure 5.10 KIN12 *continued.*

Plot the speed ($\sqrt{v_x^2 + v_y^2}$) vs. x, vs. y, and vs. t. (Use three separate plot regions.)

• •   A basketball player shoots at a basket (height 10 feet) 20 feet away (horizontally); the ball leaves his hands at a height of seven feet above the floor. (Use $g = 32$.) Let $v_o = 30$. Find the angle θ_o which results in a basket without help from backboard or rim. Solve the problem in two ways using a given-find solve block. First use the two equations $x(x_o, v_o, \theta_o, t)$ and $v(v_o, \theta_o, t)$ and then use the trajectory equation. For a given v_o, is there more than one solution? Plot the trajectories associated with the solution(s). What is the minimum v_o able to reach the basket? What is the corresponding θ_o? Is there a maximum v_o?

• •   Load KIN13, trajectory envelope (see Fig. 5.11).

Two trajectories are plotted. The trajectories start from the same point and have the same initial velocity. Only the initial angle is different.

Let $\theta 2$ successively take on the values $70°, 60°,$ and $55°$. Note the x location of the intersection as $\theta 2$ approaches $\theta 1$. Also note that any point on the 50° trajectory is reached by two trajectories (the 50° trajectory and one other). In general, any point is reached by two trajectories. The two cases correspond to the two solutions of the quadratic equation for the trajectory.

Let $\theta 1 = 45°$, $\theta 2 = 50°$, and $\theta 3 = 70°$ and process. The outer edges of the trajectories begin to define a boundary that separates accessible and inaccessible regions. Recall, in the previous problem (about the basketball), the question about minimum velocity. If the origin in this problem is the point where the ball leaves the shooter's hands, and if the basket were beyond the boundary, then the choice of angle would be irrelevant — the basket is beyond reach.

Let $\theta 1 = 45°$, $\theta 2 = 60°$, and $\theta 3° = 135°$. Add to the plot, the envelope, a second $y(\theta', x, c', x'_o)$ having defined the following values $\theta' = \pi/4$ and $v'_o = v_o\sqrt{2}$ so that $c' = 2c$ and $x'_o = -c$. Process.

For a given v_o, plot range as a function of angle and maximum height as a function of angle in the same plot region. For a given v_o, what is the ratio of maximum possible range to maximum possible height?

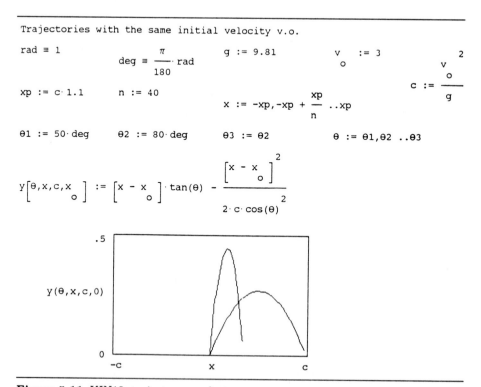

Figure 5.11 KIN13, trajectory envelope.

The boundary, an enveloping parabola, is not a trajectory. It is a line that divides the accessible from the inaccessible regions. However, being a parabola, it could be a trajectory, and in the above case it is described as one. What are the properties of the edge trajectory as compared with the trajectories issuing from the origin?

One remaining feature of trajectories is path length. This is not the range but the actual distance the projectile covers during its flight. If a length of arc along the trajectory is

$$ds = \sqrt{dx^2 + dy^2},\tag{5.24}$$

then

$$\text{pathlength} = \int ds = \int \sqrt{1 + \left(\frac{dy}{dx}\right)^2}\, dx.\tag{5.25}$$

MathCAD will perform the integration numerically, freeing us from any worry about how to integrate the specific function.

• •   Load KIN14, path length (see Fig. 5.12).

We define the range and the trajectory. The range will be used as the upper limit of integration; x goes from zero to its maximum range. We write the trajectory as a function. In the integrand, we let MathCAD perform the differentiation as well as carry out the integration. This is slow; reduce the number of angle values for a quicker response.

PathLen(θ_o) is a user-defined function. A series of values can be specified for θ_o. The integration will be performed in accordance with those values. We plot the path length as a function of angle (this plot is *not* a trajectory). The root function in combination with the derivative is used to find the maximum of the curve. The maximum path length does not occur at 45°.

Plot together the trajectories for maximum range and maximum path length.

Adapt the process to the case where the initial and final heights are not the same. In the basketball problem above, what is the path length?

Determine the time of flight as a function of angle and plot results; let initial and final heights be the same. One way to do this is to define a function in terms of a given-find solve block. For example,

$$\text{given} \qquad v_o \cdot \sin(\theta) \cdot t + 1/2 \cdot a \cdot t^2 \approx 0, \qquad T(\theta) := \text{find}(t).$$

Then, defining a sequence of angles θ_i, one can plot $T(\theta_i)$ vs. θ_i.

Determine the path length of a trajectory.

$$v_o := 10 \qquad \theta_o := 85 \cdot deg \qquad g := 9.81 \qquad a := -g \qquad rad \equiv 1$$

$$deg \equiv \frac{\pi}{180} rad$$

$$x_{max}\left[\theta_o\right] := \left[\frac{v_o^2 \cdot \sin\left[2 \cdot \theta_o\right]}{g}\right]$$

$$y\left[x, \theta_o\right] := x \cdot \tan\left[\theta_o\right] - \frac{g}{2 \cdot v_o^2 \cdot \cos\left[\theta_o\right]^2} \cdot x^2$$

The path length of the trajectory.

$$PathLen\left[\theta_o\right] := \int_0^{x_{max}\left[\theta_o\right]} \sqrt{1 + \left[\frac{d}{dx} y\left[x, \theta_o\right]\right]^2}\, dx \qquad \theta_o := .1, .3\ ..1.5$$

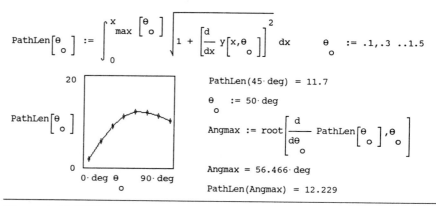

PathLen(45·deg) = 11.7

$$\theta_o := 50 \cdot deg$$

$$Angmax := root\left[\frac{d}{d\theta_o} PathLen\left[\theta_o\right], \theta_o\right]$$

Angmax = 56.466·deg

PathLen(Angmax) = 12.229

Figure 5.12 KIN14, path length.

5.3.1 Uniform Circular Motion

In the term "uniform circular motion", the word "uniform" means that the speed is constant. The acceleration associated with this motion results in the change of direction, not the change in magnitude, of the velocity vector.

• • Load KIN15, centripetal acceleration in uniform circular motion (see Fig. 5.13).

For an object to move in a circle, there must be a force toward the center. If there is force, there is acceleration, and if there is acceleration, there is change in velocity.

Acceleration is given by

$$a = \frac{\delta v}{\delta t} = \frac{v_2 - v_1}{\delta t}.$$

The diagram will help us to visualize the change in velocity, $v_2 - v_1$.

Uniform circular motion - centripetal acceleration.

We show the displacement vectors, the velocity vectors, and the velocity difference. Watch the change in direction of δv as $\delta\theta$ decreases.

$r := 1 \quad \theta := 0,.1 \ ..2\cdot\pi \qquad v := 0.5$

$$\Phi := \delta\theta + \frac{\pi}{2}$$

$x_0 := 0 \qquad x_1 := r \qquad x_2 := r$

$y_0 := 0 \qquad y_1 := 0 \qquad y_2 := v \qquad X(\theta) := r\cdot\cos(\theta) \qquad Y(\theta) := r\cdot\sin(\theta)$

$xx_0 := 0 \qquad xx_1 := r\cdot\cos(\delta\theta) \qquad xx_2 := xx_1 + v\cdot\cos(\Phi) \qquad xx_3 := xx_2$

$yy_0 := 0 \qquad yy_1 := r\cdot\sin(\delta\theta) \qquad yy_2 := yy_1 + v\cdot\sin(\Phi) \qquad yy_3 := yy_2 - v$

$xx_4 := xx_1 \qquad yy_4 := yy_1 \qquad i := 0\ ..2 \qquad j := 0\ ..4 \qquad \text{lm} := 1.15$

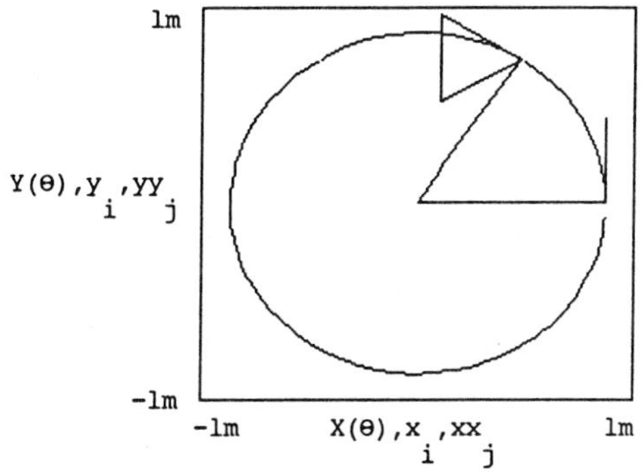

$\delta\theta \equiv 1$

Figure 5.13 KIN15, centripetal acceleration in uniform circular motion.

We want to observe the change in velocity as the object moves about the circle. To do this, we draw two displacement vectors, two radii separated by the angle $\delta\theta$. At the tips of the displacement vectors, we draw the corresponding velocity vectors. The lengths of these vectors are equal because the speed is constant. The angles between the velocity vectors are the same as those between the displacement vectors, the radii. To find the difference in velocity, we subtract the first velocity vector from the second.

We visualize the change by drawing the first velocity vector at the tip of the second. The change in velocity, δv, is represented by the line from the tangent point of the second velocity vector to the lower tip of the first vector (in its new position).

In Fig. 5.13, identify the two displacement vectors, the two velocity vectors, and the velocity difference vector.

Let the angle between successive observations of the particle in motion, $\delta\theta$, equal 1, 0.8, 0.6, 0.4, 0.2 radians successively. Observe that δv approaches the radial direction as $\delta\theta$ becomes smaller. Also note that the velocity and displacement "triangles" are similar and that at small angles the area bounded by the two radii and the short section of arc length does approximate a triangle quite well.

The velocity triangle, and the "triangle" made up of the two radii and adjoining arc δs are similar:

$$\frac{\delta v}{v} = \frac{\delta s}{r} \qquad \text{or} \qquad \delta v = \frac{v}{r}\,\delta s.$$

This yields

$$a = \frac{\delta v}{\delta t} = \frac{v}{r}\frac{\delta s}{\delta t} = \frac{v^2}{r}. \tag{5.26}$$

There are many applications involving circular motion. Keep this acceleration in mind.

> he's got all that speed and that power underneath him,
> he's comin' into the stretch and the pressure's on him
> — and he knows. Just feels, when to let go, and how much.
> So he's got everything working for him — timing, touch.
> It's a great feeling boy, it's a really great feeling
> when you're right, and you know you're right.
>
> *The Hustler*

C H A P T E R
6

Mechanics

There are five main areas in physics: mechanics, electricity and magnetism, thermodynamics and statistical mechanics, relativity, and quantum mechanics. The first three are referred to as classical theories, the last two modern. Each of these theories is rich in explanatory and predictive power. They are the cornerstones of modern science.

6.1 Newton's Laws

Mechanics can be summarized in Newton's three laws. His first law, the law of inertia, was known and expressed by Descartes (for whom Cartesian coordinates are named). Newton expressed his first law in words almost identical to those used by Descartes but did not credit their source. The first law states that a body at rest remains at rest and a body in motion continues to move with constant velocity unless acted on by an external force. The reference frame in which we observe this motion must be an inertial reference frame. An accelerating frame is not an inertial reference frame. (An object at rest in an inertial frame would not appear to be at rest or moving with constant velocity in an accelerating frame.)

Newton's second law, relating force and momentum, can be expressed as

$$\mathbf{F} = \frac{d\mathbf{p}}{dt}. \tag{6.1}$$

This law is elegant, simple, and powerful; it is applicable over a vast range.

The third law is a statement of action-reaction force pairs. It states that whenever a body, A, exerts a force on another body, B, then B exerts a force on A of equal magnitude and opposite direction.

Newton's law of universal gravitation,

$$\mathbf{F} = -\frac{Gm_1 m_2}{r_{12}^2}\hat{r}, \tag{6.2}$$

describes the force between two massive objects, m_1 and m_2, separated by a distance r_{12}. In this equation, we have a statement describing an interaction between all massive objects in the universe. We will apply Newton's laws in a variety of circumstances.

Newton's second law is a vector equation. Force and momentum are vectors. In addition, force, as used here, means the net applied external force, the sum of the external forces. So for each component, the law can be expressed as

$$\sum F_{\text{ext}} = \frac{dp}{dt}. \tag{6.3}$$

If the mass of the object remains constant, we have

$$\frac{dp}{dt} = \frac{d}{dt}(mv) = m\frac{dv}{dt} = ma$$

and

$$\sum F_{\text{ext}} = ma. \tag{6.4}$$

Remember that the latter expression is a special case and not a general statement of Newton's second law.

6.2   Constant Force Applications

A general procedure for applying Newton's second law is (1) isolate the body to which the law is to be applied, (2) indicate all applicable external forces, (3) apply the law for each component, and (4) solve for the unknowns.

MathCAD's given-find solve block is frequently useful for such problems. Once the equations are available, as in step 3, where the equations for each component have been expressed, MathCAD can be used to find the solution. Frequently, we can look at a set of solutions rather than a single solution. This feature permits us to obtain a more complete picture of the behavior of a system.

In find statements for more than one unknown, all quantities must have the same units. For example, in the case where we might wish to solve for both tension and acceleration, a solution is not possible if units are included. Either remove the units (the better idea) or temporarily adjust the units of one to match those of the other. An example is included in the following discussion.

6.2.1   Atwood Machine

The Atwood machine is frequently used to demonstrate how to apply Newton's second law. Here we consider the machine to consist of two masses suspended by massless string from a massless, frictionless pulley. Each body experiences two external forces, the gravitational force (weight) and the tension in the string supporting it. We first consider the case where the two masses m_1 and m_2 are known; we solve for the acceleration experienced by the masses and for the tension in the string.

• •   Load MECH1, Atwood machine (I) (see Fig. 6.1).

We define the fundamental and derived units, provide values for the parameters, and enter guess values for the unknowns.

As noted earlier, there is a complication involving units and the solve block. Every term in the equations must have the same units. Each quantity in the find statement must have the same units. If they do not, the simplest thing is to delete all references to units. They are included here to show the kinds of adjustments necessary if units are to be maintained.

In this case, we wish to solve for acceleration, a, and tension, T, which do not have the same units. An adjustment must be made if we are to use MathCAD's solve block. The quantity a does not appear in the solve block. Instead, we define a quantity b which has the magnitude of a and the units of T. In the equations, we express a in terms of b. In the find statement, we solve for b (and T). After solving for b, we express a in terms of b, removing the unit that had been inserted.

In this document, there are two solve blocks. In the first, we solve for acceleration and tension, assuming the values for the masses are known. (The general structure can, of course, be applied to many similar problems.) In the second solve block, we solve for the masses, accepting as true the original values for acceleration and tension. The values must,

Atwood machine 1.

m ≡ 1L   s ≡ 1T   kg ≡ 1M

$$N \equiv kg \cdot m \cdot s^{-2} \qquad g := 9.8 \cdot m \cdot s^{-2}$$

m1 := 1·kg   m2 := 0.98·kg   Guess values

$$a := 1 \cdot m \cdot s^{-2} \qquad T := 1 \cdot N$$
$$b := a \cdot kg$$

Given

$$m1 \cdot g - T \approx m1 \cdot \frac{b}{kg}$$

$$T - m2 \cdot g \approx m2 \cdot \frac{b}{kg} \qquad \begin{bmatrix} b \\ T \end{bmatrix} := Find(b,T) \qquad a := b \cdot kg^{-1}$$

check

$$a = 0.099 \cdot m \cdot s^{-2}$$

m1·g - T = 0.099·N   m1·a = 0.099·N   T = 9.701·N

T - m2·g = 0.097·N   m2·a = 0.097·N

$$a := 1 \cdot m \cdot s^{-2} \qquad T := 1 \cdot N$$

Given   m1·g - T ≈ m1·a

$$T - m2 \cdot g \approx m2 \cdot a \qquad \begin{bmatrix} m1 \\ m2 \end{bmatrix} := Find(m1,m2) \qquad \begin{array}{l} m1 = 0.114 \cdot mass \\ m2 = 0.093 \cdot mass \end{array}$$

Figure 6.1 MECH1, Atwood machine (I).

of course, correspond to a possible physical situation if there is to be a meaningful solution.

Examine the acceleration and tension as a function of the mass m_2. Change the find statement in the first solve block to a functional form. Let

$$f(m_2) := \text{find}(b, T).$$

Then add the statements

$$i := 0 \ldots 20 \qquad \text{and} \qquad m_{2i} := 0.5 \cdot kg + 0.05 \cdot i \cdot kg.$$

Plot $f(m_{2i})_0$ vs. m_{2i} and $f(m_{2i})_1$ vs. m_{2i}. The subscripts 0 and 1 refer to the zeroth and first quantities of the find statement, b and T.

As m_2 goes from $m_1 - 0.5$ kg to $m_1 + 0.5$ kg, why is the acceleration curve not symmetric about zero? Why does the tension curve continue to increase smoothly even as the acceleration changes sign? What is the tension when the acceleration is zero?

Change the find statement to $f(m_1, m_2) := \text{find}(b, T)$. Add the statements $M := 2 \cdot kg$ and $m_{1i} := M - m_{2i}$. Plot $f(m_{1i}, m_{2i})_0$ vs. m_{2i}. Similarly, plot the tension curve. Now is the acceleration curve symmetric? Why now? Explain the form of the tension curve.

If it were desired to change the masses in such a way that the acceleration is to be increased while T remains constant, how would the ratio of the masses change? Use the second solve block to examine this question. What is the upper limit on a? Find the ratio m_2/m_1 for a series of accelerations. For a given acceleration, find the ratio m_2/m_1 for a series of tensions. We could, for example, write $f(T) := \text{find}(m_1, m_2)$ and add the statements

$$i := 0 \ldots 5 \qquad T_i := 0.5 \cdot N + 0.2 \cdot i \cdot N \qquad \text{and} \qquad r_i := \frac{f(T_i)_0}{f(T_i)_1}.$$

Plot $f(T_i)_0, f(T_i)_1$ vs. T_i and r_i vs. T_i. Repeat the calculation with $a := 3 \cdot m \cdot s^{-2}$. Be sure all regions process.

• •   A mass m_1 is suspended from a pulley over the edge of a vertical cliff. The supporting string is attached to a unit that moves horizontally with constant speed, v_0. The pulley is a distance y above the plane in which the unit moves. The horizontal distance of the unit from the pulley is x (see Fig. 6.2).

Show that the tension in the string is given by

$$T(v, y, x) := m_1 \cdot g + \frac{m_1 \cdot y^2 \cdot v^2}{(x^2 + y^2)^{3/2}}.$$

For $v = 0$, what is the value of T? Explain.

For $y = 0$ and $v \neq 0$, what is the value of T? Explain.

For $y > 0$ and $x >> y$, explain the limiting value of T.

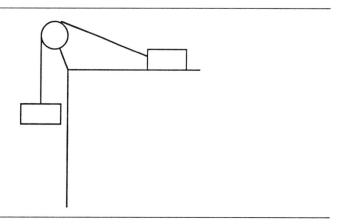

Figure 6.2 Mass, m_1, suspended; unit on horizontal surface moves at constant velocity.

Let $m_1 := 1$ and $y := 1$. Define T as above. Let $x := 0, 0.25\ldots 5.25$. Let $v := 1, 3\ldots 9$. Plot T vs. x. In the plot region, let the limits of x be 0 and 5 (why not 5.25?); let the lower limit of T be zero.

Which curve corresponds to which velocity? Explain the shape qualitatively.

Let $v := 1$ and $y := 1.5$. Plot T vs. x. In the plot region use the same limits for x, 0 and 5. Let the limits for T be 9.75 and 10.75. Change the dimensions of the plot region to (20,16). (A large vertical scale is needed to distinguish the curves.) Identify specific curves with y values. Explain the pattern qualitatively.

6.2.2 Friction

The above procedure can be adapted to a wide range of problems with interconnected masses, accelerations, and tensions. The range of problems amenable to solution can be extended even further by including friction.

The frictional force is expressed in an empirical relation,

$$F_f \leq \mu N, \tag{6.5}$$

where μ is the coefficient of friction and N is the normal force. The $\leq$ sign reminds us that the frictional force need not be constant, even though μ and N are. This resistive force depends on applied forces and cannot have a magnitude greater than the net applied force.

One procedural approach to solving problems involving frictional forces is first to solve the problem without the frictional force term and determine the sign of the acceleration. Then solve the problem again, including the friction term with its sign so as to resist the motion implied by the first solution. If both solutions yield accelerations in the same direction, then the solution including the frictional force term will yield the correct acceleration and tension. If the signs of the acceleration from the two solutions are opposite, then the acceleration is zero.

• •   Load MECH2, motion of a mass on a plane with friction (I) (see Fig. 6.3).

Mass m_1 rests on a plane tilted at $\theta = 30°$ with respect to the horizontal. A string runs from m_1 over a pulley at the top of the plane to m_2, which is freely suspended (see Fig. 6.3). Assume that the string and pulley are massless and the pulley is frictionless. The coefficient of friction between m_1 and the plane is μ.

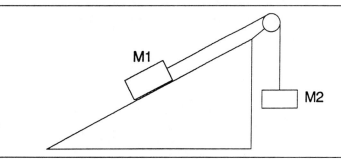

Figure 6.3 One mass on a plane, one suspended.

In the document (see Fig. 6.4), there are two given-find solve blocks. In the first, the friction term is omitted from the calculation; in the second,

```
One mass on slanted plane; one mass suspended; friction.      rad ≡ 1

Parameters                                      g := 9.8          π
                         0                  0               deg ≡ ─── · rad
θ := 30· deg   m1 := 5· 10    m2 := 1· 10    μ := 0.2          180

Guess values      a := 1        T := 1

First consider problem without friction to get direction of motion.

given    m1· g· sin(θ) - T ≈ m1· a    T - m2· g ≈ m2· a   ⎡a⎤
                                                          ⎢ ⎥ := find(a,T)
                                                          ⎣T⎦

a = 2.45      T = 12.25       sgna := if(a > 0,1,-1)     sgna = 1

Repeat calculation including friction.      N := m1· g· cos(θ)     F   := μ· N
                                                                    f
given    m1· g· sin(θ) - sgna· F   - T ≈ m1· a
                                f

         T - m2· g ≈ m2· a            ⎡a⎤
                                      ⎢ ⎥ := find(a,T)
                                      ⎣T⎦
```

The frictional force, as expressed in F.f is a maximum. If the sign of `a'
in each calculation is not the same, then the acceleration is zero.

```
         ⎡ a          ⎤              T := if(a ≈ 0,m2· g,T)     a = 1.035
a := if ⎢─── ≈ sgna,a,0⎥ ▫
         ⎣|a|          ⎦                                       T = 10.835
```

Examine sizes of various terms and check results.

```
m1· g· sin(θ) = 24.5        sgna· F   = 8.487         T = 10.835
                                  f

m1· g· sin(θ) - sgna· F   - T = 5.177           m1· a = 5.177
                       f

m2· g = 9.8            T - m2· g = 1.035         m2· a = 1.035
```

Figure 6.4 MECH2, motion of a mass on a plane with friction (1).

it is included. For this problem, we take the masses and angle as given and solve for acceleration and tension.

Because the sign of the acceleration is needed, we define *sgna* (sign of *a*). This term is ± 1, depending on whether the acceleration is greater or less than zero. *Sgna* is used in the second solve block to assure the proper sign for the frictional term. Locate the equation in the second solve block containing *sgna*. Examine the cases for $\pm a$ and convince yourself that the force acts in the desired direction, that is, that it is a resistive and not a driving force. The if statement for *a* is disabled, initially.

Observe the behavior of the two mass system. Let $\mu = 0$, $m_1 = 5$, and $m_2 = 1$; process. Check the value for the acceleration in each solve block. Note the value of *sgna*. Let m_1 keep its present value and let $m_2 = 10$. Process. Check the values of *a* and *sgna*.

Let $m_1 = 5$ and $m_2 = 1$. Let $\mu = 0.2$, 0.3, and 0.4. Note the values for *a* in each case. If the *a*'s have opposite signs, it means that the frictional force is helping to drive rather than to resist the motion — this is an unphysical condition.

Enable the if statement for *a* and process. Unless the *a*'s have the same sign, no motion is possible or allowed.

Let $\mu = 0$. Let $m_1 = 3.3$ and $m_2 = 1$. Observe the value of *a*. Now let $m_1 = 6.6$ and process. If $m_1 = 2 \times 6.6$, predict how *a* will change. Test your prediction. Let m_1 take on the values 5×10^n where $n = 1, \ldots, 5$. Explain the sequence of *a* values. Let $m_1 = 5$. Let $m_2 = 10^n$ where $n = 0, -1, \ldots, -3$. Explain the sequence of values of *a*.

• •   Examine the acceleration and tension as a function of μ. In MECH2, in the second solve block, replace the find statement with

$$f(\mu) := \text{find}(a, T).$$

Following the find statement, add the statements

$$n := 6 \quad i := 0 \ldots n \quad \mu_i := 0.2 + 0.03 \cdot i$$

$$a_i := f(\mu_i)_0 \quad T_i := f(\mu_i)_1.$$

Plot a_i vs. μ_i and T_i vs. μ_i. Explain the curves.

• •   Load MECH3, motion of a mass on a plane with friction (2) (see Fig. 6.5).

This problem is the same as that treated in MECH2. We examine the acceleration and tension as a function of m_1. However, in this case, we consider a sequence of values for m_1. The first solve block, written in functional form, is used to generate a sequence of values for the sign of the acceleration, *sgna*. The values are shown using the transpose. Normally, vectors are column vectors and appear as such in tabular form. The transpose interchanges rows and columns; the result is a row vector. Clearly, only a modest number of values can be observed in one screen.

The second solve block is also written in functional form. The arguments are the mass and the sign of the acceleration. Names without subscripts are used in the find statement. The acceleration and tension are then determined for a series of values of m (and the associated *sgna* values).

Mass, m1, on slanted plane; mass, m2, suspended; friction.   rad ≡ 1

Parameters

$$deg \equiv \frac{\pi}{180} \cdot rad$$

m1 := 5   m2 := 1   g := 9.8   θ := 30·deg   μ := 0.22

Guess values   a := 1   T := 1

First consider problem without friction to get direction of motion.

given   m1·g·sin(θ) - T ≈ m1·a   T - m2·g ≈ m2·a   f'(m1) := find(a,T)

n := 9   i := 0 ..n $m1'_i$:= 5 - i·.5 $sgna_i$:= if$\left[\left(f'\left[m1'_i \right] \right)_0 > 0,1,-1 \right]$

$sgna^T$ = (1 1 1 1 1 1 -1 -1 -1 -1)   sgnaa := 1

Repeat calculation including friction.

given   m1·g·sin(θ) - sgnaa·μ·m1·g·cos(θ) - T ≈ m1·a

T - m2·g ≈ m2·a   f(m1,sgnaa) := find(a,T)

a_i := $f\left[m1'_i, sgna_i \right]_0$ T_i := $f\left[m1'_i, sgna_i \right]_1$

The friction force, as expressed in F.f is a maximum. If the sign of `a' in each calculation is not the same, then the acceleration is zero.

a_i := $if\left[\frac{a_i}{|a_i|} \approx sgna_i, a_i, 0 \right]$ T_i := $if\left[a_i \approx 0, m2·g, T_i \right]$

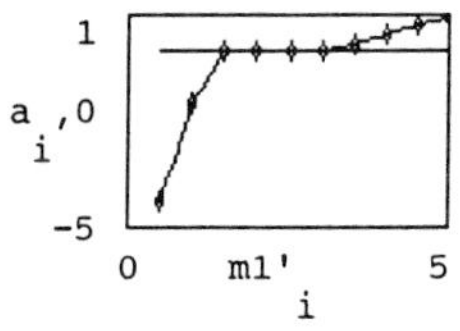

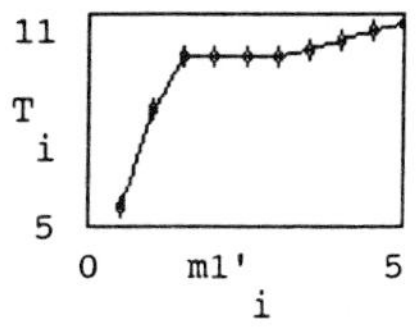

a_i	$m1'_i$
0.894	5
0.7	4.5
0.466	4
0.181	3.5
0	3
0	2.5
0	2
0	1.5
-1.516	1
-4.278	0.5

Figure 6.5 MECH3, motion of a mass on a plane with friction (2).

By requiring that the accelerations have the same sign with or without friction, in order that motion occur, we can plot the acceleration and tension, no matter whether the acceleration is positive, zero, or negative.

Interpret the graphs. Why does T take on the values that it does?

Plot the force of friction vs. m_{1_i}'.

For each of the following cases, predict how the graphs will change. Let $\theta = 40°$. Let values for (μ, m_{1_i}') equal $(0.3, 10 - i)$, $(1, 10 - i)$, $(0.6, 10 - i)$.

6.3 Velocity-Dependent Forces

The frictional force of a block sliding on a plane was treated as having no dependence on velocity. An object moving through a fluid, however, experiences a force which is dependent on velocity.

It is common practice to consider two distinct cases of velocity dependence. The two cases can be characterized as viscosity dependent and density dependent. Viscosity is associated with the resistance of fluid layers moving past each other. The coefficient of viscosity, η, is the ratio of shear stress, (force per area, F / A) and shear strain per time (change in velocity per thickness, $\delta v / th$):

$$\eta = \frac{F / A}{\delta v / th} = \frac{F\, th}{\delta v A}. \tag{6.6}$$

• •   Use dimensional analysis to find a dimensionless group that will indicate the interdependence of a velocity-dependent force and viscosity. Perform the analysis to find the relationship using the coefficient of viscosity, radius, velocity, and force. Arrange the final result so that $F \propto$ other variables.

• •   Similarly, find the dimensionless group for a velocity-dependent force where the variables are density, radius, velocity, and force. Arrange the final result so that $F \propto$ other variables.

The viscosity dimensional analysis example yields

$$F_\eta \propto \eta r v \qquad \text{or} \qquad F_\eta = C_\eta \eta r v,$$

which, in effect, is Stokes's law:

$$F_s = 6\pi \eta r v. \tag{6.7}$$

The second example yields

$$F_\rho \propto \rho r^2 v^2 \qquad \text{or} \qquad F_\rho = C_\rho \rho r^2 v^2.$$

The latter is commonly expressed as

$$F_D = \frac{1}{2} C \rho A v^2 \tag{6.8}$$

where A is the cross-sectional area (generalized from r^2 in the dimensional analysis) and C is a dimensionless drag coefficient. The net drag would be the sum of each of these contributions:

$$F_D = C_\eta \eta r v + C_\rho \rho r^2 v^2.$$

At low velocities, the viscosity term dominates; at high velocities, the density term dominates. The velocity at which these two are equal is given by

$$v = \frac{C_\eta \eta}{C_\rho \rho} \frac{1}{r}.$$

For air at standard temperature and pressure, we have

$$\eta \approx 1.8 \times 10^{-5} \frac{\text{Ns}}{\text{m}^2} \qquad \text{and} \qquad \rho \approx 1.2 \frac{\text{kg}}{\text{m}^3}.$$

The coefficients are roughly

$$C_\eta \approx 17 \qquad \text{and} \qquad C_\rho \approx 0.73.$$

Thus the velocity at which the forces are equal is given by

$$v_{\rho\eta} = 3.5 \cdot 10^{-4} \cdot \frac{1}{r}.$$

Examine the functionality by performing the following exercises.

• • Plot F_{ρ_i} and F_{η_i} vs. v_i for $i := 0 \ldots 20$ and $v_i := 0.0035 \cdot (i+1)$. Plot the two curves together. Plot both linear and log-log views.

• • Plot F_{ρ_i}/F_{η_i} vs. v_i for $v_i := 0.001 \cdot 2^i$ on a log-log scale.

• • Plot $v_{\rho\eta}$ vs. r as r goes from one micron to one meter.

• • Let $t_{\rho\eta}$ be the time for a particle in free fall, experiencing no resistance, to reach velocity $v_{\rho\eta}$. Plot $t_{\rho\eta}$ vs. r. (Let the initial velocity of the particle be zero.)

• • Plot the free fall time to reach a velocity where the ratio of F_ρ/F_η is 10:1.

These calculations help to indicate where the various contributions are relevant. For many of the velocity ranges encountered in these problems, the neglect of the F_η term is clearly justified.

The region in which the viscosity term is dominant is still of some interest. The equation of motion for a particle falling in a viscous medium is

$$mg - C_\eta \eta r v = ma = m\frac{dv}{dt}. \tag{6.9}$$

As the velocity increases, the resistive force increases. When the gravitational and resistive forces are equal and opposite, the net force is zero, the acceleration is zero, and the object has reached its terminal velocity:

$$v_t := \frac{mg}{C_\eta \eta r}. \tag{6.10}$$

The equation of motion can be rewritten as

$$\frac{dv}{dt} = g\left(1 - \frac{v}{v_t}\right). \tag{6.11}$$

The solution to this equation is

$$v = v_t(1 - e^{-t/\tau}) \tag{6.12}$$

where $\tau = v_t/g$. The constant τ is a characteristic time called the time constant. In this time, if the initial acceleration were maintained, the terminal velocity would be reached.

If the driving force is removed, the velocity decays to zero with the same time constant as did growth toward the terminal velocity. If the mg term is zero, we have

$$\frac{dv}{dt} = \frac{-g}{v_t}v = -v/\tau. \tag{6.13}$$

• •   Verify that this form follows if $mg = 0$.

The solution is

$$v = v_t\, e^{-t/\tau}. \tag{6.14}$$

• •   Verify that this solution satisfies the previous equation.

• •   Plot five time constants of growth toward terminal velocity and five returning toward zero. One way to do this would be by defining the function

$$v(t) = \; \text{if}\left[t < 5\tau, \; v_t \cdot [1 - e^{-t/\tau}], \; v_t \cdot e^{-(t-5\cdot\tau)/\tau}\right]$$

and letting t go from 0 to 10τ. Note that the velocity is always positive. The object does not return to its initial position.

• •   For the motion just described, plot position vs. time.

• •   Integrate the function $v(t)$ over time. First integrate from 0 to 5τ, then from 5τ to 10τ. Compare with the plot from the previous exercise.

6.3.1   Sky Diving

In an actual sky diving exercise, with multiple divers, one diver collided with another, hitting her in the head and knocking her unconscious. The group leader saw the woman falling limply, went into a dive, caught up with her, pulled her ripcord, and proceeded to parachute safely to earth himself. The woman sustained only relatively minor injuries (minor at least in comparison with what could have happened).

Let us model this situation approximately. As a simplification, the problem is treated as one dimensional rather than two dimensional. In this case adding the second dimension would only increase complexity and not generate any additional insight.

Before performing the rescue, we need to consider free fall with a retarding force proportional to v^2. The equation of motion is

$$mg - kv^2 = m\frac{dv}{dt} \tag{6.15}$$

where

$$k = \frac{1}{2}C_\rho\rho A.$$

As the velocity increases, kv^2 increases until $kv^2 = mg$ and $dv/dt = 0$. The terminal velocity is given by

$$v_t = \sqrt{\frac{mg}{k}}. \tag{6.16}$$

• •   Verify the above equation for the terminal velocity.

• •   We now want to make sure that we appreciate the interdependence of the various terms in the expression for terminal velocity. We can do this by looking at sequences of parameter values and looking at the families of related curves.

Examine the relationship between terminal velocity, mass, and cross-sectional area. The initial setup for this problem is in MECH4. Load the document and define two ranges: $A := 0.3 \cdot m^2, 0.5 \cdot m^2 \ldots 1.1 \, m^2$ and $m_1 := 20 \cdot kg, 40 \cdot kg \ldots 100 \cdot kg$. Plot $v_t(m_1, A)$ vs. A; set the upper limit of the abscissa at $1 \cdot m^2$.

Define $v_t'(A, m_1) := v_t(m_1, A)$. Plot $v_t'(A, m_1)$ vs. m_1. Set the lower limit on the abscissa to $21 \cdot kg$.

For the plot of $v_t(m_1, A)$ vs. A, which curve corresponds to the largest value for m_1? Which curve corresponds to the smallest? For the plot of $v_t'(A, m_1)$ vs. m_1, which curve corresponds to the largest value for A? Which curve corresponds to the smallest?

The equation of motion

$$\frac{dv}{dt} = g - \frac{k}{m}v^2 \tag{6.17}$$

can be rewritten with finite sized elements as

$$\Delta v \simeq (g - \frac{k}{m}v^2)\,\Delta t \tag{6.18}$$

where Δv and Δt are small incremental changes in velocity and time. We express the velocity difference, Δv (in the document δv), as

$$\Delta v = v_{i+1} - v_i \qquad \text{or} \qquad v_{i+1} = v_i + \Delta v_i. \tag{6.19}$$

The change in velocity for a given v_i is then

$$\Delta v_i = (g - \frac{k}{m}v_i^2) \cdot \delta t. \tag{6.20}$$

This pair of equations (6.19 and 6.20) can then be used to determine the velocity over time. One vector specifies the incremental changes; the other keeps track of the sequence of values. (It is irrelevant whether you specify the acceleration or the change in velocity. They are related by a constant factor; $a_i = v_i\,\delta t$.)

These two equations are iterated simultaneously. Once the values for v_i are known, values of x can be determined, for example, by iterating the equation $x_{i+1} = x_i + v_i\delta t$.

When equations are to be iterated simultaneously, as the v and δv equations are, the left-hand sides and right-hand sides of the equations are grouped together in one-column matrices.

[Alt]M, the create matrix command, brings to the command line a request for the size of the matrix. Enter the number of equations (in the above case, two) and press return. The form

appears. At each place marker, type the left-hand side of the coupled equations — in this case, δv_{i+1} and v_{i+1}. Use the tab button to move between place markers.

When the entries are complete, type the assignment equality, the colon. The cursor then moves to the right-hand side. Press [Alt]M again.

The value entered when the matrix for the left-hand side was created is still present. Press return and enter the corresponding right-hand sides of the equations.

The equations are then ready to be processed. Be aware that as the equations are processed, values calculated on one pass through the equations cannot be used during that pass. For example, in the pass through the equations that calculates a_2 and v_2, the value a_2 cannot be used in the calculation of v_2. The same quantities used to calculate δv_2 would need to be expressed explicitly.

Before continuing with the sky diving problem, it is interesting to note that the equation of motion with a drag force proportional to v^2 does have an analytic solution. The derivation is beyond the scope of this text, but the results can be used and compared with our iterated solution.

For the differential equation

$$m\frac{dv}{dt} = mg - kv^2, \tag{6.21}$$

the velocity is given by

$$v = v_t \tan h\left(K + \sqrt{\frac{K + kg}{m}}t\right), \tag{6.22}$$

and the position is given by

$$x = \frac{m}{k}\ln\left(\cos h(K + \sqrt{\frac{K + kg}{m}}t - \ln(\cos h(K))\right) \tag{6.23}$$

where

$$K = \tan h^{-1}(\frac{v_0}{v_t}) \quad \text{and} \quad v_t = \sqrt{\frac{mg}{k}}.$$

In these expressions, we use the sign convention where plus means down.

• •   Determine the units of each of the following: k, kg/m, $\sqrt{mg/k}$, and K.

• •   Write the position and velocity equations for the case of zero initial velocity. Start with the K equation.

• •   Explore equations 6.18, 6.19, and 6.20. Pick values for m, k, and v_0. Calculate and show v_t before entering a value for v_0. Define expressions for position and velocity and plot both vs. time. Do not use a value of v_0 greater than v_t.

Try different values of m and k. Try some values of v_0 that are negative; remember plus is down; also remember the limitations on the magnitude of v_0. For comparison, include plots of the position and velocity vs. time for similar initial conditions but with zero air resistance.

● ●   Load MECH4, resisted fall: analytic and numeric solutions (see Fig. 6.6).

We want to calculate the velocity and position of a sky diver starting a fall with zero initial velocity. A density-dependent velocity squared resistive force is included. We approach the problem in two ways. First, we iterate the difference equations as described. Then, we calculate the values using the analytical expressions above.

Plots of velocity vs. time and elevation vs. time are shown for both cases. The last two plots show the difference in velocity and difference in position as determined by the two approaches. The velocity difference, which is never large (the maximum difference is on the order of 1 mi/ hr), goes to zero as time increases. The position difference is also relatively small and becomes constant as the velocity difference goes to zero.

Considering the modest algorithm, the agreement between the analytic and numeric approaches is quite good. By examining a problem in different ways and arriving at the same result, you gain confidence in each approach.

● ●   The agreement that we note is dependent on a particular set of parameters. Vary the area and the mass. Does the shape of the velocity or elevation curve change noticeably? Does the quality of the match between the two approaches change?

Now, let's go back to the rescue. Recall that the drag force is

$$F_D = \frac{1}{2}C\rho Av^2.$$

For the speeds encountered in this problem, we ignore the viscosity-related force. We will assume that C is constant, does not depend on the orientation of the sky diver, and is the same for each diver. A is the cross-sectional area presented to the air flow. For an unconscious person, A would be reduced from the maximum. In the notation below, primes refer to maximum values. Variable names ending in 1 refer to the injured person; those ending in 2 refer to the pursuer.

Let

$$A1' = 0.65\text{m}^2 \qquad A1 = 0.5\text{m}^2 \qquad \text{(injured cross section)}$$

$$A2' = 0.7\text{m}^2 \qquad A2 = 0.35\text{m}^2 \qquad \text{(dive)}.$$

A comparison of resisted fall using an analytic and a numerical approach.

units $m := 1L$ $s := 1T$ $kg := 1M$

$$N := kg \cdot m \cdot s^{-2} \qquad ft := \frac{1}{3.28} \cdot m$$

$mi := 5280 \cdot ft$ $hr := 3600 \cdot s$

parameters

$m1 := 55 \cdot kg$ $m2 := 75 \cdot kg$ $g := 9.81 \cdot \dfrac{m}{s^2}$ $Alt := 5000 \cdot ft$

$m1 \cdot g = 539.55 \cdot N$

$C := 0.7$ $\rho := 1.2 \cdot kg \cdot m^{-3}$ $A := .75 \cdot m^2$ $k := \dfrac{1}{2} \cdot C \cdot \rho \cdot A$

$$v_t := \sqrt{\frac{m1 \cdot g}{k}} \qquad v_t = 92.556 \cdot \frac{mi}{hr} \qquad\qquad k = 0.315 \cdot \frac{kg}{m}$$

Iterative solution

$N' := 300$ $i := 0 .. N'$ $\delta t := 0.05 \cdot s$ $t_i := i \cdot \delta t$

$an_0 := g$ $vn_0 := 0 \cdot m \cdot s^{-1}$ $xn_0 := 0 \cdot m$

$$\begin{bmatrix} an_{i+1} \\ vn_{i+1} \\ xn_{i+1} \end{bmatrix} := \begin{bmatrix} \left[g - \left[\dfrac{k}{m1} \right] \cdot vn_i^2 \right] \\ vn_i + an_i \cdot \delta t \\ xn_i + vn_i \cdot \delta t \end{bmatrix}$$

Analytic solution

$$kgm := \sqrt{\frac{k \cdot g}{m1}} \qquad va_i := v_t \cdot \tanh\left[kgm \cdot t_i \right] \qquad xa_i := \frac{m1}{k} \cdot \ln\left[\cosh\left[kgm \cdot t_i \right] \right]$$

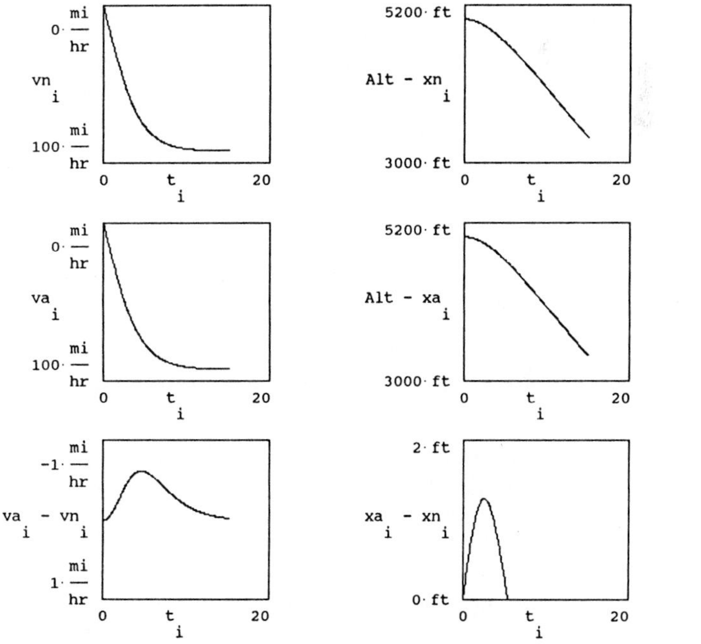

Figure 6.6 MECH4, resisted fall: analytic and numeric solutions.

This information is sufficient to define the k's and terminal velocities for the cases necessary for the problem.

In somewhat arbitrary fashion, the scenario is outlined as

$t = 0$ (diver 1 jumps)

$t = 1$ (collision occurs and diver 2 jumps)

$t = 3$ (diver 2 observes that diver 1 is injured)

$t = 3.5$ (diver 2 goes into dive).

As A changes, so does k. To incorporate those changes automatically, write k as a function of time. An if statement can select between cases. For times before the collision occurs, diver 1 has normal cross section $A1'$. For times before the dive begins, diver 2 has normal cross section $A2'$. These conditions are expressed in the equations

$$k1(t_i) := \text{if}(t_i < t_{\text{coll}}, k1', k1)$$

and

$$k2(t_i) := \text{if}(t_i < t_{\text{div}}, k2', k2).$$

• • Load MECH5, skydive: the pursuit (see Figs. 6.7 and 6.8).

Observe the sequence of areas, coefficients k, and terminal velocities. Note the numerical values of the terminal velocities. There are two sets of iterated equations; the $k(t)$ in each takes into account the sequence of events described. In this example, we do not model any additional changes in the motion which occur after the pursuer has reached the injured party.

For the pursuer to maintain his own safety, he has to reach the injured woman before she reaches an elevation of 1000 ft. If $A1 := 0.4 \cdot m^2$, could diver 2 still reach diver 1?

What value would t_{div} have, in order that the sky divers meet at an elevation of 1000 ft? Use the original values of A.

The δv curve is telling. For the first second, you see the free fall of diver 1. During the next 2.5 seconds, the velocity of diver 2 is increasing slightly faster than that of diver 1. At $t = 3.5$s, diver 2 goes into a dive by reducing his cross section. The velocity difference increases more rapidly but is not linear.

The δX curve is also informative. What is the maximum distance of separation? At what time did this maximum separation occur? At what time does diver 2 reach diver 1?

When diver 2 reaches diver 1, what is the elevation?

In a sky diving exercise, one of the sky divers was knocked unconscious. The leader of the group saw the person falling limply, went into a dive, caught up with the unconscious person, pulled that person's rip cord, and then proceeded to parachute safely to ground himself.

Let us see if we can model this problem, approximately. We refer to the falling person as 1 and the instructor as 2.

units $m := 1L$ $s := 1T$ $kg := 1M$ $N := kg \cdot m \cdot s^{-2}$ $ft := \dfrac{1}{3.28} \cdot m$

$mi := 5280 \cdot ft$ $hr := 3600 \cdot s$

parameters

$m1 := 65 \cdot kg$ $m2 := 80 \cdot kg$ $g := 9.81 \cdot \dfrac{m}{s^2}$ $C := 0.7$

$m1 \cdot g = 637.65 \, N$ $m2 \cdot g = 784.8 \, N$ $\rho := 1.2 \cdot kg \cdot m^{-3}$

Cross- sections: A1' normal, A1 injured, A2' normal, A2 dive.

$A1' := .65 \cdot m^2$ $A1 := .5 \cdot m^2$ $A2' := .75 \cdot m^2$ $A2 := .35 \cdot m^2$

$k1' := \dfrac{1}{2} \cdot C \cdot \rho \cdot A1'$ $k2' := \dfrac{1}{2} \cdot C \cdot \rho \cdot A2'$ $k1 := \dfrac{1}{2} \cdot C \cdot \rho \cdot A1$ $k2 := \dfrac{1}{2} \cdot C \cdot \rho \cdot A2$

$vt1' := \sqrt{\dfrac{m1 \cdot g}{k1'}}$ $vt1 := \sqrt{\dfrac{m1 \cdot g}{k1}}$ $vt2' := \sqrt{\dfrac{m2 \cdot g}{k2'}}$ $vt2 := \sqrt{\dfrac{m2 \cdot g}{k2}}$

$vt1' = 108.082 \cdot \dfrac{mi}{hr}$ $vt1 = 123.232 \cdot \dfrac{mi}{hr}$ $vt2' = 111.626 \cdot \dfrac{mi}{hr}$ $vt2 = 163.404 \cdot \dfrac{mi}{hr}$

$Alt := 3000 \cdot ft$ $tdiv := 3.5 \cdot s$ $tcoll := 1 \cdot s$

$N' := 300$ $i := 0 \, .. N'$ $\delta t := .05 \cdot s$ $t_i := i \cdot \delta t$ $N' \cdot \delta t = 15 \cdot s$

$k1(t) := if(t < tcoll, k1', k1)$ $k2(t) := if(t < tdiv, k2', k2)$

$v1_0 := 0 \cdot \dfrac{m}{s}$ $a1_0 := g$ $v2_{20} := 0 \cdot \dfrac{m}{s}$ $a2_{20} := g$

$x1_0 := 0 \cdot m$ $x2_{20} := 0 \cdot m$ $j := 20 \, .. N'$

$$\begin{bmatrix} a1_{i+1} \\ v1_{i+1} \\ x1_{i+1} \end{bmatrix} := \begin{bmatrix} g - \left[\dfrac{k1 \left[t_i \right]}{m1} \right] \cdot v1_i^2 \\ v1_i + a1_i \cdot \delta t \\ x1_i + v1_i \cdot \delta t \end{bmatrix}$$ $$\begin{bmatrix} a2_{j+1} \\ v2_{j+1} \\ x2_{j+1} \end{bmatrix} := \begin{bmatrix} g - \left[\dfrac{k2 \left[t_j \right]}{m2} \right] \cdot v2_j^2 \\ v2_j + a2_j \cdot \delta t \\ x2_j + v2_j \cdot \delta t \end{bmatrix}$$

Figure 6.7 MECH5, skydive: the pursuit. (See the next figure for the rest of the document.)

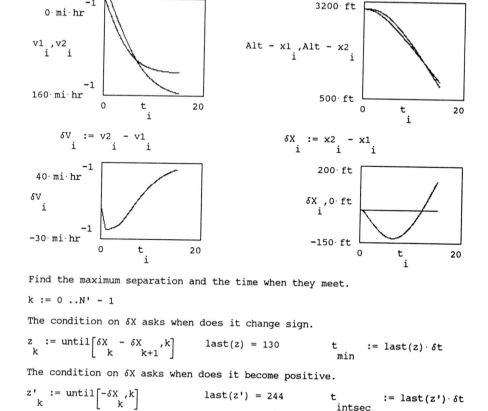

Find the maximum separation and the time when they meet.

$k := 0 ..N' - 1$

The condition on δX asks when does it change sign.

$z_k := until\left[\delta X_k - \delta X_{k+1}, k\right]$ $last(z) = 130$ $t_{min} := last(z) \cdot \delta t$

The condition on δX asks when does it become positive.

$z'_k := until\left[-\delta X_k, k\right]$ $last(z') = 244$ $t_{intsec} := last(z') \cdot \delta t$

$t_{min} = 6.5 \cdot s$ $t_{intsec} = 12.2 \cdot s$

$\delta X_{130} = -129.389 \cdot ft$ $\delta X_{244} = 1.284 \cdot ft$ $\delta X_{243} = -0.745 \cdot ft$

Figure 6.8 MECH5 *continued.*

6.3.2 *Parachute*

There are records of men in the military falling or jumping from aircraft without parachutes and suffering only minor injuries. (They landed in such a manner, for example, in deep snow, that the deceleration occurred over some distance.) The greatest recorded distance of such a fall is 23,000 ft. Still, most consider it preferable to use a parachute.

We model resisted fall without a parachute, followed by the non-instantaneous opening of the chute, and descent with the chute fully open.

The distance over which the chute opens uniformly is d_{open}. The elevation at which the chute has completely opened is d_p. The height above the earth's surface is ht.

As in the sky diving example, k varies with conditions. Let k_1 be the coefficient without the parachute, and let k_2 be the coefficient with the chute fully deployed. The term $kk(x)$ provides a linear transition between the two values. The expression $k(x)$ includes all the k conditions. These conditions are summarized as follows:

$$ht(x) \geq d_p + d_{\text{open}} \qquad (k_1)$$

$$ht(x) \leq d_p \qquad (k_2)$$

$$d_p < ht(x) < d_p + d_{\text{open}} \quad (kk(x))$$

The statement for $k(x)$ can be expressed as a nested pair of if statements. The combination permits us to express all the k conditions in one line.

● ●   Plot $k(ht(x))$ vs. $ht(x)$ for the conditions expressed in MECH6. This document contains the parameters but not all the expressions. You must write the equation for $kk(x)$, the k in the transition region. The term $kk(x)$ depends on k_1, k_2, d_{open}, d_p, and $ht(x)$. If $k_2 = 2\text{kg}/\text{m}$, is the change in velocity when the chute opens more gradual or more severe? Why?

● ●   Load MECH7, resisted fall with parachute (see Fig. 6.9).

The initial conditions are the same as in MECH6. The same iteration procedure is used here as was discussed in the sky dive example. We express k as a function of position using nested if statements as described above.

Was your expression for $kk(x)$ in the previous example essentially the same as that shown here? Verify that it is consistent with the conditions given above.

Plot v_1 vs. t. Let the upper limit of the ordinate be 0 mi/ hr, and let the lower limit be 130 mi/ hr. Plot, directly beneath the previous plot, v_1 vs. $ht(x_1)$. Use the same ordinate limits; let the lower abscissa limit be Altp; let the upper limit be 0 ft. How are the curves similar? Explain the obvious difference.

After the chute has opened, is the final velocity of the iterated solution approximately equal to the expected terminal velocity?

Plot $ht(x_1)$ vs. t. How does the first segment of the curve differ from unresisted motion?

Resisted fall with parachute.

units   m := 1L   s := 1T   kg := 1M

                                                            $ft := 3.28^{-1} \cdot m$

             mi := 5280·ft   hr := 3600·s

parameters

k1 := 0.2   k2 := 14 d_p := 500·ft d_{open} := 250·ft

Alt := 3000·ft $g := 9.81 \cdot m \cdot s^{-2}$ m1 := 55·kg

x - distance fallen, ht - elevation   ht(x) := Alt - x

The parameter, kk(x) is the k-value while the chute is opening

$$kk(x) := k2 + \frac{k1 - k2}{d_{open}} \cdot \left[ht(x) - d_p \right]$$

The general statement for k, for all x.

$$k(x) := if\left[ht(x) \geq \left[d_p + d_{open} \right], k1, if\left[ht(x) \leq d_p, k2, kk(x) \right] \right]$$

$$k(x) := k(x) \cdot kg \cdot m^{-1}$$

The iterated solution.

N := 400   i := 0 ..N   δt := .075·s $t_i := i \cdot \delta t$

$al_0 := g$ $vl_0 := 0 \cdot m \cdot s^{-1}$ $xl_0 := 0 \cdot m$

$$\begin{bmatrix} al_{i+1} \\ vl_{i+1} \\ xl_{i+1} \end{bmatrix} := \begin{bmatrix} \left[g - \left[\dfrac{k\left[xl_i \right]}{m1} \right] \cdot vl_i^2 \right] \\ vl_i + al_i \cdot \delta t \\ xl_i + vl_i \cdot \delta t \end{bmatrix}$$

Figure 6.9 MECH7, resisted fall with parachute.

Define the acceleration and plot $a1$ vs. t. To see the details of this curve, let the plot region be very wide, with dimensions 10, 60. Explain the plot. Why do the two regions have opposite signs?

Jerk is the rate of change of acceleration. Let $j := 0 \ldots N - 1$ and $jrk_j := a1_{j+1} - a1_j / \delta t$. Plot jrk vs. t directly beneath the plot of acceleration. Let the plot dimensions be the same as those of the acceleration region. Explain the shape of the jerk curve in terms of the acceleration. Change the ordinate limits to $-3 \, m \cdot s^{-3}$ and $1 \, m \cdot s^{-3}$. Why is the amplitude of the curve so much smaller at the beginning than it is near the center?

6.3.3   *Two-Dimensional Motion with Resistance*

In discussing two-dimensional motion with no resistance, we noted that the kinematic equations describing motion in the x direction were independent from the analogous equations describing motion in the y direction. With velocity-dependent forces, the two sets of equations are no longer independent. The resistive force is directed opposite to the velocity vector, and the magnitude of the resistance is proportional to some function of the velocity. As velocity typically has both x and y components, so does the resistive force. The components of the force depend on the total velocity V, which depends on both Vx and Vy. Consequently, the acceleration in each direction depends on both vx and vy, and the equations in the vertical and horizontal dimensions are not independent.

If the velocity vector v makes an angle, θ, with respect to the horizontal, the oppositely directed resistive force makes the same angle. Applying Newton's second law (constant mass),

$$\sum F_{\text{ext}} = m \frac{dv}{dt},$$

we obtain the equation of motion in the x direction,

$$-kv^2 \cos(\theta) = m \frac{dv_x}{dt} = ma_x. \tag{6.24}$$

The only force in the x direction is due to the component of the resistive force. The equation of motion for the y direction is

$$-mg - kv^2 \sin(\theta) = m \frac{dv_y}{dt} = ma_y. \tag{6.25}$$

The forces are gravitational and resistive.

Knowing that $v_x = v \cos(\theta)$ and $v_y = v \sin(\theta)$, we can eliminate the trigonometric functions from the equations of motion:

$$-kvv_x = ma_x \tag{6.26}$$

$$-mg - kvv_y = ma_y \tag{6.27}$$

or

$$a_x = \frac{-k}{m} vv_x \quad \text{and} \quad a_y = -g \frac{-k}{m} vv_y. \tag{6.28}$$

• •   Load MECH8, trajectory with resistance (see Fig. 6.10).

In the Saturday morning cartoons, the trajectories of objects or characters often seem to make an abrupt transition from motion along one

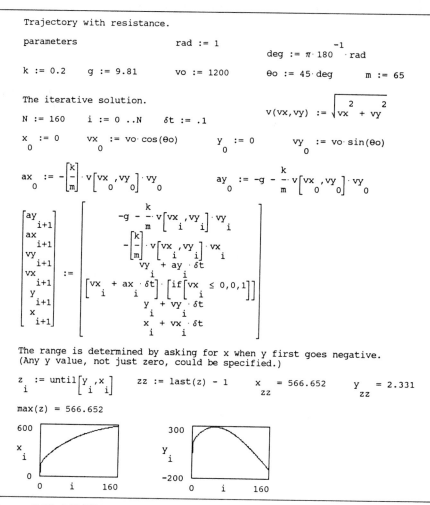

Trajectory with resistance.

parameters

rad := 1

$deg := \pi \cdot 180^{-1} \cdot rad$

$k := 0.2$ $\quad g := 9.81$ $\qquad vo := 1200$ $\qquad \theta o := 45 \cdot deg$ $\qquad m := 65$

The iterative solution.

$v(vx, vy) := \sqrt{vx^2 + vy^2}$

$N := 160$ $\quad i := 0 \;..N$ $\quad \delta t := .1$

$x_0 := 0$ $\qquad vx_0 := vo \cdot \cos(\theta o)$ $\qquad y_0 := 0$ $\qquad vy_0 := vo \cdot \sin(\theta o)$

$$ax_0 := -\left[\frac{k}{m}\right] \cdot v\left[vx_0, vy_0\right] \cdot vy_0 \qquad\qquad ay_0 := -g - \frac{k}{m} \cdot v\left[vx_0, vy_0\right] \cdot vy_0$$

$$\begin{bmatrix} ay_{i+1} \\ ax_{i+1} \\ vy_{i+1} \\ vx_{i+1} \\ y_{i+1} \\ x_{i+1} \end{bmatrix} := \begin{bmatrix} -g - \frac{k}{m} \cdot v\left[vx_i, vy_i\right] \cdot vy_i \\ -\left[\frac{k}{m}\right] \cdot v\left[vx_i, vy_i\right] \cdot vx_i \\ vy_i + ay_i \cdot \delta t \\ \left[vx_i + ax_i \cdot \delta t\right] \cdot \left[if\left[vx_i \leq 0, 0, 1\right]\right] \\ y_i + vy_i \cdot \delta t \\ x_i + vx_i \cdot \delta t \end{bmatrix}$$

The range is determined by asking for x when y first goes negative. (Any y value, not just zero, could be specified.)

$z_i := until\left[y_i, x_i\right]$ $\qquad zz := last(z) - 1$ $\qquad x_{zz} = 566.652$ $\qquad y_{zz} = 2.331$

$max(z) = 566.652$

Figure 6.10 MECH8, trajectory with resistance.

path (typically continuing in the original direction) to motion along another path (typically falling). In some medieval works, figures displaying the motion of cannon balls show a similar trajectory. The trajectory does not take the form of a triangle, but the path does trace a fairly rapid transition from motion deviating only slightly from the initial direction to motion with a greatly reduced forward velocity. We might suspect the cartoons of anything, but the medieval texts were serious. How did this point of view arise? Is it totally a misconception, or is there some basis for the trajectories as the medievals conceived them and the cartoonists represent them?

In this document, trajectories are examined where resistance to the motion proportional to the square of the velocity is included. The block of iterated equations includes calculations of acceleration, velocity, and position for both components.

Process the document. The interesting features, once again, are obtained by examining a series of plots. As the plots appear, give a qualitative explanation of each. Compare one plot with another so that all aspects of the motion become clear to you. There may be some mild surprises as you go along. Try all the plots. Keep the plot sizes small, like those included. Plots vs. i are equivalent to plotting against time because i is proportional to time.

If the document becomes too cluttered with plots, print them out and then delete them. There is no reason to keep them all resident at the same time.

Plots of x vs. i and y vs. i are shown. What would these curves look like if there were no resistance?

Plot y vs. x. Find the connections with the previous two plots. Do you find any justification for the medieval depiction of trajectories?

Plot vx vs. i, vy vs. i, vx vs. x, and vy vs. x. First predict the behaviors, then plot.

Plot ax vs. i and ay vs. i. Find the connections to the velocity plots.

Plot v, vx, vy vs. i; increase the width of this plot. Plot v vs. vx and v vs. vy.

Plot $v - vx$ vs. i and $v - vy$ vs. i. (The shape of these plots may surprise you.)

Delete all the plot regions.

For an initial speed of $v_0 = 1200$, what angle yields the maximum range? (To consider larger values for v_0, δt must be reduced proportionately; otherwise, spurious output may occur. Of course, if δt is decreased, then N will have to be increased, leading to more iterations, which means more computer time.)

If trajectories in the atmosphere carry the projectile to very high elevations, the atmospheric pressure is reduced and the resistance to the motion would be reduced as well. The effect can be approximated by permitting the coefficient to be altitude dependent.

● ●   Load MECH9, trajectory with altitude-dependent resistance (see Fig. 6.11).

Trajectory with resistance. Includes altitude dependent resistance.

parameters $rad := 1$

$deg := \pi \cdot 180^{-1} \cdot rad$

$g := 9.81$ $vo := 1200$ $\theta o := 60 \cdot deg$ $m := 65$

We compute two solutions $j := 0 \,..1$

$kc := 0.2$ $k(y) := if(j \approx 0, kc, kc \cdot exp(-a \cdot y))$

$a := 0.12 \cdot 10^{-3}$

The iterative solution

$N := 160$ $i := 0 \,..N$ $\delta t := .1$ $v(vx, vy) := \sqrt{vx^2 + vy^2}$

$x_{0,j} := 0$ $vx_{0,j} := vo \cdot cos(\theta o)$ $y_{0,j} := 0$ $vy_{0,j} := vo \cdot sin(\theta o)$

$$ax_{0,j} := -\left[\frac{k\left[y_{0,j}\right]}{m}\right] \cdot v\left[vx_{0,j}, vy_{0,j}\right] \cdot vx_{0,j}$$

$$ay_{0,j} := -g - \frac{k\left[y_{0,j}\right]}{m} \cdot v\left[vx_{0,j}, vy_{0,j}\right] \cdot vy_{0,j}$$

$$\begin{bmatrix} ay_{i+1,j} \\ ax_{i+1,j} \\ vy_{i+1,j} \\ vx_{i+1,j} \\ y_{i+1,j} \\ x_{i+1,j} \end{bmatrix} := \begin{bmatrix} -g - \frac{k\left[y_{i,j}\right]}{m} \cdot v\left[vx_{i,j}, vy_{i,j}\right] \cdot vy_{i,j} \\ -\left[\frac{k\left[y_{i,j}\right]}{m}\right] \cdot v\left[vx_{i,j}, vy_{i,j}\right] \cdot vx_{i,j} \\ vy_{i,j} + ay_{i,j} \cdot \delta t \\ \left[vx_{i,j} + ax_{i,j} \cdot \delta t\right] \cdot \left[if\left[vx_{i,j} \leq 0, 0, 1\right]\right] \\ y_{i,j} + vy_{i,j} \cdot \delta t \\ x_{i,j} + vx_{i,j} \cdot \delta t \end{bmatrix}$$

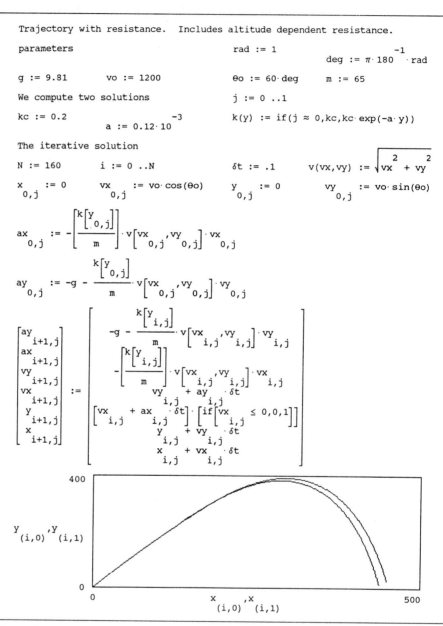

Figure 6.11 MECH9, trajectory with altitude-dependent resistance.

The procedure here is essentially the same as in MECH8. The difference is that k decreases exponentially with y. The value for the constant a is selected so as to reduce k in proportion with the decrease in atmospheric pressure as elevation increases. For modest initial velocities, the difference is not great but is clearly discernible, as you can see from the trajectory plot.

Rather than increasing the initial velocity (for example, the muzzle velocity of a cannon), which would require smaller time intervals and more iterations, increase a. In effect, this increase permits the pressure to decrease more rapidly. How does the range depend on a?

For an increased constant a, observe the trajectory for different angles. Does 45° remain the angle for maximum range (in this model)?

One way to determine the range is to find the value of x for which y first becomes negative. MathCAD's until statement can easily be used in this case. For example,

$$z_i := \text{until}(y_i,\ x_i).$$

For each value of i, z_i is assigned the value x_i until our test expression, y_i, goes negative. The range is approximately equal to the last value of z, which is also the maximum value of x. The value can be obtained in two ways. The statement $\max(z)$ returns the largest value in vector z. The statement $\text{last}(z)$ returns the index of the last value of z. This index corresponds to the first negative value for y. $\text{Last}(z) - 1$ corresponds to the last index when y is positive. The values $x_{\text{last}(z)}$ and $x_{\text{last}(z)-1}$ do not differ by much.

6.3.4 Orbits

A search for the solution to the problem of the planets was central to the development of classical mechanics. Many centuries elapsed between the time when careful observations of the planets were first made and the time when Kepler provided a precise mathematical description of planetary motion by means of his three laws. Kepler spent years performing untold numbers of calculations by hand to arrive at his solution. But now, knowing Newton's laws of motion and his law of gravitation, we can examine in detail, the properties of orbits, in minutes rather than years.

The general approach is the same as that used in the previous section on trajectories. Express the x and y components of the acceleration; write the difference equations for velocity and position. The external force is the gravitational force between the central body and the orbiting body. We shall assume that the mass, m_1, of the central body, located at the origin, is very much larger than that of the mass, m_2, of the orbiting body. The gravitational force is always attractive and along the line connecting the two bodies.

The magnitude of the force is given by

$$|F| = \frac{Gm_1m_2}{r^2}.$$

The components of this attractive force are

$$F_x = \frac{-Gm_1m_2}{r^2}\cos(\theta) \tag{6.29}$$

and

$$F_y = \frac{-Gm_1m_2}{r^2}\sin(\theta). \tag{6.30}$$

As in the case of trajectories, it is convenient to express the forces in terms of Cartesian rather than polar coordinates. Recognizing that $\cos(\theta) = x/r$ and $\sin(\theta) = y/r$, we can write the equations of motion as

$$\frac{-Gm_1m_2}{r^2} \cdot \frac{x}{r} = m_2\, a x \tag{6.31}$$

and

$$\frac{-Gm_1m_2}{r^2} \cdot \frac{y}{r} = m_2\, a y. \tag{6.32}$$

Because of MathCAD's memory limitations, we consider only one or two revolutions in these examples.

To limit the size of the equation block region, user-defined functions for distance, for the x and y components of the acceleration, and for the x and y components of the velocity are specified before the equation block is written. The value for (Gm_1m_2) is taken to be 1. These functions can then be used within the equation block. (As the equation block becomes larger, editing of the block becomes slower. The block we use here is small, so the changes are quick.)

• • Load MECH10, central force orbits (see Fig. 6.12).

Initial values are specified for the position, velocity, and acceleration. The components of acceleration and velocity are specified in terms of functions. In the equation block, we iterate the acceleration, velocity, and position for each component.

Alongside the equation block, we define the total velocity, the polar coordinates, and the distance. Once again, plots of various parameters help us to appreciate significant aspects of the motion.

Where does the orbital motion originate? Is the orbit circular? To what is i proportional (in other words, of what could i be considered a measure)?

Explain the general shapes of the x vs. i and y vs. i curves.

Examine the vx vs. i and vy vs. i curves. What is the location of the orbiting body when vx is at its minimum? At its maximum? When vy is at its maximum? At its minimum? Why are these curves nonsinusoidal?

In each of the plots for the components of acceleration and velocity that follow, include the origin. For example, plot ff, 0 vs. gg, 0. Let the plot type be lo. This choice of plot types marks the origin with an open rectangle and aids in interpretation.

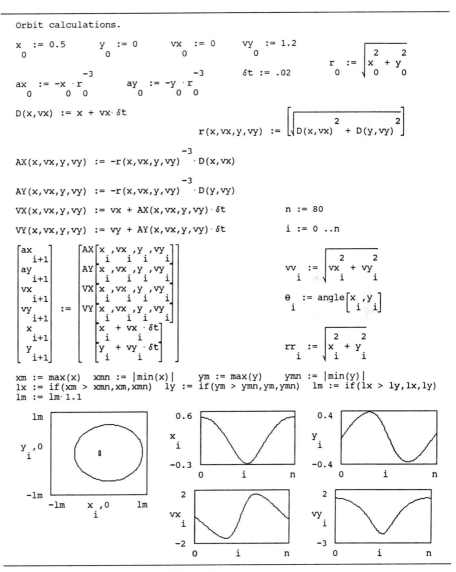

Figure 6.12 MECH10, central force orbits.

Plot vx vs. x and vy vs. y. Do the plots corroborate your answers to the previous questions?

Plot ax vs. x and ay vs. y. For each plot, carefully match where the body is in its orbit with a particular value for the acceleration. Given the initial conditions, you should know the direction in which the curves evolve in time. One quick way to check is to reduce significantly the number of iterations; then only a portion of the curve will be traced out. What is the relationship between the fact that the orbit is not circular and the shape of the ax vs. x curve?

Plot ax vs. y and ay vs. x. Explain the shapes of the curves. In which direction does the motion take them? Check by reducing the number of iterations.

Plot vv, $2 \cdot rr$ vs. i. (The factor of 2 is used to make the curvature of the rr curve more visible.) Explain the shape of this curve and the correlation between these two curves.

Examine the orbit for different initial conditions. For example, let $v_{x0} = 0$ and $v_{y0} = 1$. How would you characterize the orbit? Examine the plot regions and see what changes have occurred. Explain the changes.

Repeat for $v_{x0} = -0.5$ and $v_{y0} = 1$.

Repeat for $v_{x0} = -2$ and $v_{y0} = 1$.

One body orbiting two fixed gravitational centers is another interesting case. Very unusual orbital patterns are possible. The program may yield unphysical results if the orbiting body gets too close to an attracting center. If this happens, just change the initial conditions and try again.

• •   Load MECH11, orbits with two attracting centers (see Figs. 6.13 and 6.14).

The attractive centers are located at $(d, 0)$ and $(-d, 0)$. Observe how the previous document has been adapted. There are two distances and two components to the acceleration. The functions are similarly adapted. The algorithm remains the same.

The plots of x, y, vx, and vy vs. i are more complicated but still worth reading. For the initial case where $x_0 = 0.5$, $y_0 = 0$, $vx_0 = 0$, and $vy_0 = 1.8$, in which the orbit is not too complex, look at plots of ax vs. x and ay vs. y; the origin is included (plot type o). Interpret these plots; compare them with plots associated with a single attracting center.

Orbit. Two attracting centers.

$$x_0 := .5 \qquad y_0 := 0 \qquad vx_0 := 0 \qquad vy_0 := 1.8 \qquad \delta t := .01 \qquad d := .25$$

$$r1_0 := \sqrt{\left[x_0 - d\right]^2 + y_0^2} \qquad r2_0 := \sqrt{\left[x_0 + d\right]^2 + y_0^2}$$

$$ax_0 := -\left[x_0 - d\right] \cdot r1_0^{-3} - \left[x_0 + d\right] \cdot r2_0^{-3} \qquad ay_0 := -y_0 \cdot r1_0^{-3} - y_0 \cdot r2_0^{-3}$$

$$D1(x,vx) := (x - d) + vx \cdot \delta t \qquad D2(x,vx) := (x + d) + vx \cdot \delta t$$

$$D(y,vy) := y + vy \cdot \delta t$$

$$r1(x,vx,y,vy) := \left[\sqrt{D1(x,vx)^2 + D(y,vy)^2}\right]$$

$$r2(x,vx,y,vy) := \left[\sqrt{D2(x,vx)^2 + D(y,vy)^2}\right]$$

$$AX(x,vx,y,vy) := -r1(x,vx,y,vy)^{-3} \cdot D1(x,vx) - r2(x,vx,y,vy)^{-3} \cdot D2(x,vx)$$

$$AY(x,vx,y,vy) := -r1(x,vx,y,vy)^{-3} \cdot D(y,vy) - r2(x,vx,y,vy)^{-3} \cdot D(y,vy)$$

$$VX(x,vx,y,vy) := vx + AX(x,vx,y,vy) \cdot \delta t$$

$$VY(x,vx,y,vy) := vy + AY(x,vx,y,vy) \cdot \delta t \qquad n := 200 \qquad i := 0 \,..n$$

$$\begin{bmatrix} ax_{i+1} \\ ay_{i+1} \\ vx_{i+1} \\ vy_{i+1} \\ x_{i+1} \\ y_{i+1} \end{bmatrix} := \begin{bmatrix} AX[x_i,vx_i,y_i,vy_i] \\ AY[x_i,vx_i,y_i,vy_i] \\ VX[x_i,vx_i,y_i,vy_i] \\ VY[x_i,vx_i,y_i,vy_i] \\ x_i + vx_i \cdot \delta t \\ y_i + vy_i \cdot \delta t \end{bmatrix}$$

$$vv_i := \sqrt{vx_i^2 + vy_i^2}$$

$$\theta_i := \text{angle}\left[x_i,y_i\right]$$

Figure 6.13 MECH11, orbits with two attracting centers. (See the next figure for the rest of the document.)

Try various initial conditions. You may want to increase the number of iterations.

• • Adapt MECH8 so that the central body repels instead of attracts. Let the distance dependence remain the same. Create, examine, and interpret a set of plots similar to those in the original document.

"It's not my goddamn planet, monkey boy."

Buckaroo Banzai

```
xm := max(x)   xmn := |min(x)|   ym := max(y)   ymn := |min(y)|
lx := if(xm > xmn,xm,xmn) ly := if(ym > ymn,ym,ymn)
lm := if(lx > ly,lx,ly)·1.1
```

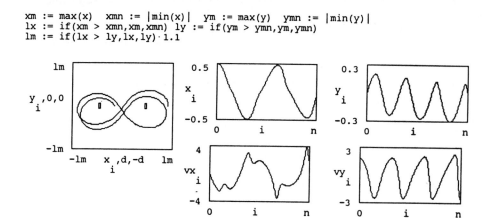

Figure 6.14 MECH11 *continued.*

C H A P T E R

7

Momentum and Collisions

The conservation laws lie at the very heart of physics. Quantities that are conserved include energy, linear momentum, angular momentum, and electric charge. For a quantity to be conserved, it means that whatever the amount of that conserved quantity is now, for some system; it will be the same in the future. For conservation to be true, the system must be closed to losses or additions. (See Richard Feynman's amusing story about energy in *The Character of Physical Law*.) Many quantities, of course, are not conserved. Kinetic energy, for example, may be conserved under particular circumstances, but, in general, it is not.

Clearly, an object's linear momentum ($\mathbf{p} = m\,\mathbf{v}$) may change. Newton's second law makes this explicit:

$$\sum \mathbf{F}_{\text{ext}} = \frac{d\mathbf{p}}{dt}.$$

However, if the sum of all external forces is zero, then $d\mathbf{p}/dt = 0$, and the momentum, $\mathbf{p}$, is a constant implying that $p_{\text{init}} = p_{\text{final}}$. For momentum, losses or additions would be in terms of external forces.

When two objects collide, each object exerts a force on the other. The forces comprise an action-reaction pair and are equal and opposite. For each individual object, the force exerted by the other is an external force and the momentum of the object changes. The momentum of neither object is conserved in a collision. However, if we say that our system consists of the two objects, then the forces are internal to the system and the total momentum of the system does not change, even though the momenta of individual parts of the system change.

Two special cases of collision are elastic and completely (or perfectly) inelastic. In an elastic collision, both kinetic energy and momentum are conserved. After a completely inelastic collision, the objects have the same final velocity.

7.1   Collisions in One Dimension

7.1.1   Inelastic Collisions

A perfectly inelastic collision is the easiest case to analyze. Momentum is conserved; initial momentum equals final momentum, and the final velocity of both masses is the same:

$$m_1 \, v_{1i} + m_2 \, v_{2i} = (m_1 + m_2) v_f. \tag{7.1}$$

• •   For a completely inelastic collision, with $v_{2i} = 0$, express the final velocity in terms of v_{1i} and the mass ratio $mr = m_2/m_1$. Define a function expressing the final velocity with the mass ratio as its argument, $v_f(mr)$. Plot $v_f(mr)$ as mr takes on the sequence of values $mr_i := 2^{i-5}$, where $i := 1, \ldots, 10$.

Ballistic Pendulum.   The ballistic pendulum is designed to permit the determination of velocities from inelastic collisions. An object such as a bullet is fired at, penetrates, and remains embedded in a pendulum bob. The rather complicated interaction process between the bullet and the pendulum bob (which, for example, could be a block of wood) can be treated as an inelastic collision. Following the collision, the pendulum swings through some angle; the maximum height or angle is determined.

After the collision occurs, mechanical energy — the sum of kinetic and potential energy — is conserved. This can be expressed as

$$KE_i + PE_i = KE_f + PE_f. \tag{7.2}$$

In this example, the potential energy is gravitational and is given by mgh. Let $h = 0$ when the pendulum is at its lowest point. As the pendulum swings following the impact, the elevation of the bob increases and the velocity of the bob decreases. The potential energy of the bob can be expressed in terms of the length, L, of the string supporting the bob and the angle through which the pendulum turns. When the pendulum is at an angle θ, the vertical distance from the pendulum bob to the support point is $L\cos(\theta)$. The height through which the pendulum has risen is the original distance from the support point, L, minus the final distance from the support point, or

$$h = L - L\cos(\theta) = L \cdot (1 - \cos(\theta)). \qquad (7.3)$$

In our statement of energy conservation, let the initial state be the instant after the collision occurs and let the final state occur when the bob reaches its maximum elevation (and has zero speed). Given these choices, $PE_i = 0$ and $KE_f = 0$. Thus, $KE_i = PE_f$.

●● Load COLL1, ballistic pendulum (see Fig. 7.1).

In COLL1, a bullet of mass m_b undergoes an inelastic collision with a ballistic pendulum of mass M. The momentum statement for an inelastic collision is expressed in terms of the final velocity. The final velocity is expressed in functional form; the initial velocity of the bullet, v_{bi}, is the parameter. The final velocity of the two masses immediately after the collision is returned by $v_f(v_{bi})$. Verify that this equation is correct.

In the solve blocks, there is only one equation. The equation is a statement of energy conservation, $KE_i = PE_f$, for the conditions mentioned above — $PE_i = 0$ and $KE_f = 0$. Is this the energy equation (where is the mass)? How can this problem be solved with an energy equation alone? Where is the momentum statement?

The maximum angle through which the pendulum can rotate can be either 90° or 180° depending on the structure of the pendulum. The initial momentum that will result in this angular displacement is a useful parameter of the system. Write an expression for that momentum, for each of the two cases. For a bullet of given mass, this translates into a maximum initial velocity.

Rewrite the equation for v_f in terms of the mass ratio $mr = M/mb$. Express v_f as $v_f(mr)$.

Determine the initial velocity that would result in the maximum possible height of the pendulum, as a function of the mass ratio. In the first solve block, change the find statement to a functional form, $vmx(mr)$. Plot the final velocity as a function of the mass ratio.

In the second solve block, a similar equation is solved. In this case, the final potential energy is not the maximum but that which occurs for a

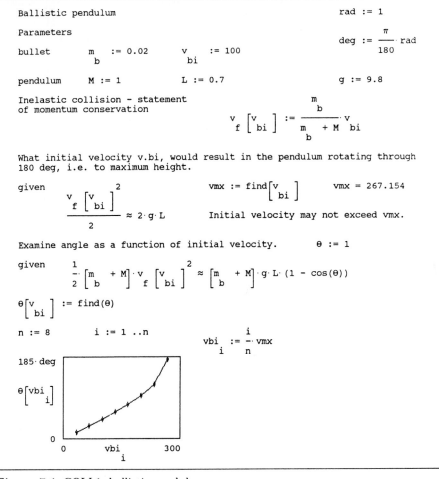

Ballistic pendulum

Parameters

bullet $m_b := 0.02$ $v_{bi} := 100$

pendulum $M := 1$ $L := 0.7$

$rad := 1$

$deg := \dfrac{\pi}{180} \cdot rad$

$g := 9.8$

Inelastic collision - statement of momentum conservation

$$v_f\begin{bmatrix} v_{bi} \end{bmatrix} := \frac{m_b}{m_b + M} \cdot v_{bi}$$

What initial velocity v.bi, would result in the pendulum rotating through 180 deg, i.e. to maximum height.

given

$$\frac{v_f\begin{bmatrix} v_{bi} \end{bmatrix}^2}{2} \approx 2 \cdot g \cdot L$$

$vmx := find\begin{bmatrix} v_{bi} \end{bmatrix}$ $vmx = 267.154$

Initial velocity may not exceed vmx.

Examine angle as a function of initial velocity. $\theta := 1$

given

$$\frac{1}{2} \cdot \begin{bmatrix} m_b + M \end{bmatrix} \cdot v_f\begin{bmatrix} v_{bi} \end{bmatrix}^2 \approx \begin{bmatrix} m_b + M \end{bmatrix} \cdot g \cdot L \cdot (1 - \cos(\theta))$$

$$\theta\begin{bmatrix} v_{bi} \end{bmatrix} := find(\theta)$$

$n := 8$ $i := 1 .. n$ $vbi_i := \dfrac{i}{n} \cdot vmx$

$185 \cdot deg$

$\theta\begin{bmatrix} vbi_i \end{bmatrix}$

0

Figure 7.1 COLL1, ballistic pendulum.

given angle θ. The solve block is written in functional form. Given a value for the initial velocity of the bullet, the function returns the maximum angle through which the pendulum swings.

How does the shape of this curve depend on the mass M?

The curve is not smooth near the 180° point. Draw a cubic spline curve through the points to show a smoother fit.

Write an expression for the ratio of kinetic energy just after the collision to the initial kinetic energy of the bullet. Express this ratio as a function of the mass ratio mr: $KEr(mr)$. Plot $KEr(mr)$ vs. mr.

Express the fractional loss in kinetic energy. Plot the fractional loss in kinetic energy as a function of the mass ratio.

•• Tarzan swings down from a tree limb, collides inelastically with Jane (saving her from impending doom), and continues to swing up to the branch of another tree. Consider this problem in three stages: (1) from the starting point to the instant just before the collision — conservation of energy, (2) during the collision — conservation of momentum, and (3) from the instant just after the collision to the final height — conservation of energy. Plot the final angle vs. the initial angle as a function of the mass ratio — $\theta_f(mr)$ vs. θ_i. If $M_T = 1.5\,M_J$, $L = 10\,m$, $\theta_o = 90°$, what is the final angle and final height? What would be the velocity of Tarzan just before the collision with Jane? Would this be a good way to save someone?

7.1.2 Elastic Collisions in One Dimension

In elastic collisions, both momentum and kinetic energy are conserved:

$$m_1 v_{1i} + m_2 v_{2i} = m_1 v_{1f} + m_2 v_{2f} \tag{7.4}$$

$$\frac{1}{2} m_1 v_{1i}{}^2 + \frac{1}{2} m_2 v_{2i}{}^2 = \frac{1}{2} m_1 v_{1f}{}^2 + \frac{1}{2} m_2 v_{2f}{}^2. \tag{7.5}$$

Three relations are commonly extracted from these two equations. One is the statement of relative velocities before and after a collision:

$$v_{1i} - v_{2i} = v_{2f} - v_{1f}. \tag{7.6}$$

This statement is readily determined using the momentum and energy equations. It shows that the rate at which particle one approaches particle two, before the collision, is the same as the rate at which particle two moves away from particle one after the collision. Verify the equation.

The other two relations express final velocities in terms of initial velocities:

$$v_{1f} = \left(\frac{m_1 - m_2}{m_1 + m_2}\right) v_{1i} + \left(\frac{2\,m_2}{m_1 + m_2}\right) v_{2i} \tag{7.7}$$

$$v_{2f} = \left(\frac{2\,m_1}{m_1 + m_2}\right) v_{1i} + \left(\frac{m_2 - m_1}{m_1 + m_2}\right) v_{2i}. \tag{7.8}$$

If $m_1 \gg m_2$ and $v_{2i} = 0$, then $v_{1f} \simeq v_{1i}$ and $v_{2f} \simeq 2\,v_{1i}$.
If $m_2 \gg m_1$ and $v_{2i} = 0$, then $v_{1f} \simeq -v_{1i}$ and $v_{2f} \simeq 0$.

•• When $v_{2i} = 0$, show that the two previous expressions are consistent with the statement about relative velocities before and after a collision.

• •   Consider an elastic collision where $m_1 := 1.5$, $m_2 = 1$, $v_{1i} = 1$, and $v_{2i} = -0.2$. Use a solve block to solve the momentum and energy relations directly for v_{1f} and v_{2f}. Using the results, evaluate the momentum and energy of the individual terms and check that the results are consistent.

Let $m_1 = m_2$. Try several different values of v_{1i} and v_{2i}. Include cases where $v_{1i} > 0$ and $v_{2i} < 0$. In each case evaluate $v_{1i} - v_{2i}$ and $v_{1f} - v_{2f}$. Compare v_{1f} with v_{2i} and v_{2f} with v_{1i}. Do the particles exchange velocities when the masses are equal? Don't base your answer on one example.

In the solve block you set up to solve for v_{1f} and v_{2f}, change the find statement from

$$\begin{bmatrix} v_{1f} \\ v_{2f} \end{bmatrix} = \text{find}((v_{1f}, v_{2f})$$

to

$$f(mr) = \text{find}(v_{1f}, v_{2f})$$

where mr is the mass ratio m_1/m_2. Of course, in this case, $v_{1f} = f(mr)_0$. Plot v_{1f} vs. mr. Plot v_{2f} vs. mr. Plot $(v_{2f} - v_{1f})/(v_{1i} - v_{2i})$ vs. mr, where $mr_i := 0.1 \cdot 2^i$, where $i := 0 \ldots 7$.

Let $v_{2i} = 0$. Plot KE_{2f}/KE_{1f} vs. mr. Plot p_{2f}/p_{1f} vs. mr. Repeat with v_{2i} negative and v_{2i} positive (but for the positive case, less than v_{1i}).

7.1.3   Displaying the Motion

The next document helps us to visualize the one-dimensional collision process in both the laboratory frame (a reference frame, stationary with respect to the laboratory in which the particles are observed) and the center-of-mass frame (a reference frame in which the center of mass is stationary).

For the case of particles of equal mass, where one particle is on a collision course with a second particle which is initially at rest. In the laboratory frame, we see the first mass move toward the second with velocity v, collide, and stop. The second mass, initially at rest, is struck and moves away with the same velocity, v. In the center-of-mass frame, motion is observed relative to the center of mass of the system. Before the collision, both particles approach the center of mass with velocity $v/2$; after the collision both move away from it with the same velocity. The particular velocities depend on the relative mass values.

• •   Load COLL2, elastic collision in one dimension (see Figs. 7.2 and 7.3).

The first plot (plot type s) shows the motion in a sequence at uniform time intervals in the laboratory frame. Time increases from top to bottom. The horizontal width of the steps is proportional to the velocity. Initially, particle one approaches particle two, which is at rest. After the collision, m_1, being larger than m_2, continues to move in the forward direction, and m_2 moves off with a larger velocity than m_1 had before the collision.

In the second plot, the same information is portrayed but in a different manner. The graph is rotated so that time proceeds from left to right, and the format style is changed. The length of the bars is proportional to velocity. The solid line shows the center-of-mass motion. Both before

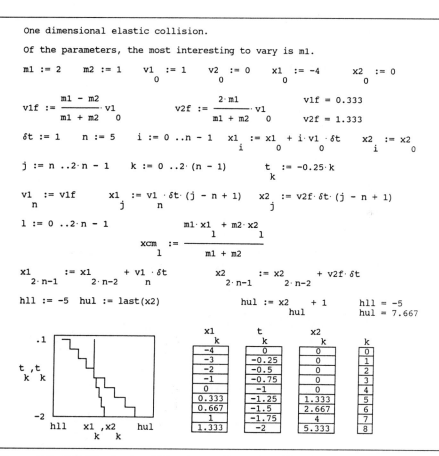

One dimensional elastic collision.

Of the parameters, the most interesting to vary is m1.

$$m1 := 2 \qquad m2 := 1 \qquad v1_0 := 1 \qquad v2_0 := 0 \qquad x1_0 := -4 \qquad x2_0 := 0$$

$$v1f := \frac{m1 - m2}{m1 + m2} \cdot v1_0 \qquad v2f := \frac{2 \cdot m1}{m1 + m2} \cdot v1_0 \qquad \begin{array}{l} v1f = 0.333 \\ v2f = 1.333 \end{array}$$

$$\delta t := 1 \qquad n := 5 \qquad i := 0 .. n - 1 \qquad x1_i := x1_0 + i \cdot v1_0 \cdot \delta t \qquad x2_i := x2_0$$

$$j := n .. 2 \cdot n - 1 \qquad k := 0 .. 2 \cdot (n - 1) \qquad t_k := -0.25 \cdot k$$

$$v1_n := v1f \qquad x1_j := v1_n \cdot \delta t \cdot (j - n + 1) \qquad x2_j := v2f \cdot \delta t \cdot (j - n + 1)$$

$$l := 0 .. 2 \cdot n - 1 \qquad \qquad xcm_l := \frac{m1 \cdot x1_1 + m2 \cdot x2_1}{m1 + m2}$$

$$x1_{2 \cdot n-1} := x1_{2 \cdot n-2} + v1_n \cdot \delta t \qquad x2_{2 \cdot n-1} := x2_{2 \cdot n-2} + v2f \cdot \delta t$$

$$hll := -5 \quad hul := last(x2) \qquad \qquad hul := x2_{hul} + 1 \qquad \begin{array}{l} hll = -5 \\ hul = 7.667 \end{array}$$

$x1_k$	t_k	$x2_k$	k
-4	0	0	0
-3	-0.25	0	1
-2	-0.5	0	2
-1	-0.75	0	3
0	-1	0	4
0.333	-1.25	1.333	5
0.667	-1.5	2.667	6
1	-1.75	4	7
1.333	-2	5.333	8

Figure 7.2 COLL2, elastic collision in one dimension. (See the next figure for the rest of the document.)

$kk := 0 \,..\, 2 \cdot n - 3$

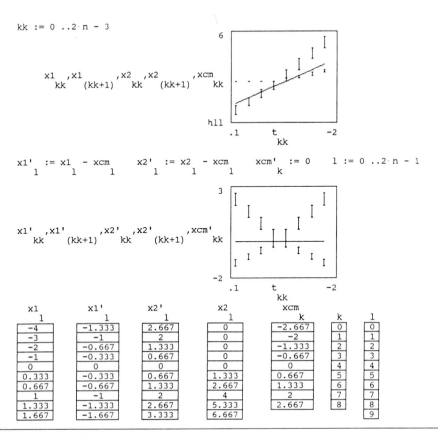

$x1_{kk} \quad ,x1_{(kk+1)} \quad ,x2_{kk} \quad ,x2_{(kk+1)} \quad ,xcm_{kk}$

hll

$.1 \qquad t_{kk} \qquad -2$

$x1'_1 := x1_1 - xcm_1 \qquad x2'_1 := x2_1 - xcm_1 \qquad xcm'_k := 0 \qquad 1 := 0 \,..\, 2 \cdot n - 1$

$x1'_{kk} \quad ,x1'_{(kk+1)} \quad ,x2'_{kk} \quad ,x2'_{(kk+1)} \quad ,xcm'_{kk}$

$.1 \qquad t_{kk} \qquad -2$

$x1_1$	$x1'_1$	$x2'_1$	$x2_1$	xcm_k	k	l
-4	-1.333	2.667	0	-2.667	0	0
-3	-1	2	0	-2	1	1
-2	-0.667	1.333	0	-1.333	2	2
-1	-0.333	0.667	0	-0.667	3	3
0	0	0	0	0	4	4
0.333	-0.333	0.667	1.333	0.667	5	5
0.667	-0.667	1.333	2.667	1.333	6	6
1	-1	2	4	2	7	7
1.333	-1.333	2.667	5.333	2.667	8	8
1.667	-1.667	3.333	6.667			9

Figure 7.3 COLL2 *continued.*

and after the collision, the motions of both m_1 and m_2 are displayed, although, in this case, before the collision the bars have zero length for mass m_2 as it has zero velocity.

The third plot shows the motion from the center-of-mass frame. Again, time increases from left to right. Before the collision, both masses approach the center of mass; after the collision, both move away from the center. The particle of lower mass has the larger velocity in each case.

Try the following values for $(m_1, m_2, v_{10} \, v_{20})$: $(2, 1, 1, 0)$, $(1, 1, 2, -1)$, $(1, 1, 2, 1)$, $(10, 1, 1, 0)$, $(0.1, 1, 1, 0)$. Try values of your own. Keep v_2 small; otherwise the collision may not occur in the region plotted. You can, of course, change the plot parameters.

7.2   Collisions in Two Dimensions

The principle to be applied, that of momentum conservation, is the same for collisions in two dimensions as for collisions in one dimension. However, since momentum is a vector, conservation applies for each component. We consider the case where $v_{2i} = 0$. The momentum statements are

$$m_1 v_{1i} = m_1 v_{1f} \cos(\theta) + m_2 v_{2f} \cos(\phi) \qquad (7.9)$$

$$0 = m_1 v_{1f} \sin(\theta) + m_2 v_{2f} \sin(\phi), \qquad (7.10)$$

where θ and ϕ are the angles that the paths of particles m_1 and m_2 make with the horizontal. Angles θ and ϕ lie on opposite sides of the horizontal, the direction of the initial path of m_1.

7.2.1   Elastic Collisions

If the collision is elastic then the statement of conservation of kinetic energy applies:

$$m_1 v_{1i}{}^2 = m_1 v_{1f}{}^2 + m_2 v_{2f}{}^2. \qquad (7.11)$$

If the masses m_1 and m_2 are equal, the following relationships can be deduced for elastic collisions:

$$\tan(\theta) \cdot \tan(\phi) = 1 \qquad (7.12)$$

$$v_{1f} = \frac{\sin(\phi)}{\sin(\theta + \phi)} v_{1i} \qquad (7.13)$$

$$v_{2f} = \frac{\sin(\theta)}{\sin(\theta + \phi)} v_{1i} \qquad (7.14)$$

• •   Examine the relationship between $\tan(\theta)$ and $\tan(\phi)$. Define $\phi(\theta)$, and plot $\phi(\theta)$ vs. θ as $\theta = 1, 5, \ldots, 90$ (degrees). Plot $\phi(\theta) + \theta$ vs. θ. How can you simplify the relationships for v_{1f} and v_{2f}?

In a grazing collision, when only a slight interaction occurs, what are the approximate values for the scattering angles of $m_1(\theta)$ and $m_2(\phi)$? What are the final velocities? In a nearly head-on collision, what are the values of θ and ϕ? What are the final velocities?

Let $v_{1i} = 1$ (we consider the case $v_{2i} = 0$). Define $v_{1f}(\theta)$ and $v_{2f}(\theta)$, and plot both vs. θ. (Note: A function $f(\theta) = \sin(\phi(\theta))$ returns the sine of ϕ, the ϕ selected being that which corresponds to the θ specified.)

Define the momenta, $p_{1fv}(\theta)$ and $p_{1fh}(\theta)$, where v and h refer to vertical/horizontal or transverse/parallel to the initial momentum. Plot the final momentum components as a function of θ. What would the corresponding curves for p_2 look like?

If the masses m_1 and m_2 are not equal, relations analogous to those given above are

$$\tan(\theta) = \frac{\sin(2\,\phi)}{m_1/\,m_2 - \cos(2\,\phi)} \tag{7.15}$$

$$v_{1f} = \frac{\sin(\phi)}{\sin(\theta + \phi)}\, v_{1i} \tag{7.16}$$

$$v_{2f} = \frac{m_1}{m_2} \cdot \frac{\sin(\theta)}{\sin(\theta + \phi)}\, v_{1i} \tag{7.17}$$

• • Show that these equations reduce to the previous set if $m_1 = m_2$.

One effect of collisions with unequal masses is that the possible range of angles for θ is affected. If $m_1 > m_2$, then θ has a maximum value less than 90°. If $m_1 < m_2$, then θ can range from 0° to 180°. However, ϕ is always restricted to the range from 0° to 90°.

• • Determine the maximum value for θ if $m_1 = 2 \cdot m_2$.

• • Load COLL3, two-dimensional collision: unequal masses (see Fig. 7.4).

Here, θ is written as a function of ϕ. The arctangent function is too restrictive in the range of angles returned and, although shown, is disabled. The angle function is used instead because it returns angles in the desired range.

For simplicity, leave the value for m_2 equal to 1 and change the value of m_1. In this document, do *not* let $m_1 = m_2$: let $m_1 > 1$ or $m_1 < 1$. Because m_1 is assigned its value using a global equality, it can be moved near plot regions farther down in the document when they are examined.

Examine the θ vs. ϕ plot ($m_1 = 0.5$, $m_2 = 1$). Interpret this case. Ask specific questions that direct you to particular cases. For example, if ϕ is approximately zero, what is θ? Why? When ϕ is $\pi/2$, what is θ? Why? How is the plot of $\theta + \phi$ different from the case where the masses are equal?

Let $m_1 = 2$ and process. This curve is quite different. In the plot of θ vs. ϕ, there are two different possible values of ϕ for a given θ. Explain how this is possible. Consider specific cases. Why should there be a maximum? How is the plot of $\theta + \phi$ different from the case where the masses are equal? How is this plot different from the case where $m_1 < m_2$?

Directly beneath the plot regions, two calculations are set up. The first, to locate the maximum of the θ vs. ϕ curve, is valid when $m_1 > m_2$. The second, to find the angle ϕ where m_1 is scattered through an angle of 90°, is valid when $m_1 < m_2$.

Let $m_1 = 1.1, 1.3, 1.5, 2, 4, 10, 50, 10^3, 10^5, 10^8$. Observe the θ vs. ϕ curve and the values of ϕ and θ at which the maximum of the curve occurs. To plot these results, after each calculation enter the values by

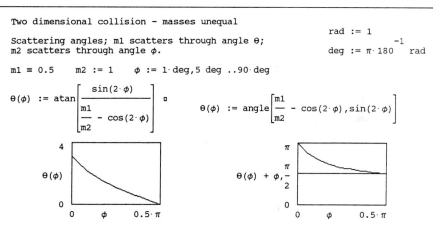

Two dimensional collision - masses unequal

Scattering angles; m1 scatters through angle θ;
m2 scatters through angle φ.

rad := 1

deg := $\pi \cdot 180^{-1}$ rad

m1 ≡ 0.5   m2 := 1   φ := 1·deg,5 deg ..90·deg

$$\theta(\phi) := \operatorname{atan}\left[\frac{\sin(2\cdot\phi)}{\frac{m1}{m2} - \cos(2\cdot\phi)}\right]□$$

$$\theta(\phi) := \operatorname{angle}\left[\frac{m1}{m2} - \cos(2\cdot\phi), \sin(2\cdot\phi)\right]$$

In the following two calculations, one to find the maximum of the θ vs φ curve, the other to determine the angle φ, at which θ equals 90 degrees, only one makes sense at a time. The first calculation, for the maximum, is valid when when m1 > m2. The second calculation, for φ(90), is valid when m1 < m2. Either the θ vs φ curve has a maximum or θ can exceed 90 degrees. Both conditions cannot be met. For the case m1 = m2 neither calculation is valid.

The maximum angle for φ and the θ which corresonds; valid when m1 > m2.

$$g := \frac{\pi}{8} \qquad \phi mx := \operatorname{root}\left[\frac{d}{dg}\theta(g), g\right] \qquad \phi mx = 1.385\cdot10^{9}\cdot deg \qquad \theta(\phi mx) = 335.462\cdot deg$$

The angle φ when m1 is scattered at 90 degrees; valid when m1 < m2.

φ90 := 1   given $\theta(\phi90) \approx \frac{\pi}{2}$ φ90 := find(φ90)   φ90 = 30·deg

v1i := 1   v2i := 0

$$v1f(\phi) := \frac{\sin(\phi)}{\sin(\theta(\phi)+\phi)}\cdot v1i \qquad v2f(\phi) := \frac{m1}{m2}\cdot\frac{\sin(\theta(\phi))}{\sin(\theta(\phi)+\phi)}\cdot v1i$$

Figure 7.4 COLL3, two-dimensional collision: unequal masses.

hand into tables. There are ten values of m_1. Let $j := 1\ldots10$. Insert values into mr_j, ϕp_j, and θp_j. Once the tables are complete, plot ϕp vs. mr and θp vs. mr; a log scale for mr is appropriate.

When the value of m_1 is less than that of m_2, m_1 can be scattered through angles greater than 90°. When m_1 is scattered through 90°, $\phi90$ is the corresponding ϕ angle.

Let m_1 take on values equal to the inverse of the previous set, that is, $1/1.1$, $1/1.3$, …. Again, put values in a table after each calculation; plot $\phi90$ vs. mr.

In the next plot region, the final velocities are plotted as a function of the scattering angle, ϕ, of m_2. Examine how these curves change as the relative mass of the two particles change. Does changing the initial velocity change the shape of the curves? Plot the momentum associated with each.

If $v_{1f} = v_{2f}$, how do the angles (θ, ϕ) vary with the mass ratio?

For a given θ $(m_1 > m_2)$, find the two values of ϕ and the corresponding velocities.

For a given mass ratio, for what angles are θ and ϕ the same? Explain.

Plot the energies of the two particles as a function of ϕ. Similarly, plot the momenta.

7.2.2 Displaying the Motion

The next two documents are an aid in visualizing the scattering process. The first provides a laboratory-frame view, the second a center-of-mass frame view. After the previous exercises, you will appreciate their content more.

• • Load COLL4, two-dimensional elastic scattering: laboratory frame (see Fig. 7.5).

The equations are familiar by now. The values for m_1, m_2, and ϕ use the global equality and are located just above the plot region. The incident mass, m_1, is represented with a series of diamonds; the target, m_2, is represented with open rectangles. The separation between adjacent plot points is a measure of the relative velocity.

For a series of values of m_1, observe the scattering process. Examine the condition of the two different scattering angles ϕ for one θ for $m_1 > m_2$.

What changes would be required to input m_1, m_2, and θ (instead of ϕ)?

• •   Load COLL5, two-dimensional elastic scattering: center-of-mass frame (see Fig. 7.6).

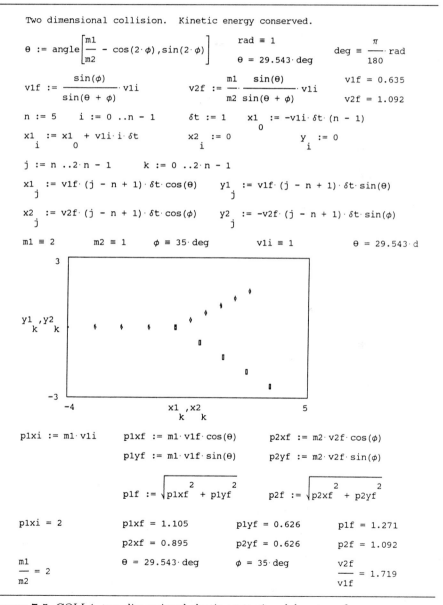

Two dimensional collision.   Kinetic energy conserved.

$$\theta := \text{angle}\left[\frac{m1}{m2} - \cos(2 \cdot \phi), \sin(2 \cdot \phi)\right]$$

$$\text{rad} \equiv 1$$
$$\theta = 29.543 \cdot \text{deg}$$
$$\text{deg} \equiv \frac{\pi}{180} \cdot \text{rad}$$

$$v1f := \frac{\sin(\phi)}{\sin(\theta + \phi)} \cdot v1i \qquad v2f := \frac{m1}{m2} \cdot \frac{\sin(\theta)}{\sin(\theta + \phi)} \cdot v1i$$

$$v1f = 0.635$$
$$v2f = 1.092$$

$$n := 5 \qquad i := 0 \ ..n - 1 \qquad \delta t := 1 \qquad x1_0 := -v1i \cdot \delta t \cdot (n - 1)$$

$$x1_i := x1_0 + v1i \cdot i \cdot \delta t \qquad x2_i := 0 \qquad y_i := 0$$

$$j := n \ ..2 \cdot n - 1 \qquad k := 0 \ ..2 \cdot n - 1$$

$$x1_j := v1f \cdot (j - n + 1) \cdot \delta t \cdot \cos(\theta) \qquad y1_j := v1f \cdot (j - n + 1) \cdot \delta t \cdot \sin(\theta)$$

$$x2_j := v2f \cdot (j - n + 1) \cdot \delta t \cdot \cos(\phi) \qquad y2_j := -v2f \cdot (j - n + 1) \cdot \delta t \cdot \sin(\phi)$$

$$m1 \equiv 2 \qquad m2 \equiv 1 \qquad \phi \equiv 35 \cdot \text{deg} \qquad v1i \equiv 1 \qquad \theta = 29.543 \cdot d$$

$$p1xi := m1 \cdot v1i \qquad p1xf := m1 \cdot v1f \cdot \cos(\theta) \qquad p2xf := m2 \cdot v2f \cdot \cos(\phi)$$

$$p1yf := m1 \cdot v1f \cdot \sin(\theta) \qquad p2yf := m2 \cdot v2f \cdot \sin(\phi)$$

$$p1f := \sqrt{p1xf^2 + p1yf^2} \qquad p2f := \sqrt{p2xf^2 + p2yf^2}$$

$$p1xi = 2 \qquad p1xf = 1.105 \qquad p1yf = 0.626 \qquad p1f = 1.271$$

$$p2xf = 0.895 \qquad p2yf = 0.626 \qquad p2f = 1.092$$

$$\frac{m1}{m2} = 2 \qquad \theta = 29.543 \cdot \text{deg} \qquad \phi = 35 \cdot \text{deg} \qquad \frac{v2f}{v1f} = 1.719$$

Figure 7.5 COLL4, two-dimensional elastic scattering: laboratory frame.

The motion of m_1, the incident particle, is represented with diamonds. The motion of m_2, the target, is represented with open rectangles. The center of mass is represented with plus signs.

In the first plot region, a laboratory-frame view of the motion of the center of mass is included. In the second plot region, the center of mass is located at the origin. Motion is described in relation to this point. The separation between adjacent plot points is a measure of the relative velocity.

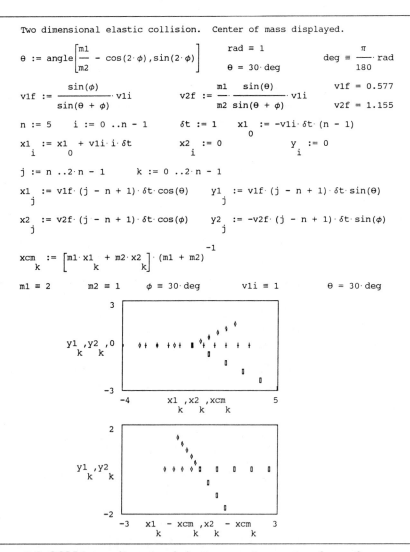

Two dimensional elastic collision. Center of mass displayed.

$$\theta := \text{angle}\left[\frac{m1}{m2} - \cos(2\cdot\phi), \sin(2\cdot\phi)\right]$$

$$\text{rad} \equiv 1 \qquad\qquad \deg \equiv \frac{\pi}{180}\cdot\text{rad}$$

$$\theta = 30\cdot\deg$$

$$\text{v1f} := \frac{\sin(\phi)}{\sin(\theta + \phi)}\cdot\text{v1i} \qquad \text{v2f} := \frac{m1}{m2}\cdot\frac{\sin(\theta)}{\sin(\theta + \phi)}\cdot\text{v1i}$$

$$\text{v1f} = 0.577$$
$$\text{v2f} = 1.155$$

$$n := 5 \qquad i := 0\ ..n - 1 \qquad \delta t := 1 \qquad x1_0 := -\text{v1i}\cdot\delta t\cdot(n - 1)$$

$$x1_i := x1_0 + \text{v1i}\cdot i\cdot\delta t \qquad x2_i := 0 \qquad\qquad y_i := 0$$

$$j := n\ ..2\cdot n - 1 \qquad k := 0\ ..2\cdot n - 1$$

$$x1_j := \text{v1f}\cdot(j - n + 1)\cdot\delta t\cdot\cos(\theta) \qquad y1_j := \text{v1f}\cdot(j - n + 1)\cdot\delta t\cdot\sin(\theta)$$

$$x2_j := \text{v2f}\cdot(j - n + 1)\cdot\delta t\cdot\cos(\phi) \qquad y2_j := -\text{v2f}\cdot(j - n + 1)\cdot\delta t\cdot\sin(\phi)$$

$$\text{xcm}_k := \left[m1\cdot x1_k + m2\cdot x2_k\right]\cdot(m1 + m2)^{-1}$$

$$m1 \equiv 2 \qquad m2 \equiv 1 \qquad \phi \equiv 30\cdot\deg \qquad \text{v1i} \equiv 1 \qquad \theta = 30\cdot\deg$$

Figure 7.6 COLL5, two-dimensional elastic scattering: center-of-mass frame.

Calculate momenta and energies in the center-of-mass frame. Try various set of values.

We have discussed only the cases of elastic or completely inelastic collisions. These are two extremes. For the case where the kinetic energy is reduced but is greater than that of the totally inelastic case, the restrictions on the scattering angles are still different. It is also possible that during the scattering process, an interaction can occur which will increase the kinetic energy. This, too, changes the relationship for the scattering angles. These cases are worth exploring if time is available.

7.3   Momentum and Rockets

You've probably heard the question about firing a high-powered rifle: which gets more momentum, the bullet or the rifle? The forces arising from the burning of gunpowder are internal, not external, to the gun-bullet system. Momentum is conserved. The momentum of the bullet and the rifle (plus whatever may be coupled to the rifle) are equal and opposite and sum to zero, the momentum before the rifle was fired.

(The problem of gunpowder is a totally separate but interesting problem. The gunpowder should not burn instantaneously; it should continue to burn as the bullet is propelled down the barrel. How should the pellets of gunpowder be shaped so that they burn at a constant rate as they are consumed and do not have a huge peak rate, as they would if they were spherically symmetric? For spherical objects, the ratio of surface area to volume is proportional to $1/r$. As spherical pellets burn, the radius is reduced and the surface area per mass of powder increases rapidly. The resulting huge rise in pressure associated with the end of the burn could be extremely dangerous. Gunpowder pellets are not spherical. One not uncommon design is that of small, hollow cylinders. In this case, the surface area and burn rate remain nearly constant as the pellets burn from both inside and outside.)

The problem of bullet and rifle can be extended to that of a machine gun and to that of a rocket. It is interesting to see how the discrete case (bullets) extends to the continuous case (ejecta from burning rocket fuel).

Consider first a machine gun or device capable of emitting units of mass. Imagine the device to be in interstellar space, far from any massive object and free to move without resistance (or free to move on a frictionless horizontal track in a vacuum). The motion is one-dimensional; there are no rotations. Also, we consider all the mass to be either in the gun or in the bullets.

A central question we would like to consider is: if the total mass of bullets is fixed, will the final velocity of the gun be any different if there is a very small number of large bullets or a large number of small bullets? Other questions can be asked. Each time a bullet is fired, the speed of the gun increases; how does the change in velocity of the rocket vary with each successive firing? Will doubling the velocity of the bullets result in doubling the final velocity of the rocket?

• •   Load COLL6, machine gun/rocket (see Fig. 7.7).

We approach the problem by specifying the mass of the rocket and the total mass and number of the bullets. We then: (1) determine the mass remaining in the rocket after each successive firing (rocket mass plus mass of unfired bullets), (2) calculate the change in velocity due to each succesive firing, (3) determine the velocity as a function of time by summing the changes.

The initial mass, M_0, is the sum of the rocket mass, $mroc$, plus the sum of all the bullets, $mbtot$. The mass of an individual bullet is specified

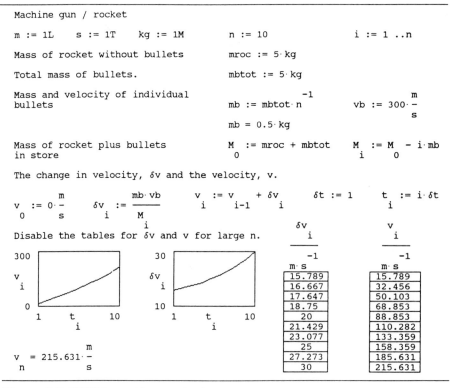

Figure 7.7 COLL6, machine gun/rocket.

in terms of the number of bullets, n, and the total mass of the bullets. The velocity of the bullets, v_b, is also specified. The rocket starts with velocity v_0. The change in velocity of the rocket each time a bullet is fired is δv, determined from the equation for momentum conservation: $m_b v_b = M_i |\delta v_i|$. The mass of the rocket decreases each time a bullet is fired; thus the change in velocity that the rocket experiences increases as each bullet is fired. We plot both the velocity and change in velocity of the rocket as a function of time; we assume the firings are regularly spaced in time.

The document is initially set with the mass of the rocket and the mass of the bullets equal. Keep the total mass constant at 10 kg. Explore the final velocity of the rocket as its mass is a decreasing fraction of the total. Let the final mass be 1/2, 1/4, then 1/8 of the total. What functional form does this suggest?

How does the final velocity depend on the total number of bullets, if the mass of bullets remains constant?

How does the final velocity depend on the velocity of the bullets? Will doubling the velocity of the bullets double the final velocity of the rocket?

If there were 100 bullets, how must *mroc* and *mbtot* be divided so that the final velocity of the rocket is ten times that of the bullet v_b?

Rewrite the document so that you can plot the final velocity as a function of the number of bullets. See COLL7 for a solution.

In COLL7, we calculate the final velocity for a series of cases. When $i = 1$, there is one bullet with a mass of 5 kg. When $i = 2$, there are two bullets with a mass of 2.5 kg; etc. The bullet mass for each successive i is specified by mb_i. The index j is associated with the sequence of firings for a given i; for example, if $i = 4$, then j goes from 1 to 4. $M_{i,j}$ gives the mass remaining after each firing; $\delta v_{i,j}$ give the changes in velocity; $v_{i,j}$ show the sequence of rocket velocities.

The final velocity is $v(i, n)$, where i is related to the number and size of the bullets. The velocity is a maximum when $i = 1$, that is, when there is one huge bullet. However, as the number of bullets increases and the mass per bullet decreases, the velocity decreases — but not indefinitely. It approaches a limit. The tables show some sequences of δv and the velocity sequences.

The equation for change in velocity,

$$\delta v_i = \frac{m_b v_b}{M_i},$$

is for bullets of finite size. Properly taking into account the direction of acceleration of the system, the change in velocity for the infinitesimal case is

$$dv = - v_b \frac{dM}{M}. \tag{7.18}$$

Integrating this expression from the initial conditions of velocity, v_o, and mass, M_o, we get

$$\int_{v_o}^{v} dv = - v_b \int_{M_o}^{M} \frac{dM}{M}. \tag{7.19}$$

The result is

$$v - v_o = v_b \ln\left(\frac{M_o}{M}\right). \tag{7.20}$$

In our example, v_o is zero. Does the final velocity of our calculation approach $v_b \ln(M_o / M)$ as the number of bullets increases and the size decreases? For the case of 10 bullets, what is the percentage difference between the associated final velocity and that determined from the continuous integrated case? A number of variations can be explored here.

"It's alright. They'll fix you. They fix everything."

Robocop

Rotational Motion

The kinematic equations for rotational motion are in one-to-one correspondence with the kinematic equations for linear motion. Linear velocity
and linear acceleration are represented by the definitions

$$v = \frac{dx}{dt} \qquad a = \frac{dv}{dt},$$

angular velocity and angular acceleration are similarly defined:

$$\omega = \frac{d\theta}{dt} \qquad \alpha = \frac{d\omega}{dt}. \tag{8.1}$$

Any linear kinematic equation has its angular analogue. If you know one
set, you know the other. For example,

$$x = x_o + v_o t + \frac{1}{2} a t^2$$

and

$$\theta = \theta_o + \omega_o t + \frac{1}{2} \alpha t^2. \tag{8.2}$$

The linear and angular relations are not independent. Consider a circle
of radius r. Measure along the circumference a linear distance, an arc

length, s, equal to one radius. The angle subtended by this arc length is, by definition, one radian. That is, when $s = r$, $\theta = 1$. As the arc length increases or decreases, the angle subtended increases or decreases proportionately; $s \propto \theta$. Combining these two concepts, you can see that

$$s = r\,\theta. \tag{8.3}$$

This relation connects linear and angular displacement and indirectly is the link to similar equations for velocity and acceleration.

For constant r, taking a derivative with respect to time yields

$$\frac{ds}{dt} = r\,\frac{d\theta}{dt}$$

or

$$v = r\,\omega, \tag{8.4}$$

where $\omega = d\theta/dt$. A second derivative with respect to time yields

$$\frac{d^2 s}{dt^2} = r\,\frac{d^2\theta}{dt^2} \quad \text{or} \quad \frac{dv}{dt} = r\,\frac{d\omega}{dt}$$

or

$$a = r\,\alpha. \tag{8.5}$$

8.1   Cycloidal Motion

When an object with a circular cross section rolls without slipping, the object translates as it rotates. The distance the center moves and the angle through which the object turns are related by $s = r\,\theta$. For example, if the object (for simplicity, we will refer to it as a cylinder) turns through one revolution as it rolls, the center moves one circumference. The velocity and acceleration of the center of the cylinder are given by $v = r\,\omega$ and $a = r\,\alpha$, respectively.

If the cylinder rolls with constant angular velocity, the center translates with constant linear velocity. The position of the center is given by

$$x_c = v_c\,t \quad \text{or} \quad x_c = R\omega\,t, \tag{8.6}$$

where R is the rolling radius. We note that each revolution corresponds to 2π radians or $\#radians = 2\pi\,\#revs$ and that

$$\omega = 2\pi\,f, \tag{8.7}$$

where f is the frequency in revolutions or cycles per second (Hz). This permits us to express the motion of the center of our cylinder as

$$x_c = R\,2\pi\,ft = C\!f t. \tag{8.8}$$

We describe the motion of any other point on the end face of the cylinder as a combination of translation and rotation. The location of any point on the cylinder, relative to the center, is given by $x = r\cos(\theta)$ and $y = r\sin(\theta)$, where θ is measured counterclockwise from the x-axis and r is the distance from the center. If the cylinder rotates uniformly with constant angular velocity, ω, then relative to the center the locations are written

$$x = r\cos(\omega t + \theta_o) \qquad y = r\sin(\omega t + \theta_o). \qquad (8.9)$$

If the cylinder is not spinning but rolling, the motions relative to the center are unchanged. The net motion is

$$x = R\omega t + r\cos(-\omega t + \theta_o) \qquad y = R + r\sin(-\omega t + \theta_o) \qquad (8.10)$$

The minus sign is present because the rotation due to the rolling motion in the positive x direction is clockwise rather than counterclockwise.

● ●   Load ROT1, cycloidal motion (see Fig. 8.1).

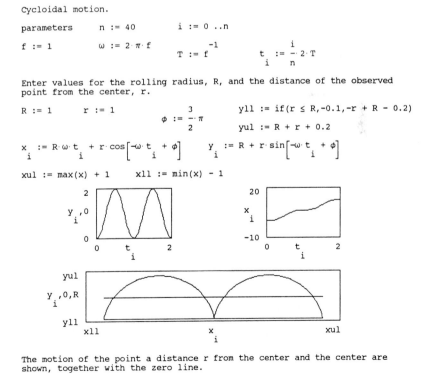

Cycloidal motion.

parameters $n := 40$ $i := 0\ ..n$

$f := 1$ $\omega := 2 \cdot \pi \cdot f$ $T := f^{-1}$ $t_i := \dfrac{i}{n} \cdot 2 \cdot T$

Enter values for the rolling radius, R, and the distance of the observed point from the center, r.

$R := 1$ $r := 1$ $\phi := \dfrac{3}{2} \cdot \pi$ $yll := if(r \le R, -0.1, -r + R - 0.2)$

$yul := R + r + 0.2$

$x_i := R \cdot \omega \cdot t_i + r \cdot \cos\left[-\omega \cdot t_i + \phi\right]$ $y_i := R + r \cdot \sin\left[-\omega \cdot t_i + \phi\right]$

$xul := \max(x) + 1$ $xll := \min(x) - 1$

The motion of the point a distance r from the center and the center are shown, together with the zero line.

Figure 8.1 ROT1, cycloidal motion.

Examine the cycloidal motion of a point on a cylinder with rolling radius R. The distance of the point from the center of the cylinder is r; r is not restricted to values less than R. The angular velocity is ω. The y vs. x plot shows two cycles of the motion; the motion of the center point is also shown. Change the number of cycles if you prefer a different combination.

Why are the plots of the x vs. t and y vs. t motions so different? Is the x motion cyclic?

The case shown is for $r = R$. Examine cases for $r < R$ and $r > R$. Before performing the computations for the different cases, predict what each of the three plots will look like. Then change r and process the document.

The initial phase angle ϕ is $3\pi/2$. Explain why this phase angle was selected. Can you make the motion start anywhere you would like within the cycle?

Imagine that there exist two coordinate systems that have a common origin. One of the systems is at rest; the other rotates with angular velocity ω. At $t = 0$, the two coordinate systems are aligned. At any time t, the relationship between the two sets is given by

$$X_i = x_i \cdot \cos(\omega t_i) + y_i \cdot \sin(\omega t_i) \tag{8.11}$$

$$Y_i = -x_i \cdot \sin(\omega t_i) + y_i \cdot \cos(\omega t_i), \tag{8.12}$$

which is a standard transformation between coordinate systems. The angle between the systems increases linearly with time (ωt).

•• Load ROT2, translation viewed from a rotating frame (see Fig. 8.2).

A simple motion in one frame, when observed from a different frame, can be complex. To appreciate how a particular pattern might develop, it is helpful to approach the final form in a series of small changes from a condition that is readily understood. For example, start with no rotation, followed by a case with a very small rate of rotation, and then continue to increase ω. If the rate becomes too large, the pattern may be unrealistic and a series of straight lines may be drawn. To see the actual pattern at high rotation rates, it is necessary to increase the number of points at which x and y are calculated and decrease the time interval.

Increase ω from 0 to 1 in small steps (for example, 0.1). Observe the development of the pattern. Back up to remind yourself how a pattern developed.

Change $t_i = (i/N) \cdot 1 \cdot \pi$. Observe when $\omega = 1, 2, 3, 4$.

Translation as seen in a rotating frame.

$N := 40$ $i := 0 .. N$ $\omega := 1$ $t_i := \dfrac{i}{N} \cdot 4 \cdot \pi$ $rad := 1$

$x_i := 2 \cdot t_i - 4 \cdot \pi$ $y_i := 1$ $deg := \dfrac{\pi}{180} \cdot rad$

$X_i := x_i \cdot \cos\left[\omega \cdot t_i\right] + y_i \cdot \sin\left[\omega \cdot t_i\right]$ $Y_i := -x_i \cdot \sin\left[\omega \cdot t_i\right] + y_i \cdot \cos\left[\omega \cdot t_i\right]$

$L := 15$

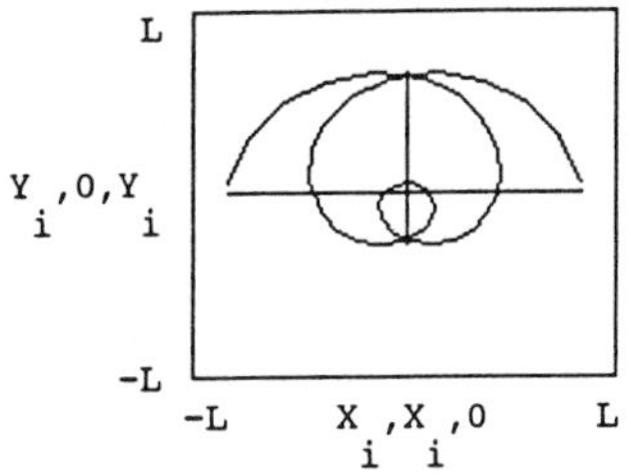

$\omega \cdot t_N = 720 \cdot deg$

$\omega \cdot t_N = 12.566$

Figure 8.2 ROT2, translation viewed from a rotating frame.

Observing motion from a rotating frame could lead the unsuspecting observer to believe that some force were acting to cause the nonrectilinear motion.

8.2   Moment of Inertia

Although the kinematic relations between linear and rotational motion are in complete correspondence, this is not true of the dynamical relationships. If a net force is applied to a mass, the mass accelerates in proportion to the force no matter how the mass is distributed within the object. If a net torque is applied to a rigid object, the object experiences angular acceleration, the rate depending on both the quantity of mass and how the mass is distributed.

The moment of inertia, I, the angular analogue to mass, includes in its expression both the quantity of mass and its distribution. For a system of discrete parts, the moment of inertia is given by

$$I = \sum_i m_i r_i^2, \tag{8.13}$$

where the r_i are measured from the axis about which the motion takes place. If the system is continuous, then the sum becomes an integral:

$$I = \int r^2 \, dm = \int \rho r^2 \, dV, \tag{8.14}$$

where ρ is the density and dV is a volume element. Again, r is a measure of the distance from the axis for which the moment is calculated. Choosing a different axis results in a different moment.

• • An estimation problem. A large icecap at the south pole of a planet melts. Assume that the planet is completely covered with water (or ice). Determine the change in the water level, which rises because of the increased volume of water caused by the melting of the ice. Ignore any change in water volume due to temperature change. How much does the moment of intertia change?

Let the planet have the radius of the earth. You must supply the other quantities, such as the thickness and density of the ice and the extent of the ice sheet. Consider the icecap to be a cylinder. Examine the size of Antarctica for an estimate of the radius of the cylinder.

• • Examine the moment of inertia of a spherical shell. A spherical shell is a sphere with a spherically symmetric cavity concentric with the outer surface. The moment of inertia of a solid sphere is

$$I_{\text{sph}} = \frac{2}{5} m r^2.$$

The moment of inertia of a spherical shell can be determined by taking the difference in moments between two solid spheres, one with the radius of the sphere and the other with the radius of the cavity.

Load ROT3, moment of inertia of a spherical shell (see Figs. 8.3 and 8.4).

The outer radius is rs; the cavity radius is rc. I_s is the moment of a sphere with radius rs; I_c is the moment of a sphere with radius rc.

Explain the form of I_s and I_c. Why is r raised to the fifth power?

When the radius of the cavity is one-half that of the sphere, by what fraction is the moment of inertia reduced? What cavity radius will result in a shell with a moment 75% of the solid value? 50%? 25%? (Try different values of rc.)

In the second part of the document, the moment of inertia of a solid sphere, $I_{ss}(r)$, is defined in functional form. The function $I(r_1, r_2)$ is the difference between the moments of two solid spheres with radii r_1 and r_2 and gives the moment of a spherical shell with outer radius r_1 and inner radius r_2. For a range of values of r_2, the ratio of the moments of inertia of the shell to that of a solid sphere with the same outer radius is shown.

Now try something slightly different. Calculate the radius of a solid sphere that has the same moment of inertia as that of a spherical shell. Let the outer radius of the shell be $r_1 = 1$. Let the radius of the cavity be r_2 and take on a range of values. Calculate the radii r_3 of solid spheres that have the same moments as the shells. Plot r_3 vs. r_2.

Determine the moment of inertia of a spherical shell. Let the radius of the sphere, rs, equal one. Specify the radius of the cavity, rc, centered about the origin. Compare the moment of inertia of the shell with that of a solid sphere with the same outside radius.

$$rs := 1 \qquad rc := 0.5 \qquad N := 24 \qquad i := 0 \text{ ..} N \qquad \theta_i := 2 \cdot \pi \cdot \frac{i}{N}$$

$$xs_i := rs \cdot \cos\left[\theta_i\right] \quad ys_i := rs \cdot \sin\left[\theta_i\right] \quad xc_i := rc \cdot \cos\left[\theta_i\right] \quad yc_i := rc \cdot \sin\left[\theta_i\right]$$

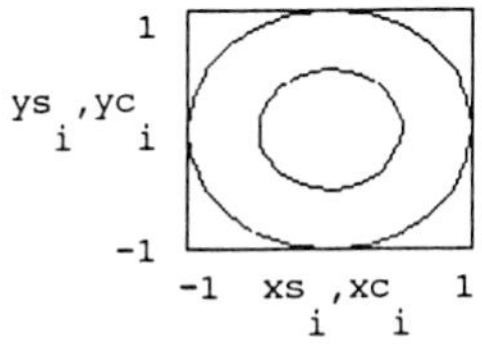

$$\rho := 1$$

$$Is := \frac{2}{5} \cdot \rho \cdot \frac{4}{3} \cdot \pi \cdot rs^5 \qquad Ic := \frac{2}{5} \cdot \rho \cdot \frac{4}{3} \cdot \pi \cdot rc^5$$

$$Is = 1.676 \qquad Ic = 0.052$$

$$\frac{Is - Ic}{Is} = 0.969 \qquad \frac{Ic}{Is} = 0.031$$

The moment of inertia of a sphere with cavity, relative to a solid sphere of the same size, is plotted as a function of the cavity radius.

$$r1 := 1 \qquad r2 := 0, .05 \text{ ...} 95 \qquad\qquad Iss(r) := \frac{2}{5} \cdot \rho \cdot \frac{4}{3} \cdot \pi \cdot r^5$$

$$I(r1, r2) := Iss(r1) - Iss(r2)$$

$$Iss(r1) = 1.676$$

$$I(r1, 0) = 1.676$$

$$\frac{I(rs, rc)}{Iss(rs)} = 0.969$$

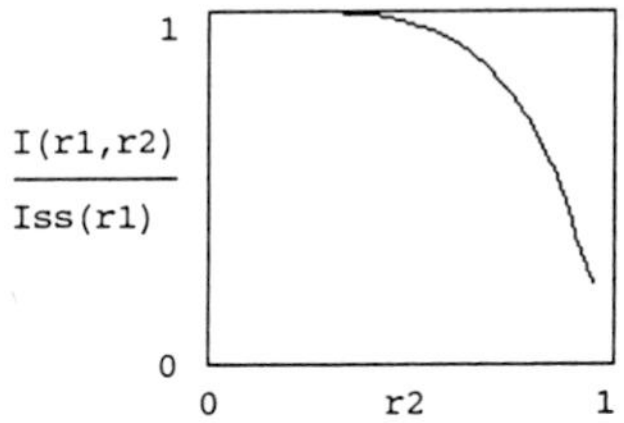

Given the moment of inertia of a shell as a fraction of the moment of a solid sphere of equal radius, find the corresponding cavity radius.

$$f := 0.5 \qquad r := 1 \qquad r' := 0.5$$

given $\qquad I(r, r') \approx f \cdot Iss(r) \qquad\qquad r' := find(r') \qquad r' = 0.871$

or to find that fraction of the moment of inertia associated with a shell of a particular cavity radius:

$$f := 0.5 \qquad r := 1 \qquad r' := 0.5$$

given $\qquad I(r, r') \approx f \cdot Iss(r) \qquad\qquad f := find(f) \qquad f = 0.969$

Figure 8.3 ROT3, moment of inertia of a spherical shell. (See the next figure for the rest of the document.)

Divide the solid sphere into a set of concentric subshells. Find the
contribution of each shell to the total moment of inertia. Perform
calculations in steps of 0.1 * r.

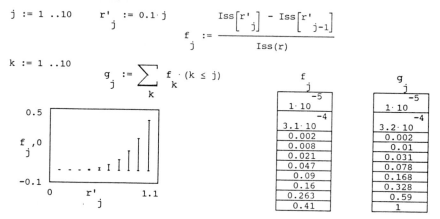

Figure 8.4 ROT3 *continued.*

In the third section of the document are two solve blocks, which provide additional means for extracting numerical information about a shell. In the first solve block, specify what fraction, f, of the total moment of the solid sphere is associated with the shell. The solve block returns the corresponding cavity radius. In the second solve block, enter the radius and return the corresponding fraction.

Start with a sphere of outer radius 1, cavity radius 0.5, and density 1. Fill the cavity with material of density ρ_c. If $\rho_c = 2$, what is the moment of the combination? What density of the core results in equal contributions to the moment from both the core and the shell?

Finally, we imagine the sphere being subdivided into ten subshells (like an onion) each with thickness equal to 10% of the outer radius (see Fig. 8.4). (The innermost shell is actually a solid sphere with radius $0.1\,r$.)

The f_j represent the individual contributions. The g_j represent the sum out to the present radius.

Note the use of the Boolean control in the sum. For each value of k, the sum is calculated for $k \leq j$. The multiplication of this condition is equivalent to a logical AND.

• • Subdivide a cylinder similar to the manner in which the sphere was subdivided above, and compute the moments of the individual cylindrical shells. For rotation about the cylindrical axis, the moment of inertia of a solid cylinder is $I_{\text{cyl}} = 1/2\,m\,r^2$.

•• Subdivide a thin rod, and compute the moments of the individual contributions. Take the axis through the center of the rod and perpendicular to it. Of the three distributions examined, which has the most extreme dependence on r? Which the least? Why?

If the moment of inertia about an axis through the center of mass, I_{cm}, is known, the moment of inertia about any other parallel axis, I_{pa}, is given by

$$I_{pa} = I_{cm} + Md^2, \tag{8.15}$$

where M is the total mass and d is the distance between the parallel axes.

•• Plot the moment of inertia of a sphere as the distance between the axis of rotation and the axis through the center of mass increases from 0 to $4R$. At what distance is the Md^2 term 99% of the total?

•• As d increases, the details of the object become less important. Given a cylinder and a sphere with the same M and the same R, calculate the moment of inertia about an axis a distance d away from the center-of-mass axis of each. At what distance d do the two moments differ by 1%? Plot the ratio of moments as a function of the distance d.

Torque. A net force is required for masses to accelerate. A net torque is required for an object to undergo angular acceleration. Torque is the angular analogue to force. Just as

$$\sum F_{\text{ext}} = m\,a,$$

the angular analogue is

$$\sum \tau_{\text{ext}} = I\,\alpha. \tag{8.16}$$

Torque is defined in terms of a cross product:

$$\tau = r \times F, \tag{8.17}$$

where r is the displacement from the origin to the point where the force acts. Choose the origin at the axis about which the torque is to be considered.

A massless, frictionless pulley requires no torque to achieve angular acceleration. Tension in the string supporting a mass on either side of such a pulley (as in massless pulley Atwood machine) is the same. The r in $(r \times F)$ for each tension force is the same. Thus the torques have the same magnitude as the r's, the F's (the tensions), and the angles (between the radius vector and the tension) are equal. The torques, however, are oppositely directed; they would accelerate the pulley in opposite directions. The torques sum to zero.

A pulley with mass requires a net torque to achieve angular acceleration. Tensions in the string on either side of the pulley cannot be the same if there is acceleration. For the case of a pulley, the law of motion is, expressing the angular acceleration α in terms of a,

$$T_1 \cdot R - T_2 \cdot R = I \cdot \frac{a}{R}.$$

•• Consider once again the Atwood machine. Before considering the case with a massive pulley, and thus before the necessity to consider torque, quickly review the earlier document, MECH1, on the Atwood machine with a massless pulley. Check the terms in the equations of motion.

Before looking at ROT4, adapt the MECH1 document to include a massive pulley. Delete the units. The tensions need to be different. A third equation is needed for the dynamics of the pulley. Let the pulley mass $M := 3$ kg. Both sides of the find statement need to be adjusted to include b, T_1, and T_2. To add a row to a matrix, move the cursor to the last item in the matrix (in this case T) and press [Alt]M. Enter +1 and press return.

•• Load ROT4, Atwood machine II (see Fig. 8.5).

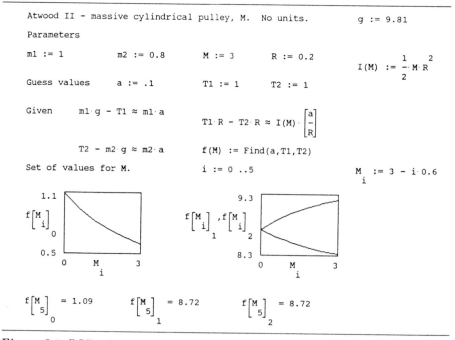

Figure 8.5 ROT4, Atwood machine II.

The suspended masses are m_1 and m_2; the mass of the pulley is M; the radius of the pulley, which is taken to be a cylinder, is R; the tensions in the string on either side of the pulley are T_1 and T_2 .

The given-find solve block is set up in functional form. There are three equations in the solve block, one associated with each of the masses. In this case, we take as the unknowns the acceleration and the tensions T_1 and T_2. In this case the find statement is set up so that on specifying a mass for the pulley, the function returns the values of a, T_1, and T_2.

In the first plot region, the acceleration is plotted as a function of pulley mass. In the second plot region, the two tensions are plotted as a function of pulley mass. Explain the shape of these curves.

Change the functional form of the find statement to depend not on the mass of the pulley but on the radius of the pulley. Let R go from 0.1 to 0.5 in steps of 0.1. It is also necessary to change the moment of inertia to depend on R rather than M. Similarly, in the plot regions, change the arguments. Explain your results.

8.3   Angular Momentum

Angular momentum of a particle relative to some axis is defined as

$$l = r \times p, \qquad (8.18)$$

where r is the radius vector from the origin to a particle with momentum p. If we take the derivative of angular momentum with respect to time, we find

$$\frac{dl}{dt} = \frac{dr}{dt} \times p + r \times \frac{dp}{dt}.$$

The first term of the right-hand side is zero because the cross product of any vector with itself is zero:

$$\frac{dr}{dt} \times p = v \times m\,v = m\,(v \times v) = 0.$$

The derivative can then be written as

$$\frac{dl}{dt} = r \times \frac{dp}{dt}. \qquad (8.19)$$

Identifying first force and then torque, the equation can be expressed as

$$\tau = \frac{dl}{dt}. \qquad (8.20)$$

This equation can readily be generalized to a system of particles, and in general

$$\sum \tau_{\text{ext}} = \frac{dl}{dt}. \tag{8.21}$$

This equation is the angular analogue of $\sum F_{\text{ext}} = dp/dt$. Just as linear momentum is conserved in the absence of external forces, so is angular momentum conserved in the absence of external torques.

Angular momentum can be conveniently expressed as the product of the moment of inertia and the angular velocity. When there are no external torques and angular momentum is conserved, we have

$$I_{\text{init}}\,\omega_{\text{init}} = I_{\text{final}}\,\omega_{\text{final}}. \tag{8.22}$$

• •   Model the change in rotation rate of a figure skater as she pulls in her arms. Keep the model very simple. Represent the body as a cylinder of mass M and radius R. Represent the arms as point masses, each of mass m, that can be moved in the range $R < r < r_{\text{max}}$.

The skater initially goes into a spin with arms extended and achieves an initial angular velocity, ω_{init}. Plot ω vs. r as the arms are pulled in. Plot rotational kinetic energy vs. r. Is the final velocity more sensitive to a decrease in R or an increase in r_{max}? Plot, for a given r_{max}, ω_{final} vs. R. Plot, for a given R, ω_{final} vs. r_{max}.

• •   Model a diver performing a dive from the high board. Keep the model simple. Let the body be a sphere of radius R. Let each of the arms be point masses with a maximum extension of $r_{a\text{max}}$ and mass m_a. Let each of the legs have length $r_{l\text{max}}$ and mass m_l.

First determine the time between the leap from the board and the moment of entrance into the water. Include an initial upward velocity in the calculation.

Then decide what is necessary in order to complete a two and one-half somersault dive. The sequence should be: dive with full extension, pull in arms and legs to increase the rotation rate, and finally extend the arms and legs to slow the rotation for a smooth entry into the water.

Plot rotation rate vs. time for the complete dive. Plot angular displacement vs. time for the dive.

> The prize at the other end is a flight education worth one million dollars, but first you have to get past me.
>
> *An Officer and a Gentleman*

Statics

An object that remains at rest (or that travels with constant linear and angular velocity) is subject to zero net force and zero net torque. The forces must sum to zero, and the torques about any axis must sum to zero. If we restrict our consideration to the x-y plane, then we reduce the number of conditions from six to three:

$$\sum F_{\text{ext,x}} = 0 \qquad \sum F_{\text{ext,y}} = 0 \qquad \sum \tau_{\text{ext,z}} = 0. \qquad (9.1)$$

(What would the other three be?)

Recall that torque is defined as $\tau = r \times F$; thus the torque is perpendicular to the plane containing the r and F (in this case the x-y plane). The torque vector is directed along axes about which the object might turn subject to the applied torque.

9.1 Simple Truss

The simple truss is a very important element in the design of rigid structures. We examine the forces along the members of a simple truss.

Construct a triangle with all acute angles and a horizontal base. Join the sides of the triangle, the members, at the corners with pins that permit rotation and that exert no torques. Forces are directed along the members. Suspend a weight from the apex of the triangle. Determine the internal forces.

• •   Load STAT1, simple truss (see Figs. 9.1 and 9.2).

Let the left end of the base of the triangle be referred to as a and the right end as b. The apex of the triangle is c. Let all the forces result from the suspended weight. Ignore the weight of the members.

Specify values for the altitude of the triangle h, the interior angles at the base of the triangle, α at point a and β at point b, and the suspended weight W.

There are no external horizontal forces. External vertical forces support the weighted triangle at the ends of the base. There are internal forces, which are directed along the lengths of the members.

First determine the external forces; there are two, F_{ay} and F_{by}. Two equations are necessary: one, the equation for vertical forces, and the other, the torque equation. An axis must be selected for the torque equation. Whatever axis is selected, any force acting along a line through that axis has no lever arm and is eliminated from consideration. In this case, the apex of the triangle is chosen; W acting along a line through that point is eliminated. The apex is at $x = 0$. In the given-find solve block are the two equations. The values F_{ay} and F_{by} are determined. These values can then be used to determine the internal forces.

The internal forces are determined using simple given-find solve blocks. You could solve these equations directly.

At the end of the document is the code that specifies the drawing of the triangle and the weight. Follow the path drawn. Which coodinates define the weight? Convenient plot limits are also specified.

Verify the equations for F_{ab}, F_{ac}, and F_{bc}.

Change the values of α and β; these values should be less than 90°. Observe the changes in F_{ay} and F_{by}. Also observe the forces F_{ab}, F_{ac}, and F_{bc}.

At corner a, what is the direction of the horizontal component of the force F_{ac}? At corner b, what is the direction of the horizontal component of F_{bc}?

Triangle supporting a weight.

$$\text{rad} \equiv 1$$

$$\text{deg} \equiv \frac{\pi}{180} \cdot \text{rad}$$

Select values for altitude of triangle, h, acute angles at base, and weight of suspended mass. (Sides of triangle have no weight.)

$$h \equiv 1 \qquad \alpha \equiv 45 \cdot \text{deg} \qquad \beta \equiv 45 \cdot \text{deg} \qquad W \equiv 10$$

$$l_{ac} \equiv \frac{h}{\sin(\alpha)} \qquad l_{bc} \equiv \frac{h}{\sin(\beta)} \qquad x_{ac} \equiv l_{ac} \cdot \cos(\alpha) \qquad x_{bc} \equiv l_{bc} \cdot \cos(\beta)$$

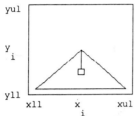

α is the interior angle at lower left
β is the interior angle at lower right.

The lower left corner is referred
to as, a
the lower right corner, b,
the peak, c.

The peak is located at x = 0, y = h.

Guess values for the external vertical forces at points a and b, where the triangle is supported.

$$F_{ay} := 5 \qquad F_{by} := 5$$

Solve the static force and torque equations for the vertical forces.

$$\text{Given} \qquad F_{ay} + F_{by} - W \approx 0 \qquad F_{ay} \cdot x_{ac} - F_{by} \cdot x_{bc} \approx 0$$

$$\begin{bmatrix} F_{ay} \\ F_{by} \end{bmatrix} := \text{Find}\begin{bmatrix} F_{ay}, F_{by} \end{bmatrix} \qquad \begin{aligned} F_{ay} &= 5 \\ F_{by} &= 5 \end{aligned}$$

Solve for forces in individual beams.

$$F_{ac} := \frac{F_{ay}}{\sin(\alpha)} \qquad F_{ab} := F_{ac} \cdot \cos(\alpha) \qquad F_{bc} := \frac{F_{by}}{\sin(\beta)}$$

$$F_{ac} = 7.071 \qquad F_{ab} = 5 \qquad F_{bc} = 7.071$$

Figure 9.1 STAT1, simple truss. (See the next figure for the rest of the document.)

Which of the beams, *ab, ac, bc*, are under compression and which under tension (force tending to pull them apart)?

If α equals β, at what angle does F_{ab} have the same magnitude as W?

• • Invert the triangle, with point *c* beneath the line *ab*. The weight is suspended from *c*. Analyze the forces. Determine which members are under compression and which are under tension.

Details for plot region and limits.

$i \equiv 1 \, ..10$ $z \equiv 0.15$

$x_2 \equiv -x_{ac}$ $x_3 \equiv x_{bc}$ $x_6 \equiv \dfrac{-z}{2}$ $x_8 \equiv \dfrac{z}{2}$ $x_7 \equiv x_6$ $x_9 \equiv x_8$ $x_{10} \equiv 0$

$y_1 \equiv h$ $y_4 \equiv h$ $y_5 \equiv \dfrac{h}{2}$ $y_6 \equiv y_5$ $y_7 \equiv y_6 - z$ $y_8 \equiv y_7$ $y_9 \equiv y_6$

$y_{10} \equiv y_9$ $hr \equiv 1_{ac} \cdot \cos(\alpha) + 1_{bc} \cdot \cos(\beta)$ $vr \equiv h$ $r \equiv if(vr > hr, vr, hr)$

$xll \equiv -\left[x_{ac} + z \right]$ $xul \equiv r + xll + 2 \cdot z$ $yll \equiv -z$ $yul \equiv r + z$

Figure 9.2 STAT1 *continued.*

• •   Extend this concept to a truss structure of three triangles and seven members. Analyze the forces. Determine which members are under compression and which are under tension.

9.2   Center of Mass

The center of mass of an object is, in effect, the average location of the mass. If a massive object is placed in a uniform gravitational field, each element of the object, dm, is acted on by the gravitational force. All those individual forces are equivalent to a single force, equal to Mg, acting at the center of mass of the body.

We can calculate the location of the center of mass of an object by a weighted average calculation. If the masses are discrete, then the location of the center of mass is given by

$$x_{cm} = \frac{\sum m_i x_i}{\sum m_i} \quad \text{and} \quad y_{cm} = \frac{\sum m_i y_i}{\sum m_i}. \tag{9.2}$$

If the distribution is continuous, then the summation goes over to an integration:

$$x_{cm} = \frac{\int x \, dm}{\int dm} \quad \text{and} \quad y_{cm} = \frac{\int y \, dm}{\int dm}, \tag{9.3}$$

where dm could be expressed in terms of a volume density, a surface density, or a line density:

$$dm = \rho dV \quad \text{or} \quad dm = \sigma dA \quad \text{or} \quad dm = \lambda dl,$$

where the units of ρ, σ, and λ are mass per volume, mass per area, and mass per length, respectively.

If we apply these concepts to determining the center of mass of a triangle, the mass elements are those of area (see Fig. 9.3). Let the elements of area be thin rectangles of height y and width dx. The y-location of the center of mass of each of these strips is $y/2$. The area associated with the mass element dm is $y\,dx$, so we integrate:

$$y_{cm} = \frac{\int_0^1 (y(x)/2)\,y(x)\,dx}{\int_0^1 y(x)\,dx}.$$

The corresponding x-value is obtained by replacing the y-value, $y/2$, with x.

• •   Four particles are located at the corners of a square. The masses have the values 1, 2, 3, and 4. Find the center of mass.

• •   Calculate the center of mass of the area enclosed by $y(x) = \sin(x)$ in the range from zero to π. Determine the center of mass of the area enclosed by the cosine curve over the same range.

• •   Determine the center of mass of the semicircular region bounded by $x^2 + y^2 = 2$ with $x > 0$. Try repeating for a hemisphere.

9.3   A Tower of Blocks

The pressure between stacked blocks is similar to the static pressure in a fluid. The process is interesting because we can analyze a simple system consisting of a small number of blocks and describe approximately the pressure in either a compressible or an incompressible fluid.

Construct a tower by neatly stacking one block on top of another. Each block added increases the weight of the column and increases the pressure between the blocks beneath it. The problem is to determine the pressure between the blocks. If we know the density, volume, and cross-sectional area, we can determine the weight per area or pressure.

• •   Load STAT3, incompressible blocks (see Fig. 9.4).

The blocks are 1 m by 1 m by 10 cm. The blocks have the density of aluminum. We calculate the weight of one block and the associated pressure. Descending from the top of the stack, the pressure between successive blocks, P_i, increases in proportion to the number of blocks. The distance from the top of the stack is h_i; the elevation above the ground is el_i.

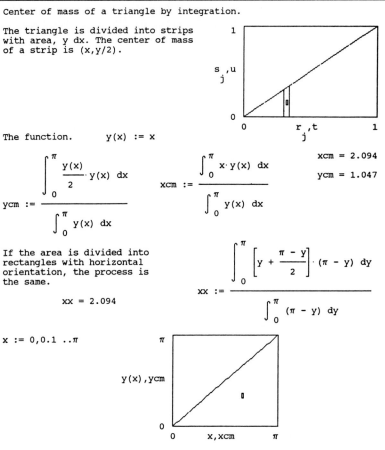

Center of mass of a triangle by integration.

The triangle is divided into strips with area, y dx. The center of mass of a strip is (x,y/2).

The function. $y(x) := x$

$$ycm := \frac{\displaystyle\int_0^{\pi} \frac{y(x)}{2} \cdot y(x)\ dx}{\displaystyle\int_0^{\pi} y(x)\ dx}$$

$$xcm := \frac{\displaystyle\int_0^{\pi} x \cdot y(x)\ dx}{\displaystyle\int_0^{\pi} y(x)\ dx}$$

$xcm = 2.094$

$ycm = 1.047$

If the area is divided into rectangles with horizontal orientation, the process is the same.

$$xx := \frac{\displaystyle\int_0^{\pi} \left[y + \frac{\pi - y}{2}\right] \cdot (\pi - y)\ dy}{\displaystyle\int_0^{\pi} (\pi - y)\ dy}$$

$xx = 2.094$

$x := 0, 0.1\ ..\ \pi$

$y(x), ycm$

The coordinates to draw the first triangle and show a representative area.

$r_1 \equiv 0$ $s_1 \equiv 0$ $r_2 \equiv 1$ $s_2 \equiv 1$ $r_3 \equiv 1$ $s_3 \equiv 0$

$r_4 \equiv 0$ $s_4 \equiv 0$ $r_5 \equiv 0.3$ $s_5 \equiv 0.3$ $r_6 \equiv 0.3$ $s_6 \equiv 0$

$r_7 \equiv 0.34$ $s_7 \equiv 0$ $r_8 \equiv 0.34$ $s_8 \equiv 0.34$ $t \equiv .32$ $u \equiv \dfrac{t}{2}$

$j \equiv 1\ ..\ 8$

Figure 9.3 STAT2, center of mass of a triangle.

Plots show that the pressure increases linearly as one descends through the pile of blocks, just as the pressure increases linearly as one descends into an incompressible fluid, such as water.

The equation for pressure in a fluid is given by

$$P = \rho g h, \tag{9.4}$$

A tower consists of a series of solid rectangular blocks stacked neatly on top of one another. Let the number of blocks be n. Determine the pressure between each of the blocks. (The lowest index corresponds to the maximum height.

$$n := 10 \qquad i := 1 \, .. \, n \qquad\qquad g := 9.81 \cdot m \cdot s^{-2} \qquad\qquad m \equiv 1L$$

Properties of the blocks $\qquad\qquad\qquad\qquad\qquad\qquad\qquad\qquad s \equiv 1T$

$$l := 1 \cdot m \quad w := 1 \quad th := 10 \cdot cm \qquad\qquad\qquad\qquad kg \equiv 1M$$
$$\rho := 2.7 \cdot 10^3 \cdot kg \cdot m^{-3}$$

$$V_b := l \cdot w \cdot th \qquad A_b := l \cdot w \qquad\qquad\qquad\qquad N \equiv kg \cdot m \cdot s^{-2}$$

$$m_b := \rho \cdot V_b \qquad w_b := m_b \cdot g \qquad\qquad\qquad\qquad Pa \equiv N \cdot m^{-2}$$

$$cm \equiv .01 \cdot m$$

el is the elevation above the ground.

$$P_i := \frac{w_b}{A_b} \cdot i \qquad h_i := th \cdot i \qquad el_i := th \cdot n - h_i$$

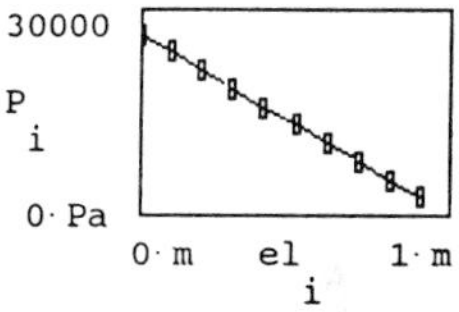

Figure 9.4 STAT3, incompressible blocks.

where h is the depth beneath the surface. Can the equation for P_i be rewritten to resemble this fluid equation?

In fact, the change of pressure with height (we use the same notation as in the document) is

$$\frac{\Delta P}{\Delta h} = \frac{w_b / A_b}{th} = \frac{m_b \cdot g}{A_b \cdot th} = \rho_c \frac{A_b \, th \, g}{A_b \, th} = \rho g.$$

This is identical to the standard expression for the pressure in a fluid, although if y is positive upward, a minus sign is introduced. Upon integration, this yields

$$P = P_{\text{surf}} + \rho g \, h. \tag{9.5}$$

P_{surf} is the pressure at the upper surface of the liquid, which could be atmospheric pressure. This pressure presses down on our tower of bricks as well.

In the previous example, the bricks were incompressible, as was the water in our fluid comparison. Now permit the blocks to be compressible. An effect of compression is that the density is no longer a constant. In the case of the tower of bricks, the density is greater near the ground than at the top of the tower.

Assume that each block has constant density, but different blocks have different densities. The compressibility of each block is proportional to the force applied, a Hooke's law behavior that can be expressed as $(F = -k\,x)$.

• •   Load STAT4, compressible blocks (see Fig. 9.5).

A tower consists of a series of solid rectangular blocks stacked neatly on top of one another. Let the number of blocks be n; what is the pressure between each of the blocks. (Let the lowest index correspond to the maximum height.

$n := 10$ $i := 1 .. n$ $g := 9.81 \cdot m \cdot s^{-2}$ $m \equiv 1L$

Properties of the blocks $s \equiv 1T$

$l := 1 \cdot m$ $w := l$ $th := 10 \cdot cm$ $kg \equiv 1M$

$\rho := 2.7 \cdot 10^{3} \cdot kg \cdot m^{-3}$

$V_b := l \cdot w \cdot th$ $A_b := l \cdot w$ $N \equiv kg \cdot m \cdot s^{-2}$

$m_b := \rho \cdot V_b$ $w_b := m_b \cdot g$ $Pa \equiv N \cdot m^{-2}$

$cm \equiv .01 \cdot m$

Let the blocks be compressible. Let each block have constant density. The density of each block will increase as the distance from the top increases. The compressibility of each block is proportional to the force applied, a Hooke's law behavior. δth is the change in thickness of a block when another block is set on top of it.

$F := w_b$ $k_H := 3 \cdot 10^{5} \cdot \dfrac{N}{m}$ $\delta th := \dfrac{-F}{k_H}$ $F = 2.649 \cdot 10^{3} \cdot N$

$\delta th = -0.883 \cdot cm$

$\delta V := w \cdot l \cdot \delta th$

$th'_i := th + (i - 1) \cdot \delta th$ Thickness of individual blocks.

$k := 1 .. n$ $ir_i := n - i + 1$

$el_k := \displaystyle\sum_i th'_i \cdot (i \geq k)$

el - elevation above ground of the tops of individual bricks.

$\dfrac{el_i}{m}$	$\dfrac{th'_i}{m}$
0.603	0.1
0.503	0.091
0.412	0.082
0.329	0.074
0.256	0.065
0.191	0.056
0.135	0.047
0.088	0.038
0.05	0.029
0.021	0.021

$P_i := \dfrac{w_b}{A_b} \cdot i$

The pressure as a function of the elevation.

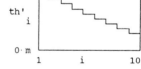

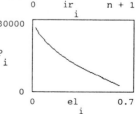

Figure 9.5 STAT4, compressible blocks.

The density of the top block is the uncompressed density. The block below it is compressed, the change in thickness being determined by Hooke's law. The change in thickness results in a change of volume and a change of density. Descending from the top of the tower, for each successive brick, the density increases and the thickness decreases. The elevation is determined by summing the thickness of the compressed blocks. Pressure between bricks is the same as in the previous case. The mass of the bricks, of course, has not changed.

The thickness of the blocks, th'_i, is plotted. The spring constant for the bricks is intentionally made quite small. By the time the bottom of the pile has been reached, a significant change in total thickness has occurred. Summing the individual heights starting from the bottom, the elevation is determined (notice the Boolean condition in the sum) and plotted. The elevation and individual thicknesses are shown in tabular form. Finally the pressure is plotted as a function of the elevation.

Plot the pressure vs. elevation on a semi-log plot. Does this curve resemble (somewhat) an exponential? (Let plot type be V.)

Explore the pressure curve for a variety of values of k. Note that k is very nearly at its lower limit. Why might there be a lower limit? How would it manifest itself? By what factor would k need to increase before the pressure vs. elevation curve begins to look like the incompressible case?

In an analysis of the compression, the previous equation still applies:

$$\frac{dP}{dy} = -\rho g. \tag{9.6}$$

In addition, we recognize that a consequence of the compression is that the density is proportional to the pressure, or

$$\frac{\rho}{\rho_o} = \frac{P}{P_o}. \tag{9.7}$$

Solving the second equation for ρ and substituting in the first yields

$$\frac{dP}{dy} = -cP \qquad \text{where} \qquad c = \frac{\rho_o}{P_o} g.$$

Upon integration, we obtain

$$P = P_o e^{-cy}. \tag{9.8}$$

This relation expresses, for example, the approximate pressure in the earth's atmosphere as a function of elevation.

The ratio ρ/P for an ideal gas, $PV = nRT$, can be expressed in terms of atomic weight, M, and the temperature, T. Thus we have

$$\rho = \frac{m}{V} = \frac{mP}{nRT}$$

and

$$\frac{\rho}{P} = \frac{M}{RT}.$$

In this case,

$$c = \frac{\rho_o}{P_o} g = \frac{Mg}{RT}. \tag{9.9}$$

Plot the pressure associated with the compressible tower together with the pressure associated with a gas, $P' = P'_o \cdot \exp(-c\,el_i)$ vs. el_i. Pick values for P'_o and c. Note that c has units of $1/m$. Two very simple values with P'_o in even thousands and c an integer give an excellent fit. Why would you expect a deviation at the very highest elevation? If k is increased, the curvature diminishes. How are k and c related?

> I wanted to tear my teeth out,
> I didn't know what I wanted to do.
> And I want to remember it, I never want to forget it.
> I never want to forget. And then I realized — like I was *shot* ...
> like I was *shot* with a diamond ...
> a diamond bullet right through my forehead. And I thought,
> "My God, the genius of that".
>
> *Apocalypse Now*

CHAPTER

10

Oscillations

10.1 Simple Harmonic Motion

Motion that repeats in a periodic way is called harmonic. Harmonic motion is called simple harmonic if the motion is described by a sinusoidal function. For example, the equation

$$x = A \sin(\omega\, t + \phi) \tag{10.1}$$

describes the displacement of a particle as a periodic function of time. The motion is simple harmonic. A is the amplitude of the motion, the maximum displacement; $|x| \leq A$. When the sine function is at its maximum $x = A$. At the time $t = 0$, $x = A \sin(\phi)$. The angular frequency, as noted above, is

$$\omega = 2\,\pi\, f. \tag{10.2}$$

The period is related to the frequency by

$$T = \frac{1}{f}. \tag{10.3}$$

For example, if a motion has a frequency of ten cycles per second (10Hz), the period is one-tenth of a second per cycle. Note that f, T, and ω are interchangeable and all contain the same basic information:

$$\omega = 2\pi f = \frac{2\pi}{T} \qquad f = \frac{1}{T} = \frac{\omega}{2\pi} \qquad T = \frac{1}{f} = \frac{2\pi}{\omega}.$$

The equations for velocity and acceleration are readily obtained by taking derivatives with respect to time:

$$x = A\sin(\omega t + \phi) \qquad\qquad x_{max} = A \qquad\qquad (10.4)$$

$$v = \frac{dx}{dt} = A\omega\cos(\omega t + \phi) \qquad\qquad v_{max} = A\omega \qquad\qquad (10.5)$$

$$a = \frac{dv}{dt} = \frac{d^2 x}{dt^2} = -A\omega^2\sin(\omega t + \phi) \qquad a_{max} = A\omega^2 \qquad (10.6)$$

• •   What is the average velocity? What is the root-mean-squared velocity?

The first two documents in this chapter include these basic expressions, reminding us of the role of the phase angle and relating uniform circular motion with simple harmonic motion.

• •   Load HM1, simple harmonic motion (see Fig. 10.1).

The position, velocity, and acceleration are plotted as a function of time. Notice the relative starting point and sign of each curve. Over what portion of the position curve are the signs of position and velocity the same? Are the signs of velocity and acceleration the same?

When the acceleration is at its maximum, what is the position? Process for values of ϕ equal to 0, $\pi/4$, $\pi/2$, $3\pi/4$, and π. Express the function $y = A\sin(\omega t + \pi/4)$ in terms of the cosine function. Plot to verify.

Simple harmonic motion and uniform circular motion are interrelated. Consider a radius vector of length A oriented at an angle θ counterclockwise from the positive x-axis. The projection of the vector on the y-axis is $A\sin(\theta)$. If θ increases uniformly in time, $\theta = \omega t$, the tip of the vector traces out a circle. The projection, $A\sin(\omega t)$, executes simple harmonic motion.

In the second part of HM1, we show a radius vector and its projection at $t = 0$. We see the y-projection of the vector as a function of time. We also show the exponential representation with a complex argument to display the radius vector at its initial position. (Refer to Chapter 2.) If the angle increased in time, the vector would rotate.

Let ϕ take on a series of values. Note that the y-component of the vector and the tip of the projection vector give the initial y-value for the

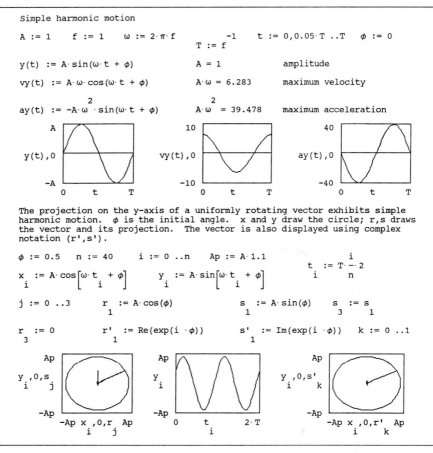

Figure 10.1 Simple harmonic motion.

The text and equations shown in the figure:

Simple harmonic motion

$A := 1$ $f := 1$ $\omega := 2 \cdot \pi \cdot f$ $T := f^{-1}$ $t := 0, 0.05 \cdot T \ldots T$ $\phi := 0$

$y(t) := A \cdot \sin(\omega \cdot t + \phi)$ $A = 1$ amplitude

$vy(t) := A \cdot \omega \cdot \cos(\omega \cdot t + \phi)$ $A \cdot \omega = 6.283$ maximum velocity

$ay(t) := -A \cdot \omega^2 \cdot \sin(\omega \cdot t + \phi)$ $A \cdot \omega^2 = 39.478$ maximum acceleration

The projection on the y-axis of a uniformly rotating vector exhibits simple harmonic motion. ϕ is the initial angle. x and y draw the circle; r,s draws the vector and its projection. The vector is also displayed using complex notation (r',s').

$\phi := 0.5$ $n := 40$ $i := 0 \ldots n$ $Ap := A \cdot 1.1$ $t_i := T \cdot \frac{i}{n}$

$x_i := A \cdot \cos[\omega \cdot t_i + \phi]$ $y_i := A \cdot \sin[\omega \cdot t_i + \phi]$

$j := 0 \ldots 3$ $r_1 := A \cdot \cos(\phi)$ $s_1 := A \cdot \sin(\phi)$ $s_3 := s_1$

$r_3 := 0$ $r'_1 := Re(\exp(i \cdot \phi))$ $s'_1 := Im(\exp(i \cdot \phi))$ $k := 0 \ldots 1$

Figure 10.1 HM1, simple harmonic motion.

y vs. t curve. If we could display the rotating vector in time, its projection would always correspond to the y-value in the y vs. t curve.

• • Load HM2, rotating vector (see Fig. 10.2).

MathCAD cannot display animation, but if several curves are plotted you can see them laid down in time.

In the document, we show four locations of the rotating vector and the four corresponding y vs. t curves. Imagine the radius vector rotating counterclockwise; the four positions lead us through the maximum and most of the way to a minimum. From our fixed vantage point, the y-curve appears to step to the left.

Rotating vector and simple harmonic motion. Connect the projection of
the rotating vector on the y-axis of the first plot region with the y-t
curves in the second.

$A := 1$ $f := 1$ $\omega := 2 \cdot \pi \cdot f$ $T := f^{-1}$ $t := 0, 0.05 \cdot T \, .. \, T$ $Ap := A \cdot 1.1$

$n := 40$ $i := 0 \, .. \, n$ $m := 3$ $k := 0 \, .. \, m$ $l := 0 \, .. \, 2 \cdot m$

$\phi o := \dfrac{\pi}{3}$ $\phi_k := 0.15 \cdot k \cdot 2 \cdot \pi + \phi o$ $t_i := \dfrac{i}{n} \cdot 2 \cdot T$

$y_{k,i} := A \cdot \sin\left[\omega \cdot t_i + \phi_k\right]$ $r_{2 \cdot k} := A \cdot \cos\left[\phi_k\right]$ $s_{2 \cdot k} := A \cdot \sin\left[\phi_k\right]$

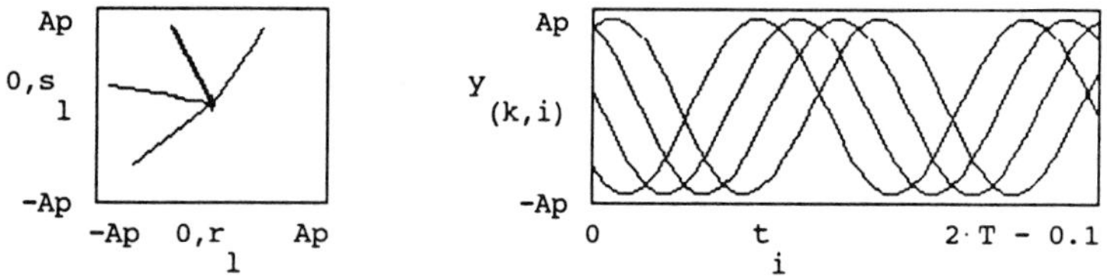

Figure 10.2 HM2, rotating vector.

Try different phase angles; ϕ_0 gives the starting position. In the expression for ϕ_k, changing the value from 0.15 will change the size of the
angle between successive pictures. Increasing m will increase the number
of snapshots of the motion. However, the y vs. t graph becomes cluttered
as m increases.

10.1.1 Spring-Mass Oscillation

The force exerted by a spring that obeys Hooke's law is

$$F = -k\,x. \tag{10.7}$$

The force is a linear (proportional to x), restoring $(-)$ force. If x is positive, F is negative; if x is negative, F is positive; thus the tendency is
always to return to the equilibrium position. A linear restoring force results in simple harmonic motion.

Attach a mass to the end of a spring, stretch the spring, and release it.
Let the force exerted by the spring be the only force acting. On applying
Newton's second law, we obtain

$$-k\,x = m\frac{d^2 x}{dt^2}. \tag{10.8}$$

If $x = A \sin(\omega t)$, $d^2 x / dt^2 = -\omega^2 x$. After substituting, we obtain an expression relating the angular frequency, the spring constant, and the mass.

$$-kx = -m\omega^2 x \quad \text{or} \quad \omega = \sqrt{\frac{k}{m}} \tag{10.9}$$

Every frequency equation for oscillatory motion includes a ratio of terms of a similar nature; k is a measure of the restoring force, and m is a measure of inertia. If there is a large restoring force and a small mass, the resulting oscillation will have a high frequency; if there is a small restoring force and a large mass, the resulting oscillation will have a low frequency.

As the spring is stretched, work is performed:

$$W = \int F \cdot dx. \tag{10.10}$$

If we stretch the spring, the force we apply is directed opposite to that of the spring, $F_{\text{appl}} = kx$:

$$W = \int_0^x kx\, dx = \frac{1}{2} k x^2. \tag{10.11}$$

The energy is stored as potential energy. In an oscillating system the energy flows back and forth between kinetic and potential energy.

● ●   Load HM3, simple harmonic motion in a spring-mass system (see Fig. 10.3).

In the specification of parameters, the spring constant, k, and the angular frequency, ω, are defined. Given these choices, the value for the mass is determined. Change the description so that you input k and m, rather than k and ω.

Plot ω vs. k for a mass of 0.1 kg; let successive values of k increase by factors of $10^{n/2}$. Plot a corresponding curve that shows how much the spring would stretch if the mass of 0.1 kg were simply hung from the spring (with no oscillation taking place). At 10^3 Hz, how much does the spring stretch in supporting the mass? At 10^5 Hz?

In the next section, the work done in stretching the spring is calculated. The work is defined as a function and then calculated for a series of x-values and plotted. A comparison is then made with the expression $(1/2)\,k x^2$, which was determined by integrating the Hooke's law force analytically.

Next, we examine the kinetic and potential energy as a function of time. The position is specified as a function of time. The velocity is obtained by taking the derivative of position with respect to time. The

Work and potential energy associated with stretching a `Hooke's law` spring.

Spring-mass system

k := 1   m := 0.1

$F(x) := -k \cdot x$ $\omega := \sqrt{\dfrac{k}{m}}$ $f := \dfrac{\omega}{2 \cdot \pi}$ $T := f^{-1}$

The work

A := 2.5   x := 0, 0.2 .. A $W(x) := \int_{0}^{x} -F(x)\ dx$ W(A) = 3.125

$\dfrac{1}{2} \cdot k \cdot A^2 = 3.125$

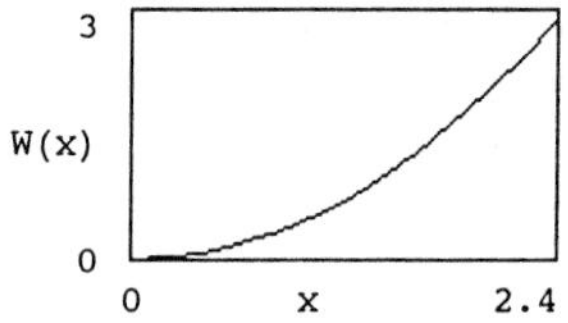

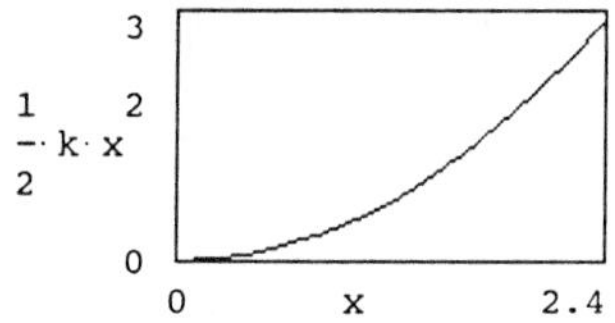

Potential and kinetic energy.

$\phi := 0$ t := 0, .05 \cdot T .. 2 \cdot T $x(t) := A \cdot \sin(\omega \cdot t + \phi)$ $v(t) := \dfrac{d}{dt} x(t)$

$U(t) := \dfrac{1}{2} \cdot k \cdot x(t)^2$ $K(t) := \dfrac{1}{2} \cdot m \cdot v(t)^2$ $Umax := \dfrac{1}{2} \cdot k \cdot A^2$

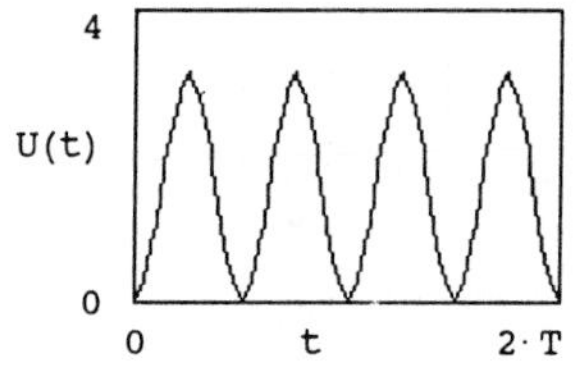

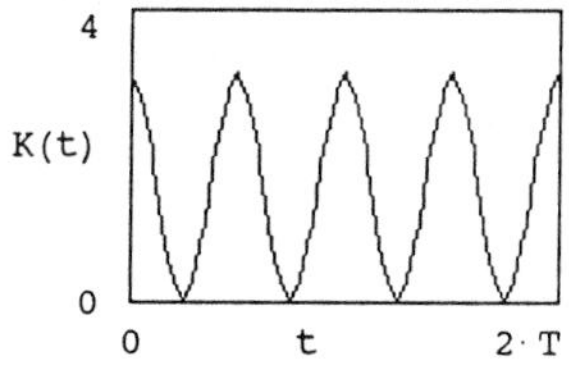

Figure 10.3 HM3, simple harmonic motion in a spring-mass system.

potential energy, $U(t)$, and kinetic energy, $K(t)$, are then defined and plotted. Note that $U(t)$ and $K(t)$ are out of phase. What is the phase difference? Is it the same phase difference as between position and velocity? The maximum value of the potential energy, U_{max}, is given. Determine the maximum value of the kinetic energy, K_{max}.

Plot in one region $U(t), K(t), U(t) + K(t)$ vs. t. Let the upper limit of the ordinate be $1.1 \cdot U_{max}$.

What information does the constant ϕ supply about the initial division of energy between potential and kinetic? At $t = 0$, what values would ϕ take on to make the energy all kinetic? All potential? Half kinetic and half potential? Verify by plotting. If you know U_{max} and a specific value of U, do you know the position of the object? Do you know its velocity?

Time in the plots of $U(t)$ and $K(t)$ ranges from zero to $2 \cdot T$. Why are there four cycles of $U(t)$ and $K(t)$? Correlate precisely with the motion. Define $C_1 := (1/2) \cdot k \cdot A$ and $C_2 := (1/2) \cdot m \cdot A \cdot \omega$. Plot $U(t)$, $C_1 \cdot x(t)$ vs. t and plot $K(t)$, $C_2 \cdot v(t)$ vs. t. These plots should assist in the previous question. How are C_1 and C_2 related to maximum potential energy and maximum kinetic energy?

• •   Load HM4, Lissajous figures: phase diagrams and phase differences (see Fig. 10.4).

Lissajous figures provide an amusing way to compare harmonic curves. They are also useful in the interpretation of the phase difference between two signals.

Lissajous figures are created by driving both the x and y motions harmonically. That is, x and y may both vary sinusoidally in time but they are not plotted against time but against each other. For example,

$$x_j := \sin(2 \cdot \pi \cdot f1 \cdot \frac{j}{N}) \qquad y_j := \sin(2 \cdot \pi \cdot f2 \cdot \frac{j}{N} + \phi).$$

Here, time is not specifically mentioned; the curves are simply evaluated at a series of points that could be sequential in time. Both x and y are harmonically driven. Each is driven at its own frequency, f_1 and f_2; these frequencies could be the same or they could be different. A phase shift can be introduced.

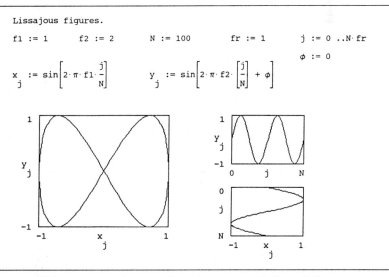

Figure 10.4 HM4, Lissajous figures: phase diagrams and phase differences.

There are three plot regions. The plot region y vs. x displays the Lissajous figures. The regions y vs. j and j vs. x show the individual oscillations of the two components. Note that the last plot is tipped on its side to correspond to the direction in which the x-aspect of the Lissajous motion takes place.

Start with $f_1 := 1$ and $f_2 := 2$. Look at the small plots. We see that x goes through one cycle while y goes through two cycles. Starting at the center of the larger plot, follow the motion. As x goes from zero to one (one-quarter of its cycle), y goes from zero to one and back to zero. As x returns from one to zero, y goes from zero to minus one and back to zero. Therefore, x has gone through one-half of a cycle, when y has gone through one complete cycle.

In a separate plot region, plot x_j, y_j vs. j for one cycle of the smaller frequency. Do this in a general way so that if, for example, f_1 is 98 and f_2 is 34, it still works.

Near the beginning of the document is the quantity fr, for fraction. When fr is 1, all N points are plotted. As fr is reduced, so is the upper limit. By plotting the figure with a reduced value for fr, you can determine where the Lissajous figure begins and in what direction it evolves. This is helpful, as the starting point and initial direction are not always clear in such plots, especially if $\phi \neq 0$.

Let $f_1 := 1$ and $f_2 := 1$ and process. In the small plots, we see the individual motions. In the y vs. x plot, where does the trace begin? In what direction does it move? Let $fr = 0.1$ and process. Next let $fr = 0.25$ and 0.5. Why are the 0.25 and 0.5 plots the same on the y vs. x graph?

Let $fr := 1$ and $\phi := 0.1$ and process. Watch both the y vs. j and j vs. x curves. Let ϕ take on the values $\pi/4$, $\pi/2$, and π. A Lissajous figure displaying a straight line, ellipse, or circle corresponds to equal frequencies for the x and y motions. Only the relative phase is different. What is the phase difference between the straight line and the circle? Does changing the sign of the phase difference change the appearance of the curve?

Let $\phi := 0$ and $f_2 := 1.1$ and process. Why doesn't the curve close?

Try the following values for (f_1, f_2, fr): (1, 2, 1), (1, 2, 0.2), (1, 2, 0.45), (2, 1, 0.45).

Change the y-function to cosine. Try (1, 2, 0.2), (1, 2, 0.5), (1, 2, 1).

With the functions for the x and y directions sine and cosine, try (1, 3, 1), (1, 3, 0.45), (1, 2, 0.45), (2, 1, 0.45).

Finally, with the functions for the x and y directions sine and sine, and for the case (2, 1, 1), let $\phi := 0$, 0.2, 0.6, 0.785. These curves are analogues to the circle, ellipse, and straight line when the frequency ratio was 1:1.

For some fun, try (20, 19, 1), (40, 39,1), (40, 47, 1), (48, 49, 1)($\phi = 0$ or $\phi = 0.785$).

10.2 Pendulums

The simple pendulum consists of a point mass suspended from a massless rod or string. A physical pendulum consists of a rigid body suspended from an axis. Pendulum motion is generally different for small and large amplitude oscillations. The motion can be damped or driven.

10.2.1 *Simple Pendulum*

When the pendulum bob is displaced from its equilibrium position, there is a restoring force equal to $m\,g\sin(\theta)$. The equation of motion is

$$- m\,g\sin(\theta) = m\,a = m\frac{d^2s}{dt^2}, \tag{10.12}$$

where s is the displacement, the arc length as measured from the equilibrium point. From the discussion of the spring-mass system, we know that a linear restoring force is necessary for simple harmonic motion. For the pendulum, there is a restoring force but it is not linear because the sine function is nonlinear. However, for small angles, $\sin(\theta) \approx \theta$. (Check the comparisons made in Chapter 2 for the range of accuracy of the small-angle approximation.) Using this approximation and substituting for s, the value $L\theta$, the equation becomes

$$- m\,g\,\theta = m\,L\frac{d^2\theta}{dt^2}. \tag{10.13}$$

This equation is of the same form as that for the spring-mass system,

$$-k\,x = m\frac{d^2x}{dt^2},$$

which yielded

$$\omega = \sqrt{\frac{k}{m}}.$$

Similar analysis for the angular frequency of the pendulum yields

$$\omega = \sqrt{\frac{g}{L}} \qquad \text{or} \qquad T = \frac{2\pi}{\omega} = 2\pi\sqrt{\frac{L}{g}}. \qquad (10.14)$$

In a static situation, with the pendulum bob at the equilibrium point, the weight of the bob and the tension in the supporting string are equal and opposite. If a second string is attached to the bob and the bob is pulled slowly to the side (the pull being tangential to the circle in which the bob moves), the tension in the supporting string is reduced. The static tension in the string, T_s, is given by

$$T_s = mg\cos(\theta). \qquad (10.15)$$

When the bob swings freely along its circular path, there must be a force that constrains the bob to move in a circle. The dynamic tension, T_d, is given by

$$T_d = m\frac{v^2}{L} = mL\left(\frac{d\theta}{dt}\right)^2. \qquad (10.16)$$

The total tension, T_n, is the sum of these contributions.

• •   Load HM5, simple pendulum (see Fig. 10.5).

(Information from the Lissajous figure example will be used in this example. Examine that document first.)

To speed the calculations, the equations for the angular displacement, velocity, and acceleration are written explicitly rather than using derivatives. Maximum values of θ' and θ'' are θ'_m and θ''_m.

Note that ω refers to the frequency of the oscillations and depends on g and L. Do not confuse this ω with $d\theta/dt$, which represents the angular velocity of the pendulum bob. In this pendulum problem, ω is a constant while $d\theta/dt$ varies with angle.

The first plot region shows the curves $\theta(t)$, $\theta'(t)$, and $\theta''(t)$; the curves are scaled to have the same amplitude. Identify each curve.

Position the document on the screen so that the set of three phase diagrams (θ' vs. θ, θ'' vs. θ', and θ'' vs. θ) are visible. The diagrams are reminiscent of Lissajous figures. However, the motions are not independent but related by derivatives.

What is the phase relation between θ' and θ? Between θ'' and θ'? Between θ'' and θ?

Knowing the initial conditions, where does the θ' vs. θ curve begin? The θ'' and θ' curve? The θ'' and θ curve?

As time increases, predict the directions the paths take in each of the three cases. Let $fr = 0.1$ and process. (As before, fr is a control on the range of the calculations. Note assignment for t.) Were you right?

Now examine the tension statements and the last tier of three plots. Why does the total tension T_n go through two cycles in one period, T? The T_n vs. θ curve should be reminiscent of curves in the Lissajous figure study. What frequency ratio is associated with curves of this shape? What is the difference between the T_n vs. θ and T_n vs. θ' curves? Express your answer

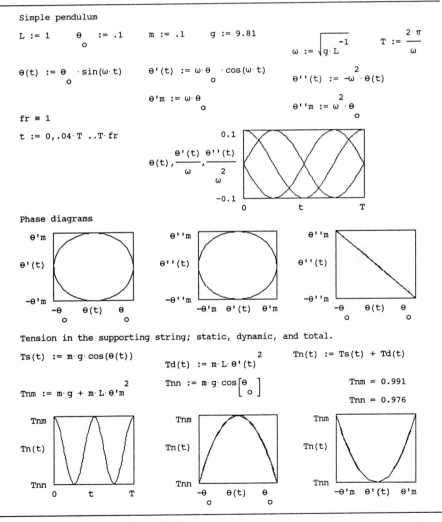

Figure 10.5 HM5, simple pendulum.

in terms of frequencies and phase relationships. Refer to the definitions of θ, θ', and θ''. In what direction do these T_n curves evolve? Let fr equal 0.1, 0.25, 0.5. What happened to the T_n vs. θ curve? Why? In the T_n vs. θ' curve, at what point is $\theta = 0$?

Plot T_s vs. t and T_s vs. θ. Plot T_d vs. t and T_d vs. θ. Notice the range of values, both maximum and minimum. Plot T_s vs. T_d. Plot T_d/T_s vs. t.

Damped Motion.   If the amplitude of the oscillation decreases exponentially with time, the displacement is given by

$$\theta = \theta_o \sin(\omega t)\, e^{-t/\tau},$$

where τ is a time constant.

• •   Take the first and second derivatives of θ with respect to time. Express the results, as far as possible, in terms of θ and θ'. Note that you have more than one time-dependent term when you take the derivatives.

• •   Load HM6, simple pendulum with damping (see Fig. 10.6).

This document has the same general structure as the previous one; the difference is that exponential damping has been included. To save time, the derivatives are written explicitly. Check your values for the derivatives against θ' and θ''.

Identify θ, θ', and θ'' in the first plot. Are these curves precisely in phase as they were in the case with no damping? How do you decide? Are the peaks an appropriate measure?

Compare each of the three phase diagrams with those of the previous example. Follow the motion of the first, for example, thinking about velocity and displacement. How are these plots similar? How are they different? Why? What happens if $\tau = 10$?

Compare the tension plots with those of the previous example. In the T_n vs. t plot, the tension seems to be approaching some value. What value might that be?

In the T_n vs. θ plot, what point corresponds to the end of one cycle? Are the maxima in the T_n vs. θ curve all at $\theta = 0$?

Plot $T_s(t)$ vs. t. Plot $T_d(t)$ vs. t. Plot $T_s(t)$ vs. $T_d(t)$.

Examine different rates of damping. Let τ equal 4, 2, 1, 0.5. As the damping decreases, the plots should approach those of the previous example.

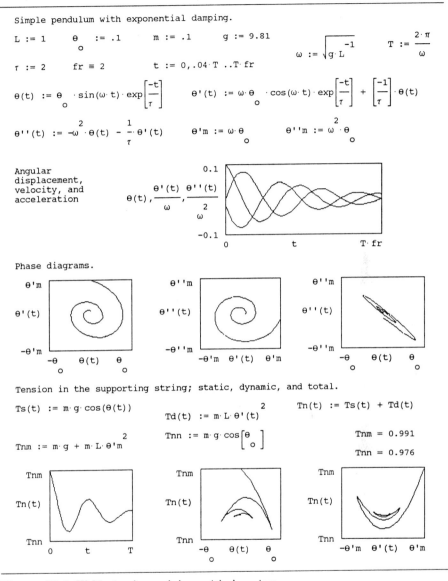

Simple pendulum with exponential damping.

$$L := 1 \qquad \theta_o := .1 \qquad m := .1 \qquad g := 9.81 \qquad \omega := \sqrt{g \cdot L^{-1}} \qquad T := \frac{2 \cdot \pi}{\omega}$$

$$\tau := 2 \qquad fr \equiv 2 \qquad t := 0, .04 \cdot T \,..\, T \cdot fr$$

$$\theta(t) := \theta_o \cdot \sin(\omega \cdot t) \cdot \exp\left[\frac{-t}{\tau}\right] \qquad \theta'(t) := \omega \cdot \theta_o \cdot \cos(\omega \cdot t) \cdot \exp\left[\frac{-t}{\tau}\right] + \left[\frac{-1}{\tau}\right] \cdot \theta(t)$$

$$\theta''(t) := -\omega^2 \cdot \theta(t) - \frac{1}{\tau} \cdot \theta'(t) \qquad \theta'm := \omega \cdot \theta_o \qquad \theta''m := \omega^2 \cdot \theta_o$$

Angular
displacement,
velocity, and
acceleration

$$\theta(t), \frac{\theta'(t)}{\omega}, \frac{\theta''(t)}{\omega^2}$$

Phase diagrams.

Tension in the supporting string; static, dynamic, and total.

$$Ts(t) := m \cdot g \cdot \cos(\theta(t)) \qquad Td(t) := m \cdot L \cdot \theta'(t)^2 \qquad Tn(t) := Ts(t) + Td(t)$$

$$Tnm := m \cdot g + m \cdot L \cdot \theta'm^2 \qquad Tnn := m \cdot g \cdot \cos\left[\theta_o\right] \qquad Tnm = 0.991$$

$$Tnn = 0.976$$

Figure 10.6 HM6, simple pendulum with damping.

Large Amplitude Motion. As the amplitude of the pendulum's motion increases, the precision with which the values are determined decreases as the small-angle approximation becomes less suitable. In this example, we explore the motion using a numerical technique; the Euler algorithm is used in the simulation of the motion. The method applies for all angles from 0° to 180°. (For the large-amplitude case, consider the pendulum bob to be supported by a massless rod.)

The algorithm is essentially the same as that used in previous one-dimensional cases. The difference is that the acceleration, velocity, and displacement are expressed in angular, rather than linear, variables. Otherwise, the correspondence is complete.

The restoring force is $-m g \sin(\theta)$; the corresponding acceleration is $-g \sin(\theta)$. The angular acceleration $\alpha = a/r$. In this case, the angular acceleration is given by

$$\alpha = -\left(\frac{g}{L}\right) \sin(\theta). \tag{10.17}$$

• • Load HM7, simple pendulum with large amplitude (see Fig. 10.7).

After the initial values and the range of iteration are specified, the equations to be iterated are written into an equation block. Verify that the algorithm is indeed the Euler algorithm. How would it be different if it were the Euler-Cromer algorithm?

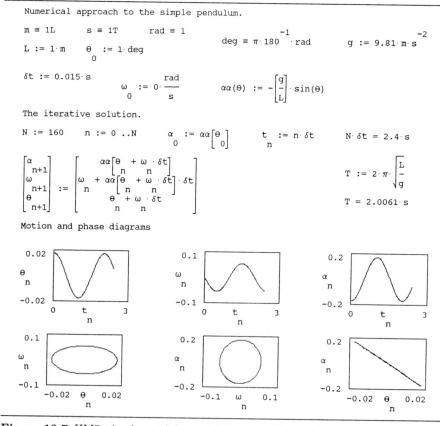

Figure 10.7 HM7, simple pendulum with large amplitude.

In this application, $\omega_n = d\theta_n / dt$.

We examine the results for small angles and compare with known results before considering cases for which the outcome is unknown (to us). Process the document. Examine the plot regions and compare with the undamped case.

Estimate the period. How does this value compare with the value for the simple pendulum, which, of course, has no dependence on θ.

Specify the tension and plot it vs. time, displacement, and angular velocity. Compare the results with the undamped case.

Next consider an initial angle of 90°. Before performing the calculation, anticipate the changes that will occur. When the pendulum is near 90°, is the magnitude of the acceleration large or small? Does the pendulum spend equal times for equal angular differences? Or does it spend more time in certain regions? Considering these questions will permit you to anticipate changes in the α vs. t or α vs. θ plots and in other associated plots.

Let $\theta = 90°$ and $N = 180$ and process. Estimate the period. Save this value.

The α vs. t curve shows regions that are nearly constant, as would be expected from the questions in the previous paragraph. These regions of nearly constant acceleration occur when $\theta \approx 90°$. That region of constant acceleration implies linear change in velocity, which is observed in the ω vs. t plot. Changes in the θ vs. t curve are subtle, and the curve does not look much different from that associated with the small-amplitude motion of a simple pendulum. The phase diagrams show a significant change as well.

Now anticipate the changes that will occur if the pendulum starts at 175°. This is difficult to do. Perhaps the easiest curve to predict is that of α vs. θ. At large angles, is the acceleration large or small? At the 90° positions, is the acceleration large or small? At the 0° positions, is the acceleration large or small? What phase diagram does this suggest?

Let $\theta = 175°$ and $N = 300$ and process. Estimate the period. Save this value.

Now the differences in each of the curves are quite striking. Did you guess the α vs. θ curve approximately? Spend a few minutes to see if you can, in your head, match the curves with the motion. To see where the phase diagrams begin and in what direction they evolve, reduce N to a fraction of that necessary for a complete cycle and process. Become familiar with these diagrams; they are a very useful way of presenting information.

• •   There are analytic solutions for large-amplitude oscillations. The period, for example, for oscillations with maximum amplitude θ_{max} is given by the series

$$T = 2\pi \sqrt{\frac{l}{g}} \left(1 + \frac{1}{2^2}\sin^2\left(\frac{\theta_{max}}{2}\right) + \frac{1}{2^2}\frac{3^2}{4^2}\sin^4\left(\frac{\theta_{max}}{2}\right) + \cdots \right). \qquad (10.18)$$

How do your estimates of the period for the three cases above compare with the formula? The next term is given by

$$\frac{1}{2^2}\frac{3^2}{4^2}\frac{5^2}{6^2}\sin^6\left(\frac{\theta_{max}}{2}\right).$$

In integral form the period is given by

$$T = 4\sqrt{\frac{l}{2g}} \int_0^{\theta_{max}} \frac{d\theta}{\sqrt{\cos(\theta) - \cos(\theta_{max})}}. \qquad (10.19)$$

Compare values with the series, with the integral, and with your estimated results.

With a numerical algorithm in place, it is relatively straightforward to introduce changes. However, changes may have unexpected consequences, and one should always explore to make sure that the behavior of the algorithm is appropriate to the situation being modelled.

For example, an oscillatory system with exponential damping was considered earlier. If damping is to be introduced in the numerical model, how might that be done? What would happen if, after each iteration, the magnitude of the acceleration term were reduced by some factor, and after many iterations the acceleration were reduced to zero? Would this result in damped oscillatory motion?

• •   Define the function

$$f(n) = \frac{N - n}{N}.$$

As n increases from zero to N, f decreases monotonically. Multiply the $\alpha\alpha$ term by $f(n)$. That is, on the right-hand side of the α, ω, θ equation block in HM7 change the top term to

$$\alpha\alpha(\theta_n + \omega_n \cdot \delta t) \cdot f(n).$$

Process this algorithm and look at the plot regions. Did the acceleration term reduce as we had anticipated? What about the θ and ω curves? Are they damped? Explain.

The function $g(n) = f(n)/f(n-1)$ decreases much more slowly than $f(n)$. Define the function $g(n)$. Remove the $f(n)$ term that was entered in the previous step from the top right-hand side of the expression. Multiply the $(\omega + \alpha\alpha)$ term, the center term on the right-hand side by $g(n)$ and process. Does this result in damped motion? Explain.

Remove $g(n)$ from the center term. Multiply the $(\theta + \omega)$ term, the lower term on the right-hand side, by $g(n)$. Does this result in damped motion?

The purpose of the previous examples was simply to show the effect of changing certain terms. The examples were not based on a physical model for damping. What if the resistance to the motion is proportional to the square of the velocity of the pendulum bob?

The angular acceleration

$$\alpha = a/r = -k\,\frac{v^2}{r} = -k\,r\,\omega^2.$$

Let $\mathrm{sgn}(\omega) = \omega/|\omega|$. Then replace both the $\alpha\alpha$ (one each in the α and ω equations) terms with

$$\alpha\alpha\left(\theta_n + \omega_n \cdot \delta t\right) - k \cdot L \cdot \omega_n^2 \cdot \mathrm{sgn}(\omega_n)$$

Don't forget the δt term in the ω equation. Let $k = 0.25 \cdot m^{-1}$.

Changing the sign of the $k \cdot L \cdot \omega^2$ terms would result in driven motion, though one would be hard pressed to imagine a driving force proportional to the square of the velocity. The motion would be driven at the natural frequency of the system.

A driving force can be added to the $\alpha\alpha$ terms. Leave the damping term in place and add a third term,

$$15 \cdot \sin(2 \cdot \pi \cdot t_n).$$

Note that the motion rather quickly responds to the driving force and the frequency of the motion changes within a few cycles. The rate at which this happens depends both on the strength of the driving force and on the damping term.

Physical Pendulum. The physical pendulum, in contrast to the simple pendulum, does not have its mass concentrated at one point; the mass is distributed. The pendulum is suspended from an axis about which it is free to rotate. (For pendulum motion, why should the axis not pass through the center of mass?) In equilibrium, the center of mass is directly beneath the center of support, the axis. If the object is displaced from equilibrium through some angle θ, there is a restoring torque. If the distance from the center of mass of the object to the axis is d, the torque $(r \times F)$ is $d \cdot \sin(\theta) \cdot (m \cdot g)$. Applying $\sum \tau_{\text{ext}} = I\,\alpha$, the equation of motion is

$$-m\,g\,d\sin(\theta) = I\,\alpha = I\,\frac{d^2\theta}{dt^2}, \tag{10.20}$$

where I is the moment of inertia of the pendulum about the selected axis. The equation is similar to that of the simple pendulum. Limiting the amplitude to small angles permits us to use the small-angle approximation and the equation simplifies to

$$\frac{d^2\theta}{dt^2} = -\left(\frac{m\,g\,d}{I}\right)\theta \tag{10.21}$$

which yields

$$\omega = \sqrt{\frac{m\,g\,d}{I}}. \tag{10.22}$$

Again, in the expression for ω the numerator tells us of the restoring force, the denominator of the inertia.

• • Describe and plot the motion of a meter stick suspended 10 cm from one end.

The motion of a physical pendulum depends on the choice of axis. If the axis is selected far from the center of mass, the restoring torque is large, but so is the moment of inertia. Right at the center of mass, the moment of inertia is least but the restoring torque is zero. In the next document, we examine the effect of choosing different axes in the movement of a long, thin rod, in this case, a meter stick.

• • Load HM8, physical pendulum: changing the axis location (see Fig. 10.8).

We examine the properties of the meter stick as a physical pendulum in 1-cm steps from the center, at 50 points. We plot the restoring torque for some arbitrary angle and the moment of inertia (making use of the parallel axis theorem) as a function of distance from the center of mass. Even though neither curve has a maximum or minimum, their ratio does. We then plot the angular frequency, the period, and the frequency. We use the root function with the derivative of the period to find the location of the axis for which the period is a minimum.

Use the root function to find the maximum of the frequency curve.

Imagine that you can measure the period to an error of ±0.06 s. What is the largest range of location of the pivot point that could yield values within this error range?

The last two plots in the document are of two different lengths. One is that of a simple pendulum; the other is that of a physical pendulum.

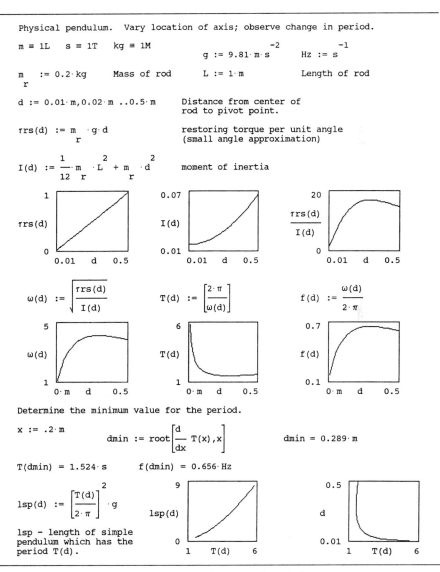

Figure 10.8 HM8, physical pendulum: changing the axis location.

The length of the physical pendulum is plotted vs. its period. The length of the simple pendulum is that length which has the same period as the physical pendulum.

• •   Perform a similar set of calculations for a thin cylinder pivoting about an axis parallel to the cylinder axis. Is there a minimum period? If yes, what is the ratio of the distance of the minimum axis from the axis of symmetry to the radius of the cylinder?

• •   Explore the large-amplitude motion of a meter stick. Compare the phase diagrams with that of a simple pendulum.

• •   A thin rod of length 15 cm and mass 10 g has a 25-g point mass attached to one end. The rod pivots about an axis 3 cm above the weighted end. A second point mass is attached above the pivot point and can be moved to any location on the rod above the pivot point. Find the period of the system as a function of the location of the upper mass. (Select magnitude of second point mass to suit.) This system is a kind of metronome.

You're trying to seduce me, right?

The Graduate

Waves

The oscillating systems that we have considered so far are of the form

$$x(t) = A \sin(\omega t),$$

where x describes displacement as a function of time. A wave is a disturbance that propagates through space. It requires the same sort of time description but also needs an analogous description for its movement through space. In our discussion of oscillatory systems, we used the variables period T, angular frequency ω, and frequency f. A spatial description requires similar quantities. The first two time-domain variables have common analogues; they are the wavelength λ and the wavenumber k. The wavelength is the distance between successive crests of a wave or the distance from any point in the wave to the corresponding point in the next wave with the same phase. The wave number k is a spatial frequency:

$$k = \frac{2\pi}{\lambda};$$

it is the analogue of

$$\omega = \frac{2\pi}{T}.$$

These descriptions are especially convenient for harmonic waves in which 2π corresponds to one cycle.

11.1 Wave Properties

Let the shape of a wave be represented by the function $y = f(x)$ at time $t = 0$. As the wave propagates, it maintains its original shape. We can follow any point on a wave that is travelling with velocity v, and after a time t, we will find that the point has moved a distance x equal to vt. In general, we can describe the displacement at any point x at time t in the form

$$y(x, t) = f(x - vt). \tag{11.1}$$

Given the displacement of the wave at any point x_o at time $t = 0$, at any time t at a distance $x = vt$ the displacement will be the same. Because the point is arbitrary, y describes the displacement of the wave in terms of position and time.

As a wave travels one wavelength, λ, in one period, T, the wave's velocity is given by

$$v = \frac{\lambda}{T} \qquad \text{or} \qquad v = \frac{\lambda}{2\pi}\frac{2\pi}{T} = \frac{\omega}{k}. \tag{11.2}$$

Consequently,

$$y(x, t) = f(x - \frac{\omega}{k}t). \tag{11.3}$$

If the waveform is sinusoidal, then $y(x, 0) = A \sin B(x - (\omega/k)t)$, where A and B are constants; A is the amplitude of the wave. As x goes from x_o to $x_o + \lambda$, the function must go through one cycle. Thus the constant B must be the wavenumber k and

$$y(x, 0) = A \sin(2\pi\frac{x}{\lambda}). \tag{11.4}$$

In general,

$$y(x, t) = A \sin 2\pi\left(\frac{x}{\lambda} - \frac{t}{T}\right). \tag{11.5}$$

The latter form reminds us that if x moves through one λ or if t moves through one T, y moves through one cycle.

Two different movements occur in wave propagation. For example, as waves move down a string, individual bits of the string oscillate back and forth but undergo no net movement. The combined motion of all the particles, the wave, propagates through the medium. We must distinguish between the displacement of particles from equilibrium and the distance the wave travels. We must distinguish between the velocity of the oscillating particles and the velocity of the wave as a whole.

• •   Load WAVE1, motion at a point vs. motion of the wave (see Fig. 11.1).

We define the parameters and give a description of the wave, $y(x, t) = \sin(k\,x - \omega\,t)$ and the derivative of y with respect to time, $y'(x, t)$. This velocity is the velocity of the individual particles along the string or within the medium through which the wave is propagating. Recall that the wave velocity, a constant, is given by ω/k or λf.

The plots are in the form of vertical bars rather than a continuous line. For the present example, this helps to emphasize that we are discussing the motion of individual points along the path. The plots permit us to see the displacements and velocities of the particles both as they are distributed in space and as they are spread out in time.

In the first group of three plots, the time t is zero. Imagine the wave frozen in time. At our leisure, we can walk alongside the wave and examine the displacement. The second plot shows the velocities of the particles at that instant. Compare these two curves carefully. At $x = \lambda/4$, at the peak of the curve, what is the velocity of those particles that have maximum displacement? At $x = \lambda/2$, where the displacement changes sign, what is the velocity of the particles? Pay special attention to the direction of the motion. How might the phase diagram contribute to your explanation? Over what portion of the wave is the particle velocity upward? Downward? Explain the overall propagation process in terms of the displacement and the velocities at particular points. If time is increased from zero, how will the y vs. x curve shift? Change the t value and verify.

In the next group of three plots, the position x is zero. Imagine standing at the $x = 0$ point and observing the displacement of the wave as time increases. The y vs. t plot shows the displacements that we would observe at $t = 0$, $t = 0.05$, $t = 0.1$, etc. At $x = 0$, $t = 0$, the displacement is zero. The descending portion of the wave is approaching. Distinguish carefully between the y vs. x and y vs. t curves. At $t = 0$, what is the velocity of the particle at $x = 0$? Why? Over what portion of the y vs. t curve is the particle velocity upward? Downward? If x is increased from zero, how will the y vs. t curve shift? Verify your answer.

The phase diagrams appear to be identical; yet the conditions $t = 0$, where x takes a range of values and $x = 0$ where t takes a range of values would appear to be different. Do the curves start in the same place? Do they evolve in the same direction?

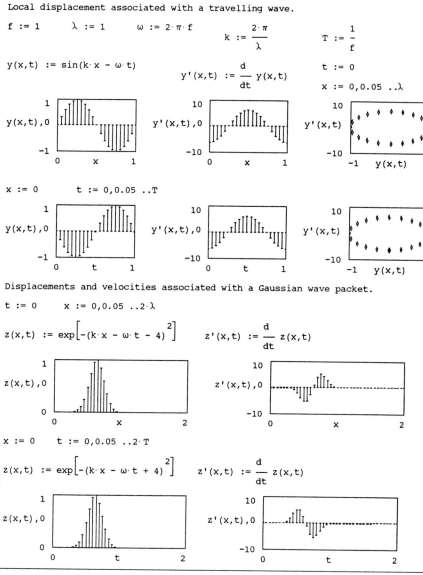

Figure 11.1 WAVE1, motion at a point vs. motion of the wave.

To see that the same concepts apply for nonharmonic waves, consider a travelling Gaussian waveform. Explain the z' vs. x and z' vs. t curves. What parameter should be changed to permit the z waveform to advance to larger x values? To larger t values? Verify your answer.

11.2 Superposition

Imagine two different waves passing through the same space at the same time. What happens when the waves overlap? Will each wave lose its individual identity? Will the collision change the properties (amplitude, frequency, phase) of the individual waves? We restrict our consideration to waves of small amplitude. Shock waves from an explosive event, such as a stroke of lightning or the beam from a high-intensity laser, are not of small amplitude and we exclude them. Given this caveat, the waves that we discuss obey linear wave equations. This means that when waves combine, the process is a linear one. In effect, each wave is independent of the other. Each wave contributes to the net amplitude of the resultant wave as if the other wave were not there. The principle of superposition tells us that the resultant wave is a linear combination of the contributing waves. At a given instant, the resultant wave is literally a sum of the components:

$$\Phi = \sum_i a_i\, \phi_i. \qquad (11.6)$$

• •　Plot the sum of two waves. Let $v := 331$ and $f := 5 \cdot 10^2$. For the case $x = 0$, plot 2.5 cycles of the sum of $y_1 + y_2$ vs. time, where $y_1 := A_1 \cdot \sin(k\,x - \omega\,t)$ and $y_2 := A_2 \cdot \sin(k\,x - \omega\,t + \phi)$. Let $A_1 = 1$ and $A_2 = 0.8$. Let $\phi = 0$. Why do these curves look like $-\sin(\theta)$? What is the frequency of the sum? How will these curves change if $x = \lambda/4$? Is the frequency of the sum dependent on x? Dependent on ϕ? Predict the answer, then verify.

Let $t = 0$ and plot 2.5 cycles of the sum of $y_1 + y_2$ vs. position. Let ϕ equal $0°$, $30°$, $90°$, and $180°$. For $A_1 = 1$ and $A_2 = 1$, what phase angle results in a maximum value for the sum, $y_1 + y_2$, of one?

• •　Load WAVE2, a sum of nonharmonic waves travelling in opposite directions (see Fig. 11.2).

Two nonharmonic waves, $f(t, x)$ and $g(t, x)$ are defined. Notice that they both use the form $(x - v\,t)$ (although g describes a wave travelling in the negative x-direction and includes a plus sign). Note that f and g are Gaussian wave forms. Starting points are chosen for convenience.

The first plot region shows a sequence of wave forms depicting the wave as it progresses in the positive x-direction. The second region shows a similar sequence, starting from x_0 and propagating in the negative x-direction. The sum of those functions is $h(x)$; the factors a and b permit

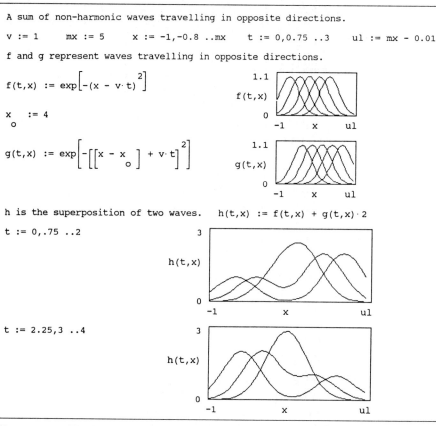

A sum of non-harmonic waves travelling in opposite directions.

$v := 1$ $mx := 5$ $x := -1, -0.8 \ ..mx$ $t := 0, 0.75 \ ..3$ $ul := mx - 0.01$

f and g represent waves travelling in opposite directions.

$$f(t,x) := \exp\left[-(x - v \cdot t)^2\right]$$

$$x_o := 4$$

$$g(t,x) := \exp\left[-\left[\left[x - x_o\right] + v \cdot t\right]^2\right]$$

h is the superposition of two waves. $h(t,x) := f(t,x) + g(t,x) \cdot 2$

$t := 0, .75 \ ..2$

$t := 2.25, 3 \ ..4$

Figure 11.2 WAVE2, a sum of nonharmonic waves travelling in opposite directions.

an adjustment of amplitude. The wave moving in from the right is given a larger amplitude to help distinguish the two curves. The first h-plot shows the sum at the times $t = 0$, 0.75, and 1.5. The second h-plot continues the sequence, showing the sums at the times $t = 2.25$, 3.0, and 3.75. As the sequence progresses, the waves, initially separate, gradually merge and pass on. Viewing the sum only at time $t = 2.1$, you might conclude that there is only one wave pulse.

To see only one or two curves or for a different spacing, change the t statements. Too many curves result in a cluttered diagram.

To check the details of the procedure, consider various values of (a, b). What would be expected for the cases $(1, 0)$ or $(0, 1)$? Try $(1, 1)$ and $(1, 0.2)$. Also try $(1, -1)$, $(1, -0.5)$, and $(1, -2)$.

• •   Load WAVE3, sum of nonharmonic waves travelling in opposite directions, II (see Fig. 11.3).

This example is similar to the previous one, except that the two waves summed are a triangle and a trapezoidal wave. Examine the various sums. To observe any intermediate value, enter a value for t at the statement "The sum at time t." How does the spike occur? Adjust "The sum at time t" to examine. Change the values in the t-sequences to see other groupings.

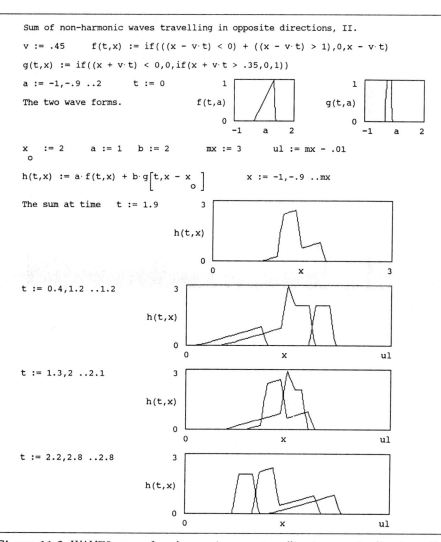

Figure 11.3 WAVE3, sum of nonharmonic waves travelling in opposite directions, II.

11.2.1 *Standing Waves*

The summing of two waves with equal amplitude, frequency, and wavelength, travelling in opposite directions results in a standing wave. The expression

$$y = A \sin(k\,x - \omega\,t) + A \sin(k\,x + \omega\,t) \tag{11.7}$$

is an example.

With the aid of the identity for the $\sin(\theta \pm \phi)$, this sum can be expressed as

$$y = 2A \sin(k\,x) \cos(\omega\,t). \tag{11.8}$$

The amplitude $2A$ is not surprising because the sum is of two waves with amplitude A. In the second expression for y, there is a separation of variables, $y = f(x) \cdot g(t)$; both the x and t portions of the curve can be described independently from the other. The $\cos(\omega\,t)$ term indicates that every element described has the same harmonic motion in time. The $\sin(k\,x)$ term indicates that the amplitude is dependent on position; in fact, there are some points where the amplitude is always zero. The zero points are determined according to the following condition, which makes the sine function zero:

$$k\,x = n\,\pi \quad n = 0, 1, 2\ldots \tag{11.9}$$

or

$$x = \frac{n\,\lambda}{2} \quad n = 0, 1, 2\ldots. \tag{11.10}$$

The points of zero amplitude are referred to as nodes, the points with maximum amplitude, antinodes.

• •   Load WAVE4, standing wave (see Fig. 11.4).

Two waves are defined: y_1 travels in the plus x-direction, and y_2 travels in the minus x-direction. For the present, ignore the factor B in the definition of y_2. There are two regions defining t. Only one should be enabled at a time.

Why are the functions written $f(t, x)$ and not $f(x, t)$?

Let fr take on the values 0, 0.3, 0.5, 0.8, and 1. Carefully explain the results when $fr = 0.5$ or $t = T/4$.

Disable the $t = fr \cdot T/2$ region; enable the t-range region. Process. Watch the sequence of cases. To repeat, press [Ctrl]r. Which points are nodes? Which are antinodes?

What happens if the amplitudes of the two waves are not equal? Try changing A in small steps and see how the sum changes.

```
Standing wave

λ := .1                      f := 1                        ω := 2·π·f
              2·π                         1
         k := ───                   T := ─
               λ                         f
```

One can look at either a single sum or a set of sums. First consider invidual sums, then disable the second t region and enable the first.

```
       λ                  T    T              fr := 1                  T
x := 0, ── ..λ    t := 0, ── ..─                           t := fr·─ □
        20                16    2             A := 1                   2

y1(t,x) := sin(k·x - ω·t)          y2(t,x) := A·sin(k·x + B·ω·t)     B ≡ 1

              y1(t,x) + y2(t,x)
y(t,x) :=  ────────────────────
                     2
```

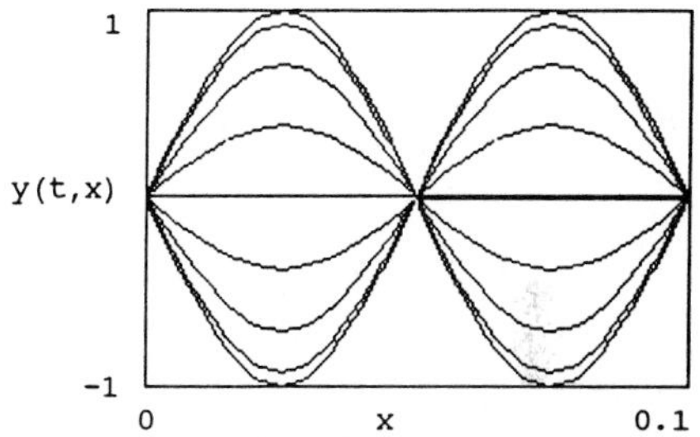

Figure 11.4 WAVE4, standing wave.

What happens if the frequencies of the two waves are not equal? Try changing B in small steps and see how the sum changes.

11.2.2   Shock Waves

Consider a sound source that emits short blasts of noise once per second. Sound waves propagate outward radially. After one second there is another blast, and another wave propagates outward. If the source is stationary, the spherical waves are concentric; if the source moves, they are not.

If the velocity of the source exceeds that of the outgoing wave, the successive spheres are spread out in such a manner that their edges form a cone-shaped wave front.

• •   Load WAVE5, shock wave (see Fig. 11.5).

The velocity of the waves is c; the velocity of the source is v. Examine the equation for $r(t)$. How are the radii of successive curves determined?

Let $v = 0$, corresponding to a stationary source. Process. Let v take on the values 4, 8, 10, 12, 16, and 20.

The angle of the cone for $v > c$ is given by $\sin(\theta) = c/v$. Plot θ vs. v as v goes from $1.5\,c$ to $10\,c$.

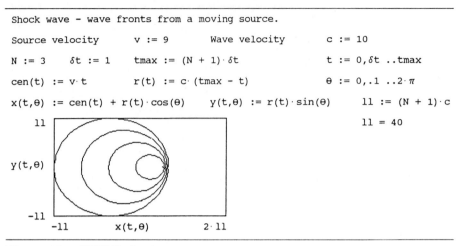

Shock wave - wave fronts from a moving source.

Source velocity $\quad$ v := 9 $\quad$ Wave velocity $\quad$ c := 10

N := 3 $\quad$ δt := 1 $\quad$ tmax := (N + 1)·δt $\quad$ t := 0,δt ..tmax

cen(t) := v·t $\quad$ r(t) := c·(tmax - t) $\quad$ θ := 0,.1 ..2·π

x(t,θ) := cen(t) + r(t)·cos(θ) $\quad$ y(t,θ) := r(t)·sin(θ) $\quad$ ll := (N + 1)·c

ll = 40

Figure 11.5 WAVE5, shock wave.

11.3 Fourier Series

Hook a microphone up to an oscilloscope and talk, sing, or play your favorite musical instrument. The microphone, a transducer, converts the pressure variations of the sound waves into voltage variations. This electrical signal, appropriately amplified, is used to drive the electron beam of the oscilloscope vertically, while at the same time an internally generated voltage drives the beam from left to right. That is, while the beam is moving from left to right with a uniform speed, the beam moves up and down according to the sound variations. The net effect is that the input signal (y-axis) is displayed as a function of time (x-axis).

If you display a single clear note — for example, from a flute or picolo — on the oscilloscope, it will look almost identical to a sinusoidal wave. However, if you sing the same note, the display will not be so smooth. The dominant oscillation will have the same overall motion, but the curve will not be smooth the way it was with a "clean" note. The difference is associated with the fact that the sounds that were sung are more complex; they contain more information than does the single pitched sound.

The clean signal is a sound wave that consists of essentially one frequency. The voice signal is more complex. It can be constructed from a combination of several different waves, each with its own frequency and amplitude. The oscilloscope display presents the information of the composite wave in the time domain, the time-amplitude composition.

The procedure known as harmonic analysis deals with the problem of finding the components of a signal such as that seen on the oscilloscope screen. Jean Baptiste Joseph Fourier, the French mathematician who accompanied Napoleon to Egypt, carried out the first extensive treatment of this process while developing the mathematical theory of the conduction of heat in solids. He was not the first to treat harmonic problems, but his name, his work, and his theorem have resulted in the generic name, Fourier series.

Fourier's idea is that some arbitrary function or signal can be expressed as an infinite series of certain prescribed functions. Here, those functions will be sines and cosines. Some examples of Fourier series will be presented below.

11.3.1   Ancient Astronomy

Long before Fourier, a less sophisticated form of harmonic analysis flourished. The ancients sought to describe the motion of the planets with various schemes. The most powerful idea, and one they exploited, was uniform circular motion.

Ptolemy's synthesis of planetary motion includes the system of epicycles, and this system of spheres is fundamentally the same as that used by Copernicus. (The present description will be in terms of circles rather than spheres.) A brief description of epicyclic motion follows.

A circle rotates uniformly about its center; refer to this circle as the deferent. At a fixed point on the circumference of the deferent, a second circle is attached; refer to this circle as an epicycle. The epicycle moves with the deferent as it rotates about its center; at the same time, the epicycle revolves about its own center (on the circumference of the deferent) with its own frequency. The combination of two circles moving with two frequencies can describe a variety of paths.

• •   Load WAVE6, epicycles (see Figs. 11.6 and 11.7).

The deferent rotates uniformly with angular velocity ω_a. The x and y motions of a point on the circle are given by $x_i := A \cdot \cos(\omega_a t_i + \phi_a)$ and $y_i := A \cdot \sin(\omega_a t_i + \phi_a)$, respectively. The resulting motion is circular. If the motion of the epicycle is added in, the equations become

$$x_i := A \cdot \cos(\omega_a t_i + \phi_a) + B \cdot \cos(\omega_b t_i + \phi_b)$$
$$y_i := A \cdot \sin(\omega_a t_i + \phi_a) + B \cdot \sin(\omega_b t_i + \phi_b).$$

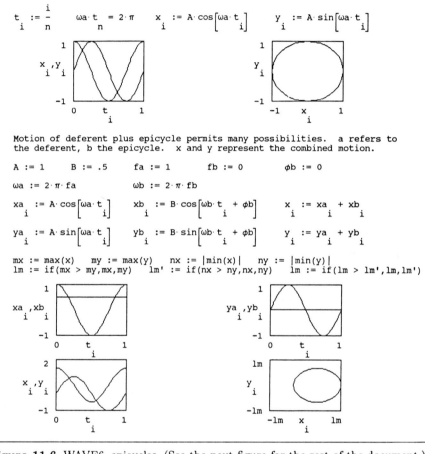

Deferent, epicycle, and equant.

Motion associated with a deferent is uniform circular motion.

$n := 36$ $i := 0 .. n$ $A := 1$ $fa := 1$ $\omega a := 2 \cdot \pi \cdot fa$ $\phi a := 0$

$t_i := \dfrac{i}{n}$ $\omega a \cdot t_n = 2 \cdot \pi$ $x_i := A \cdot \cos\left[\omega a \cdot t_i\right]$ $y_i := A \cdot \sin\left[\omega a \cdot t_i\right]$

Motion of deferent plus epicycle permits many possibilities.   a refers to the deferent, b the epicycle.   x and y represent the combined motion.

$A := 1$ $B := .5$ $fa := 1$ $fb := 0$ $\phi b := 0$

$\omega a := 2 \cdot \pi \cdot fa$ $\omega b := 2 \cdot \pi \cdot fb$

$xa_i := A \cdot \cos\left[\omega a \cdot t_i\right]$ $xb_i := B \cdot \cos\left[\omega b \cdot t_i + \phi b\right]$ $x_i := xa_i + xb_i$

$ya_i := A \cdot \sin\left[\omega a \cdot t_i\right]$ $yb_i := B \cdot \sin\left[\omega b \cdot t_i + \phi b\right]$ $y_i := ya_i + yb_i$

$mx := \max(x)$ $my := \max(y)$ $nx := |\min(x)|$ $ny := |\min(y)|$
$lm := \text{if}(mx > my, mx, my)$ $lm' := \text{if}(nx > ny, nx, ny)$ $lm := \text{if}(lm > lm', lm, lm')$

Figure 11.6 WAVE6, epicycles. (See the next figure for the rest of the document.)

The first two plots show the motion associated with x and y individually, and the associated circular motion. (At the beginning of the next section in the document, some of the quantities are redefined to keep them close to the plot regions of interest. These values can then be varied without having to move to the top of the document.) Note carefully what is plotted in each of the four plots — x-motion of the deferent and epicycle, y-motion of the deferent and epicycle, total x and y motions, all vs. t, and finally y vs. x. Explore the possible motions of the deferent-epicycle system. Observe all four plots regions as you proceed.

The equant. Motion takes place at constant distance from the center. The rate at which the position changes is constant about an offset point.

r := 1   ofs := .4   radius of circular orbit and offset from center.

n := 8   i := 0 ..n $t_i := \dfrac{i}{n}$ $\Phi_i := asin\left[\dfrac{ofs}{r} \cdot sin\left[\omega a \cdot t_i\right]\right]$

$\theta_i := \omega a \cdot t_i - \Phi_i$

$d_i := \sqrt{r^2 + ofs^2 - 2 \cdot r \cdot ofs \cdot cos\left[\theta_i\right]}$

$x_i := cos\left[\omega a \cdot t_i\right] \cdot d_i + ofs$ $y_i := sin\left[\omega a \cdot t_i\right] \cdot d_i$ ll := 1.2

j := 0 ..2·n

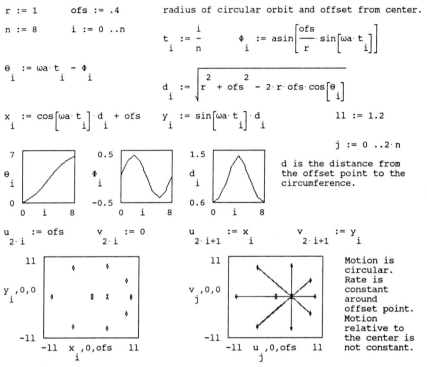

d is the distance from the offset point to the circumference.

$u_{2 \cdot i} := ofs$ $v_{2 \cdot i} := 0$ $u_{2 \cdot i+1} := x_i$ $v_{2 \cdot i+1} := y_i$

Motion is circular. Rate is constant around offset point. Motion relative to the center is not constant.

Figure 11.7 WAVE6 *continued.*

Let (A, B, f_a, f_b, ϕ_b) equal $(1, 0.5, 0, 1, 0)$. (Position the document so that the line starting $A := 1$ is at the top of the screen.) Explain this and each ensuing motion as it is displayed. Try $(1, 0.5, 1, 0, 0)$ and $(1, 0.5, 1, 1, 0)$. Be sure that you understand the difference between the previous two cases. Try $(1, 0.5, 1, -1, 0)$. What affect will changing ϕ_b have on this figure?

Next consider examples with different frequency ratios and different phase angles. Try $(1, 0.5, 1, 2, 0)$, $(1, 0.5, 1, 2, \pi/4)$, and $(1, 0.5, 1, 2, \pi/2)$. Try $(1, 0.5, 2, 1, 0)$ and $(1, 0.5, 2, 1, \pi/4)$.

Finally, increase the frequency of the epicycle. Try $(1, 0.5, 1, 3, 0)$ and $(1, 0.5, 1, 4, 0)$.

Systems like these were used to fit the observational data of planetary motion. Other features included the minor epicycle, another circle, attached to a fixed point on the epicycle and free to turn at its own frequency; the eccentric, in which motion was still uniform about the center, but the earth was offset rather than at the center; and the equant, where uniform motion occurred relative to a point offset from the center but with the earth still located at the center. This last example is included as the last section of this document. You may wish to explore the other features on your own.

Motion associated with the equant has the effect of changing the velocity of the planetary motion relative to the center (see Fig. 11.7). The planet moves along a circle centered on the earth. The planet moves through equal angles in equal times about a point that is offset from the center. For which part of the orbit is the velocity the greatest relative to the central point? For which part is it the least? Make a crude plot of the velocity vs. time.

As elegant as the methods of Ptolemy and the ancients were, gradually the system failed for both philosophical and scientific reasons. Ultimately, it was Kepler who solved the problem of the planets. His three laws state that (1) planetary orbits are ellipses, and the sun is at one focus of the ellipse; (2) a line from the sun to the planet sweeps out equal areas in equal times; and (3) $T^2 \propto R^3$ where T is the period of planetary motion and R is the mean distance between planet and sun. (We examined the third law previously, in the section on dimensional analysis.) The second law is true because there is a central force. The first and third laws depend on the attractive force varying inversely with the square of the distance.

• •   Load WAVE7, Kepler's second law (see Figs. 11.8 and 11.9).

We explore Kepler's second law by generating an orbit and noting the location of the orbiting body at uniform time intervals.

An equation for the ellipse is

$$r = \frac{a \cdot (1 - \epsilon^2)}{1 - \epsilon \cdot \cos(\theta)}. \tag{11.11}$$

Draw two radius vectors, from one of the foci to two different points of the orbital path that are near each other. Let the separation between the vectors be a small angle $\delta\theta$. The area contained between these radius vectors and the orbit is given approximately by

$$\delta A = \frac{1}{2} r^2 \delta\theta. \tag{11.12}$$

This can be determined by considering the area as that of a triangle with base r and height, the arc length, $r\,\delta\theta$.

The set of four plots shows the orbit (an ellipse with large eccentricity was selected to emphasize the changes), a subset of the individual radii (equal areas between each), and a plot of the radius and angle as a function of time. By increasing f, the number of vectors plotted is reduced, g changes the subsets for a given f; f and g are both integers; $g < f$.

In the final plot area, we show points spaced according to the equal area law. The values of k_1 and k_2 selected determine which triangles are to be outlined. The values should not exceed n. The equal area law has clear implications about the velocity at various points in the orbit. Compare this example with that of the equant above. Is the eccentric a better

---

Kepler's second law.   A radius vector from the attracting body to the orbiting body sweeps out equal areas in equal times.

The ellipse: semimajor axis, a, eccentricity ϵ.

$$a := 1 \qquad \epsilon := .7 \qquad \theta_0 := 0 \qquad r_0 := a \cdot (1 + \epsilon) \qquad \delta\theta_0 := 0.1 \cdot r_0^{-2}$$

$$n := 45 \qquad\qquad i := 0\ ..n$$

Generate the set of equally spaced points.

$$\begin{bmatrix} r_{i+1} \\ \delta\theta_{i+1} \\ \theta_{i+1} \end{bmatrix} := \begin{bmatrix} \dfrac{a \cdot \left[1 - \epsilon^2\right]}{1 - \epsilon \cdot \cos\left[\theta_i + \delta\theta_i\right]} \\ \dfrac{\left[1 - \epsilon \cdot \cos\left[\theta_i + \delta\theta_i\right]\right]^2}{\left[a \cdot \left[1 - \epsilon^2\right]\right]^2} \cdot .1 \\ \theta_i + \delta\theta_i \end{bmatrix}$$

$$x_i := r_i \cdot \cos\left[\theta_i\right]$$

$$y_i := r_i \cdot \sin\left[\theta_i\right]$$

$$\sum_i \delta\theta_i = 2.015 \cdot \pi$$

$$ll := 1.8 \qquad j := 0\ ..2 \cdot n \qquad u_{2 \cdot i} := 0 \qquad v_{2 \cdot i} := 0 \qquad f := 2 \qquad g := 1$$

$$u_{2 \cdot i+1} := x_i \cdot \text{if}(\text{mod}((i + g),f) \neq 0,0,1)$$

$$v_{2 \cdot i+1} := y_i \cdot \text{if}(\text{mod}((i + g),f) \neq 0,0,1)$$

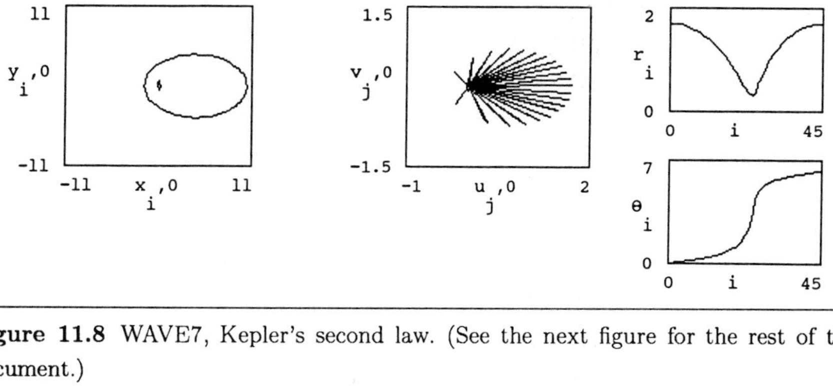

Figure 11.8 WAVE7, Kepler's second law. (See the next figure for the rest of the document.)

$k1 := 10$ $k2 := 25$ $n = 45$ $0 \leq k1 \leq n - 1$ ▫ $0 \leq k2 \leq n - 1$ ▫

$k := 0 \, ..6$ $c_k := 0$ $d_k := 0$ $c_1 := x_{k1}$ $c_2 := x_{k1+1}$

$d_1 := y_{k1}$ $d_2 := y_{k1+1}$ $c_4 := x_{k2}$ $c_5 := x_{k2+1}$

$d_4 := y_{k2}$ $d_5 := y_{k2+1}$

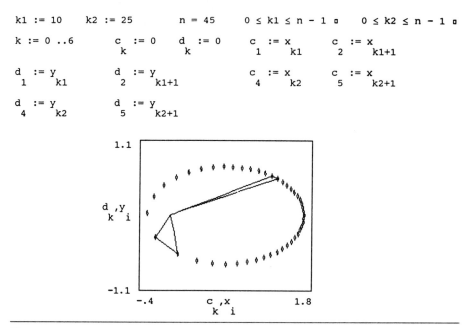

Figure 11.9 WAVE7 *continued.*

model to describe this phenomenon? Is the rotation approximately constant relative to the actual center of the ellipse? Plot the different motions in the same region for comparison.

The apparent epicyclic motion of a planet as seen from the earth is, of course, the result of viewing the motion of one orbiting body from another. The motion can be obtained from two elliptic orbits. First determine the elliptical motion; then determine the location of one body relative to the other.

• • Load WAVE8, elliptical orbits and retrograde motion (see Figs. 11.10 and 11.11).

The orbits for two different planets are calculated and plotted. The distance between the two planets is determined using the law of cosines. The angle as measured from the horizontal is determined using the law of sines and some basic geometry.

In the first plot region (see Fig. 11.11) are the two elliptical orbits of the planets about the central attracting body; we assume no interaction between the planets. In the second plot region is the orbit of the outer

Determine two elliptical orbits. Then specify the location of the outer orbiting body relative to the inner orbiting body.

$a1 := 1$ $\epsilon1 := .1$ $a2 := 1.2$ $\epsilon2 := .15$

$n := 66$ $i := 0 .. n$

$$aa := \left[\frac{a1}{a2}\right]^{1.5}$$

$\theta1_0 := 0$ $r1_0 := a1 \cdot (1 + \epsilon1)$ $\delta\theta1_0 := 0.1 \cdot r1_0^{-2}$ $\delta\theta2_0 := aa \cdot \delta\theta1_0$

$\theta2_0 := 0$ $r2_0 := a2 \cdot (1 + \epsilon2)$

$$\begin{bmatrix} r1_{i+1} \\ \delta\theta1_{i+1} \\ \theta1_{i+1} \end{bmatrix} := \begin{bmatrix} \dfrac{a1 \cdot \left[1 - \epsilon1^2\right]}{1 - \epsilon1 \cdot \cos\left[\theta1_i + \delta\theta1_i\right]} \\ \dfrac{\left[1 - \epsilon1 \cdot \cos\left[\theta1_i + \delta\theta1_i\right]\right]^2}{\left[a1 \cdot \left[1 - \epsilon1^2\right]\right]^2} \cdot .2 \\ \theta1_i + \delta\theta1_i \end{bmatrix}$$

$x1_i := r1_i \cdot \cos\left[\theta1_i\right]$

$y1_i := r1_i \cdot \sin\left[\theta1_i\right]$

$$\sum_i \delta\theta1_i = 4.212 \cdot \pi$$

$$\begin{bmatrix} r2_{i+1} \\ \delta\theta2_{i+1} \\ \theta2_{i+1} \end{bmatrix} := \begin{bmatrix} \dfrac{a2 \cdot \left[1 - \epsilon2^2\right]}{1 - \epsilon2 \cdot \cos\left[\theta2_i + \delta\theta2_i\right]} \\ \dfrac{\left[1 - \epsilon2 \cdot \cos\left[\theta2_i + \delta\theta2_i\right]\right]^2}{\left[a2 \cdot \left[1 - \epsilon2^2\right]\right]^2} \cdot .2 \cdot aa \\ \theta2_i + \delta\theta2_i \end{bmatrix}$$

$x2_i := r2_i \cdot \cos\left[\theta2_i\right]$

$y2_i := r2_i \cdot \sin\left[\theta2_i\right]$

$$\sum_i \delta\theta2_i = 2.207 \cdot \pi$$

$$rd_i := \sqrt{r1_i^2 + r2_i^2 - 2 \cdot r1_i \cdot r2_i \cdot \cos\left[\theta2_i - \theta1_i\right]}$$

$\theta3_i := \text{asin}\left[\dfrac{r1_i}{rd_i} \cdot \sin\left[\theta2_i - \theta1_i\right]\right]$ $\theta e_i := \theta2_i - \theta1_i + \theta3_i$

$\theta_i := \theta1_i + \theta e_i$ $x_i := rd_i \cdot \cos\left[\theta_i\right]$ $y_i := rd_i \cdot \sin\left[\theta_i\right]$

Figure 11.10 WAVE8, elliptical orbits and retrograde motion. (See the next figure for the rest of the document.)

planet as seen by the inner planet. From an earth-centered frame, the system of epicycles invented by ancient astronomers was quite a reasonable way to account for planetary motion.

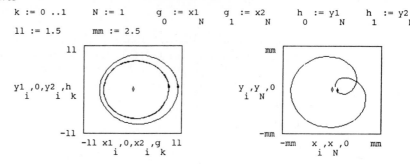

$$k := 0 \ ..1 \qquad N := 1 \qquad g_0 := x1_N \qquad g_1 := x2_N \qquad h_0 := y1_N \qquad h_1 := y2_N$$

$$ll := 1.5 \qquad mm := 2.5$$

Every fifth point is plotted in the d vs c region and written to file in xcoor and ycoor.

$$j := 0 \ ..\frac{n}{5} \qquad c_j := x_{5 \cdot j} \qquad d_j := y_{5 \cdot j}$$

$$WRITE\!\left[xcoor\right] := c \qquad WRITE\!\left[ycoor\right] := d$$

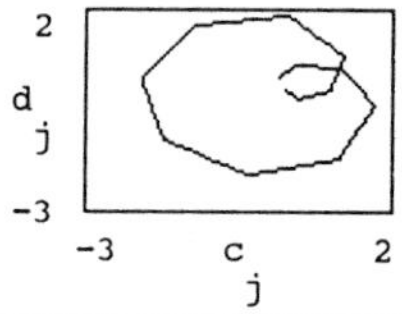

Figure 11.11 WAVE8 *continued.*

The c[i and d[i data read in are the coordinate values from WAVE8 specifying the location of one orbiting body relative to another. See how well you can fit this data with a system consisting of a deferent and an epicycle.

$$n := \frac{66}{5} \qquad i := 0 \ ..n \qquad c_i := READ\!\left[xcoor\right] \qquad d_i := READ\!\left[ycoor\right]$$

$$mm := 2.5$$

Supply values for A, B, f, α, and ϕ. A "fair" match is possible with relatively simple values.   A is the radius of the deferent, B the epicycle, f is the frequency of the epicycle relative to that of the deferent, α and ϕ are phase angles.

$$A := 0 \qquad B := 0 \qquad f := 0 \qquad \alpha := 0 \qquad \phi := 0$$

$$\theta_i := 2 \cdot \pi \cdot \frac{i}{n} \qquad\qquad \Phi_i := 2 \cdot \pi \cdot \frac{i}{n} \cdot f$$

$$x_i := A \cdot \cos\!\left[\theta_i + \alpha\right] + B \cdot \cos\!\left[\Phi_i + \phi\right] \qquad y_i := A \cdot \sin\!\left[\theta_i + \alpha\right] + B \cdot \sin\!\left[\Phi_i + \phi\right]$$

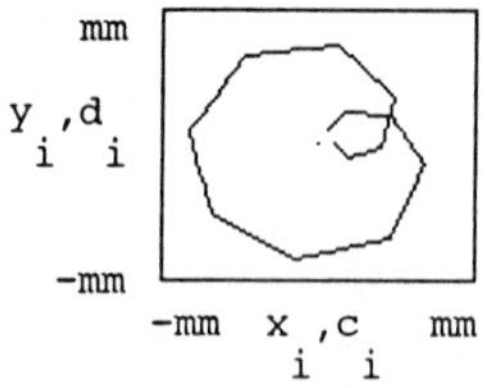

Figure 11.12 WAVE9, fitting the data with an epicycle.

• •   Load WAVE9, fitting the data with an epicycle (see Fig. 11.12).

At the end of the last document, a small number of data points are written to disk. Those data points are read in at the beginning of this document. Supply values for A, B, f, α, and ϕ and see how well you can fit the data. The data can be fit with relatively simple values. The ancient astronomers were very clever. See how well you can fit this realistic data with a system composed of a deferent and a single epicycle. (Every once in a while, it is useful to stop and think of the kinds of computations that are being performed, computations of such magnitude that they would simply be prohibitive if you attemped them on your own.)

11.3.2 Harmonic Series

The complex motion of the planets can be described by various combinations of circular motion. In general, any periodic motion or function can be described by summing a series of sine and/or cosine functions, with different amplitudes and different frequencies:

$$f(t) = \frac{a_0}{2} + a_1 \cdot \sin(\omega\, t) + a_2 \cdot \sin(2\,\omega\, t) + \cdots$$
$$b_1 \cdot \cos(\omega\, t) + b_2 \cdot \cos(2\,\omega\, t) + \cdots \tag{11.13}$$

To gain some insight into the process, examine the sum of a small number of sine functions of different frequencies.

• •   To perform a sum, two indices are necessary, one for the different frequency terms and the associated amplitudes, and one for the points at which the function is to be evaluated. For example, the frequencies might be f_j and the times t_i.

Try something simple. Create a sum with a maximum of four terms. Let $j = 1, \ldots, M$, where M can take on any integral value from one to four. (M could, in fact, be larger, but at this point four terms is enough.) Type $a_j :=$, and enter four simple values for amplitude. For example, let each successive amplitude be half the previous one. Type $f_j :=$, and enter four values for frequency — start for example, with the values one to four.

Observe the sum for two cycles of the lowest frequency. Let the number of points at which the sum is computed be N; start with $N = 60$. If processing is too slow, reduce N to 50 or 40. Let the time be specified as

$$t_i = \frac{i}{N} \cdot \frac{2}{f_1} \qquad \text{where} \qquad i = 0, \ldots, N.$$

Why does this statement correspond to two cycle of the lowest frequency?

Then define the y-values,

$$y_{i,j} = a_j \cdot \sin(2\,\pi\,f_j\,t_i).$$

To observe the individual sine curves, define $y'_{j,i} = y_{i,j}$. Plot $y'_{j,i}$ vs. t_i. M sine curves with different frequencies should be displayed.

Sum the individual curves. Define the sum

$$z_i := \sum_k y_{i,k} \qquad \text{where} \qquad k := 1, \ldots, 1.$$

By limiting the range of k, we limit the number of terms in the sum. Plot z_i vs. t_i. Increase the upper limit of k to four in steps of one. Observe how the shape develops.

If the series were a cosine series instead of a sine series, would the shape be the same? If sine and cosine differ by a fixed phase, won't the curve just be shifted? If not, why not? Define

$$yy_{i,j} := a_j \cdot \cos(2\,\pi\,f_j\,t_i) \qquad \text{and} \qquad zz_i := \sum_j yy_{i,j}.$$

Plot zz_i vs. t_i.

• •   Examine some sine and cosine series. Look at a limited region to minimize the computation time. Examine each sum with one, then two, and then three terms. Then, if you wish, increase the number of terms in the sum in larger steps.

For the region from $x = 0$ to $x = \pi$, explore in pairs the following sine and cosine series:

$$s_j := 2 \cdot \sum_i \frac{(-1)^{i+1}}{i} \sin(i \cdot x_j).$$

$$t_j := \frac{8}{\pi} \cdot \sum_i \frac{\cos((2 \cdot i - 1) \cdot x_j)}{(2 \cdot i - 1)^2},$$

and

$$s_j := \sin(x_j)$$

$$t_j := \frac{2}{\pi} - \frac{4}{\pi} \cdot \sum_i \frac{\cos(2 \cdot i \cdot x_j)}{4 \cdot i^2 - 1}.$$

For the range $x = 0$ to $x = \pi/2$, examine

$$s_j := \frac{2}{\pi} \cdot \sum_i \left(1 - \cos\left(\frac{i \cdot \pi}{2}\right)\right) \cdot \frac{\sin(i \cdot x_j)}{i}$$

$$t_j := \frac{1}{2} - \frac{2}{\pi} \cdot \sum_i (-1)^i \cdot \frac{\cos((2 \cdot i - 1) \cdot x_j)}{2 \cdot i - 1}.$$

These series should approximate the functions x, $\sin(x)$, and 1.

• •   Load WAVE10, Lanczos smoothing (see Fig. 11.13).

After a finite number of terms are summed in a Fourier series, a certain amount of "ringing" remains. This is representative of the higher frequency elements that have been omitted. This ringing can be greatly

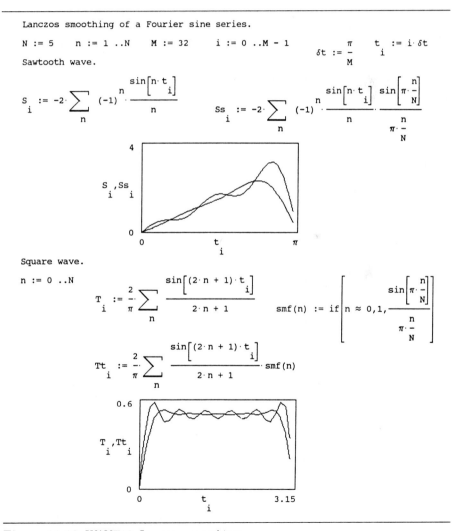

Figure 11.13 WAVE10, Lanczos smoothing.

reduced by including the Lanczos smoothing factor, or convergence factor. The factor is

$$\frac{\sin(n\pi/N)}{(n\pi/N)}.$$

As an example, the first sine series of the last problem is calculated and plotted. The same series is then plotted with the Lanczos convergence factor. The effect is to remove much of the high-frequency ringing that still remains after a finite number of terms are summed. A second example for a square wave series is shown. Change values of N in each case and observe the ringing and the quality of the smoothing.

The Fourier series is observed here in connection with the concept of superposition. Later, when we deal with optics, we will examine the fast Fourier transform, which provides a means of transforming the description of a signal between two different domains, for example, time and frequency.

> But sir, that's Charlie's beach,
>
> Charlie don't surf.
>
> *Apocalypse Now*

12

Heat and
Thermodynamics

The concept of heat is approached in this chapter through a variety of problems: an asteroid striking the earth; Newton's law of cooling; the wind chill effect; and heat flow from the interior to the exterior of a house, where both convective and conductive effects are considered. The ideal gas law is compared with the van der Waals equation of state. The Carnot cycle is explored as we look first at the problem of isothermal and adiabatic transitions connecting two arbitrary points. We then explore the area enclosed by the cycle and ask questions about the conditions for maximum work and maximum efficiency.

12.1   Heat

Temperature is a subtle concept. For the present, consider temperature to be the quantity that a thermometer measures. The unit is the Kelvin or degree Celsius. Heat is the transfer of energy resulting from temperature differences. The energy unit, the Joule, is already familiar. The calorie, an energy unit used frequently in heat computations, is defined as the amount of energy necessary to raise the temperature of one gram of water from $14.5°$ C to $15.5°$ C. Before the concept of energy was clearly understood, it was a major revelation to realize that heat and work had similar measures. The relative sizes of the units are $1\,\text{cal} = 4.186\,\text{J}$.

When heat is transferred to or from an object, the temperature or the phase of the object changes. The relation between the heat flow and the change in temperature is

$$\Delta Q = m\,c\,\Delta T, \tag{12.1}$$

where m is the mass of the object and c is the specific heat. The units of specific heat are energy per mass per temperature, for example, calories per gram per degree Celsius. Note that the specific heat depends on whether the substance is held at constant volume during the heat transfer process or, for example, the transfer takes place at constant pressure.

During a phase transition, a substance either absorbs or gives off heat without changing its temperature. The equation

$$\Delta Q = m\,L \tag{12.2}$$

expresses this relationship. L is the latent heat associated with the phase transition; ΔQ is the energy exchanged; and m is the mass. For water, the latent heat of fusion is 80 calories per gram. That is, 80 calories are required to change one gram of ice at $0°C$ to one gram of water at $0°C$. Similarly, the latent heat of vaporization for water is 540 calories per gram.

• •   One postulated mechanism for the end of all life on earth is that a huge asteroid will collide with the earth. Let the asteroid land in the ocean, and make the assumption that half of its kinetic energy ultimately goes into evaporating water. (The actual energy transfer process is quite complex. See, for example, *Science*, 250, 1078 (1990).) When the evaporation has taken place, by how many meters will the combined oceans of the world have dropped?

Try several sizes of asteroid. An enormous asteroid could be 400 to 500 km in diameter. Assume it has a density similar to that of iron and an incoming velocity of $10^4\,\text{m/s}$. Plot the depth of ocean evaporated vs. the diameter of the asteroid.

12.1.1 *Newton's Law of Cooling*

An object warmer than its surroundings will cool; an object cooler than its surroundings will warm. Newton's law of cooling, an empirical statement, suggests that the rate of temperature change is proportional to the temperature difference between an object and its surroundings (s). That is,

$$\frac{dT}{dt} = -k\,(T - T_s),\qquad(12.3)$$

where k is a positive constant. If $T > T_s$, dT/dt is negative and the temperature, T, decreases with time. If $T < T_s$, the reverse is true.

We will treat this problem both numerically and analytically. The analytic solution can be obtained by integration (the integration is not complicated; it involves the equivalent of integrating $dx/\,x$):

$$\int_{T_{\text{init}}}^{T} \frac{dT'}{T' - T_s} = -k \int_{0}^{t} dt'.$$

Integrating, we find

$$T = T_s + (T_{\text{init}} - T_s)\,e^{-k\,t}.\qquad(12.4)$$

• •   Take the derivative of T with respect to time and verify that the differential form of the equation is returned.

• •   Verify that at $t = 0$, $T = T_{\text{init}}$. Evaluate T when t is very large.

• •   Plot T_s and $(T_{\text{init}} - T_s)\exp(-k\,t)$ vs. t. Plot T vs. t. Verify that the expected behavior occurs whether $T_{\text{init}} > T_s$ or $T_{\text{init}} < T_s$. Remove the minus sign in the exponential. What happens to the solution?

• •   Load HEAT1, Newton's law of cooling (see Fig. 12.1).

The analytic solution is presented and the temperature is plotted as a function of time. What is the effect of changing k?

Data associated with the cooling of an object are presented in tabular form. The data are taken at one-minute intervals. The temperature, TT_i, is in degrees Celsius.

Find the best value for the cooling constant, k, using the general curve fitting technique. First, plot TT_i and $T(k, T_i)$ with various values of k; when a reasonable fit is found, use this k value as the guess value in a fitting process. Minimize the sum of the squares of the differences between TT_i and $T(k, T_i)$.

Plot TT_i and $T(k, T_i)$ vs. t_i. Plot the difference, $TT_i - T(k, T_i)$ vs. t_i.

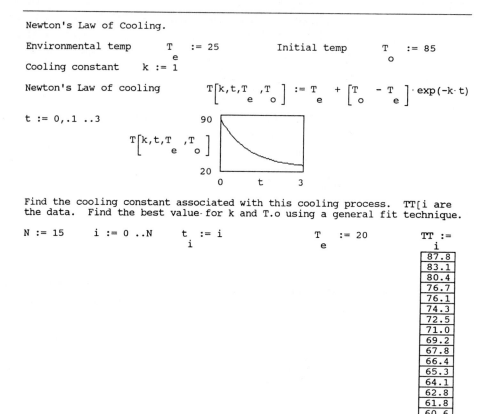

Newton's Law of Cooling.

Environmental temp $T_e := 25$ Initial temp $T_o := 85$

Cooling constant $k := 1$

Newton's Law of cooling $T[k, t, T_e, T_o] := T_e + [T_o - T_e] \cdot \exp(-k \cdot t)$

$t := 0, .1 .. 3$

Find the cooling constant associated with this cooling process.   TT[i are
the data.   Find the best value for k and T.o using a general fit technique.

$N := 15$ $i := 0 .. N$ $t_i := i$ $T_e := 20$ TT :=

TT_i
87.8
83.1
80.4
76.7
76.1
74.3
72.5
71.0
69.2
67.8
66.4
65.3
64.1
62.8
61.8
60.6

Figure 12.1 HEAT1, Newton's law of cooling.

Try slightly larger and slightly smaller values for k and observe how
the fit changes. From the appearance of the curves, if the data were available for 30 minutes instead of 15, would you guess that k would be the
same, larger, or smaller? Why?

Now consider the problem using a numerical approach. Rewrite the
differential equation for Newton's law of cooling in difference form. First
express the equation in terms of finite steps:

$$\delta T = -k(T - T_s)\, \delta t \qquad (12.5)$$

and identify $\delta T = T_{j+1} - T_j$. The difference equation is (new temperature
equals former temperature plus change)

$$T'_{j+1} = T'_j - k\,(T'_j - T_s)\, \delta t. \qquad (12.6)$$

To keep the various temperatures distinct, let the temperature, time and
index for the analytic case be T, t, i and for the numeric case T', t', j.

What is the initial value T_0'? Let $N = 1$, let $j = 0 \ldots, N$ and let $\delta t = 15/N$. Let t_j' represent the elapsed time; define t_j'. Plot T_j' vs. t_j' and TT_i vs. t_i in the same plot region. Make the region of moderate size, say 10, 20, and process. In this example, for the particular value of k and for the limited period of time, the cooling, as determined analytically, does not deviate markedly from a straight line. Consequently, the error, on taking one 15-minute time step, is not as large as you might think. Let $N = 2$ and process. The kink between the 0 to 1 and 1 to 2 steps is noticeable. Try $N = 5$, 10, and 100.

How does the quality of the numerical fit compare with that of the analytical fit? How does the numerical result compare with the analytical result? Plot them together in the same plot region.

12.1.2   *Wind Chill*

As an object's temperature changes, there is a corresponding heat flow. As we noted earlier in our discussion of temperature change,

$$\Delta T = \frac{1}{mc} \Delta Q.$$

Writing this change per time and multiplying the right-hand side by A/A where A is the surface area, we get

$$\frac{\Delta T}{\Delta t} = \frac{A}{mc} \frac{\Delta Q}{A \Delta t} = \frac{A}{mc} H, \tag{12.7}$$

where H is the heat transfer per area per time. The wind chill temperature involves the quantity H.

Imagine that you are standing naked in the open on a windy winter day. How does the heat loss from your body depend on the air temperature, T, and wind velocity, v? Data (hot water bottle hung from a telephone pole) suggest that $H(v, T) = f(v) \cdot g(T)$.

• •   Load HEAT2. Some data are presented on the velocity-dependent part of H, the function $f(v)$. Plot f_i' vs. v_i. From your knowledge of curve shapes (review your plots from the section on families of curves in Chapter 2), decide on an appropriate function with a maximum of three terms. Define a function $f(v)$ incorporating these terms; include in the argument list the parameters that are to be adjusted, for example, $f(a, v) = a v^2$. This particular velocity-dependent term, however, is not one that you should select. Guess the coefficients and plot the defined function together with the data. Adjust the values of the coefficients by hand until a decent fit is obtained. Change the terms in your function if necessary.

To improve on the values just obtained for the coefficients, use these values as guess values in a general fit method (see Chapter 2). This approach determines the coefficients using a given-minerr solve block. Recall that we want to minimize the sum of the squared errors, the errors being the differences between the data f_i' and the function that is to fit the data. (MathCAD is somewhat unpredictable here. Solutions may be slow, and the results depend somewhat on guess values. The results are not determined to the precision shown.) Plot the fitted curve and data in the same plot region.

There are a number of ways to fit the data in HEAT2. After you have obtained a fit, open HEAT3 for a brief examination of the data.

Whatever the specific form that f and g may take, the concept behind the wind chill temperature is that different sets of conditions (different temperatures and wind velocities) may result in the same heat flow. To determine the wind chill temperature, we equate the heat flows from two different cases: one case is based on the actual temperature and wind velocity, $H(T, v)$; the other, $H(T', v')$, is for a velocity $v' = 1.79$ m/s and T', the wind chill temperature. The question is what temperature T' will make

$$H(T', v') = H(T, v).$$

• • Let $H(T, v) = f(v) \cdot (T_s - T)$. Express $H(T', v')$ similarly. Solve for T' in terms of $f(v)$, $f(v')$, T_s, and T. (Leave the $f(v)$'s in functional form; do not substitute an expression for $f(v)$.)

• • Load HEAT4, wind chill (see Figs. 12.2 and 12.3).

We are now ready to explore the heat loss and the wind chill temperature as a function of the velocity and the temperature of the environment, T_s. The function $f(v)$ has the form described above in HEAT3; $g(T)$ is simply the temperature difference $(T_s - T)$. The wind chill temperature, in terms of the air temperature and velocity, $T'(T, v)$, is the form determined from the previous exercise.

For plotting purposes, the heat loss function H and the wind chill temperature function T' are defined twice with the order of the variables interchanged. For example, we write the heat loss function as $H(T, v)$ and $H'(v, T)$; they are equated to the same expression. This is necessary to plot the family of curves as a function of v (where we want the (T, v) form) or as a function of T (where we want the (v, T) form). To avoid a line connecting individual curves when we plot a family of curves, either the first or last point of each curve must lie outside the plot region. In

Wind chill – after H. R. Crane.

A range of wind velocities.   v := 2,4 ..20

A range of temperatures.   T := 10,0 ..-20

Parameters in the heat loss equation.   a := 10.45   b := 10   c := -1

T.s is the temperature of the exposed surface. T_s := 33

The empirical heat loss equation.

$$f(v) := \left[a + b \cdot \sqrt{v} + c \cdot v \right] \qquad H(T,v) := f(v) \cdot \left[T_s - T \right]$$

Equate two heat losses - one at actual wind speed and temperature, H(T,v), the other at the nominal speed v', 1.79 m/s and wind chill temperature T'. H(T,v) = H(T',v') Solve for T'.

v' := 1.79

$$T'(T,v) := T_s - \left[T_s - T \right] \cdot \frac{f(v)}{f(v')}$$

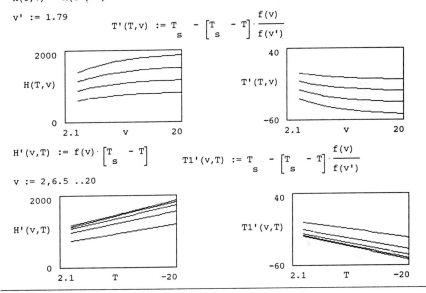

$$H'(v,T) := f(v) \cdot \left[T_s - T \right] \qquad T1'(v,T) := T_s - \left[T_s - T \right] \cdot \frac{f(v)}{f(v')}$$

v := 2,6.5 ..20

Figure 12.2 HEAT4, wind chill. (See the next figure for the rest of the document.)

this case, we set the lower ordinate limit to exclude either the smallest v or largest T values.

For each family of curves, identify the specific value of the parameter that goes with a particular curve.

In the function $f(v)$, the coefficients a, b, and c all have different units. In the last portion of the document, units are attached to the coefficients and all other necessary quantities. The procedure was to multiply in the complete set of desired units and divide out the units of the multiplying terms. G is equivalent to H. Including units in the expressions has the advantage that units can be specified in any system.

The coefficient a, b, and c all have different units. Units are included
in the expression below.

$m := 1L$ $s := 1T$ $kg := 1M$ $C := 1Q$

$h := 3600 \cdot s$ $J := kg \cdot m^2 \cdot s^{-2}$ $cal := 4.186 \cdot J$

$kcal := 1000 \cdot cal$

$epat := kcal \cdot h^{-1} \cdot m^{-2}$ epat - energy per area per time.

$vel' := \dfrac{m}{s}$

$a' := a \cdot \dfrac{epat}{C}$ $b' := b \cdot \dfrac{epat}{C} \cdot \dfrac{1}{\sqrt{vel'}}$ $c' := c \cdot \dfrac{epat}{C \cdot vel'}$

$v' := 2 \cdot \dfrac{m}{s}, 4 \cdot \dfrac{m}{s} \,.. 20 \cdot \dfrac{m}{s}$ $T' := 10 \cdot C, 0 \cdot C \,.. -20 \cdot C$ $T'_s := 33 \cdot C$

$G(T', v') := \left[a' + b' \cdot \sqrt{v'} + c' \cdot v' \right] \cdot \left[T'_s - T' \right]$

$G\left[0 \cdot C, 5 \cdot \dfrac{m}{s} \right] = 917.752 \cdot epat$ $H(0,5) = 917.752$

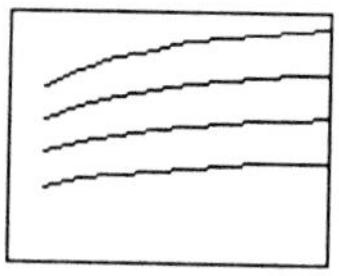

Figure 12.3 HEAT4 *continued.*

Define the Fahrenheit degree, F, miles, mi, and hours, hr. Plot the
wind chill family of curves; show two sets of curves. Plot wind chill temperature vs. wind speed for several air temperatures and wind chill temperature vs. air temperature for several wind speeds. Put the limits of
the abscissa in miles per hour and the limits of the ordinate in degrees
Fahrenheit.

• •   One of the estimation problems near the beginning of the text
was to determine the surface area of your body. Use this value in H.
Instead of kcal per square meter per hour, use kcal per your body area
per hour. If your entire body were exposed and if the body were able
to maintain its surface area at a constant temperature (this is totally
unreasonable), determine your weight loss as a function of time. (One
pound is approximately 3500 kcal.) Plot your weight loss per hour for
$0° C$ as a function of wind speed.

• •   Assume your body is able to produce heat at the rate of 150 kcal per
hour. For a given external temperature and wind velocity, at what rate
would the body temperature decrease? Plot body temperature vs. time
for a person with a mass of 65 kg.

12.1.3   *Heat Flow*

If an insulator is placed between reservoirs at two different temperatures,
the heat flow depends directly on the cross-sectional area of the insulator

A and on the temperature gradient across the insulator, that is, $\Delta T/\Delta x$. The constant of proportionality k, is the thermal conductivity of the material selected:

$$\frac{\Delta Q}{\Delta t} \simeq -k\,A\,\frac{\Delta T}{\Delta x}. \tag{12.8}$$

For an object of constant cross section A, under steady-state conditions (where temperatures throughout the insulator do not change with time), the flow of heat across any cross section must be the same. (If this were not true, the temperatures would change or something would melt or boil.) If $\Delta Q/\Delta t$ is constant, then for constant A $\Delta T/\Delta x$ must be constant. Just as constant dx/dt implies $x \propto t$, constant $\Delta T/\Delta x$ implies $T \propto x$. Therefore, the temperature change across the conductor is linear. The conduction heat flow equation becomes

$$\frac{\Delta Q}{\Delta t} = -k\,A\,\frac{T_2 - T_1}{L} \qquad \text{where} \qquad T_2 > T_1.$$

T_1 and T_2 are the temperatures of the reservoirs; L is the thickness of the insulator. This equation can be rewritten, where $H = \Delta Q/\Delta t$ (area is not included in this H, as it was in the wind chill example) as

$$(T_2 - T_1) = H \cdot \frac{L}{k\,A}.$$

The heat flow equation is analogous to Ohm's law for current flow in electrical circuits:

Quantity	Heat	Electrical
Flow	H	I
Condition that drives flow	$T_2 - T_1$	$V_2 - V_1$
Resistance to flow	$\Re = L/(k\,A)$	$R = L/(\sigma\,A)$
Ohm's law	$T_2 - T_1 = H\,\Re$	$V_2 - V_1 = I\,R$

(In engineering practice, the R-value of an insulator is just L/k; the A is not included.)

Heat flow through a wall, for example, might pass through drywall, then insulation, and finally the outer wall. Such a configuration is referred to as a compound slab when the cross section of each material is the same, but there is a sequence of materials with different resistances. The materials are in series and the resistances add:

$$T_2 - T_1 = H \cdot (\Re_1 + \Re_2 + \cdots).$$

Convective heat flow is complex and not easy to characterize quantitatively. For a vertical plate—for example, a wall—an empirical expression for the heat flow is

$$H_{conv} \propto (T_{air} - T_{wall})^{1.25}. \qquad (12.9)$$

For horizontal plates—for example, the ceiling—the relation is similar. The constant varies depending on whether the heat is being exchanged with the air above or the air below and whether the plate is warmer or cooler than its surroundings.

Objects also lose or gain heat through radiation. All objects radiate electromagnetic energy continuously. How such energy is radiated depends on the temperature, the area radiating, and the properties of the surface. The relationship is

$$P = e \sigma A T^4, \qquad (12.10)$$

where P is the power radiated in watts (Joules per second) (the same units as H); T is the temperature in Kelvin, unless otherwise stated; σ is the Stefan-Boltzmann constant $(5.67 \cdot 10^{-8} \, J/(s \, m^2 \, K^4))$; and e, which varies from zero to one, is the emissivity of the surface. For a perfect absorber e is 1. Perfect absorbers are black when cool (the sun approximates an ideal black body radiator). For a perfect reflector e is 0. (A good reflector at one wavelength may be a good absorber at a different wavelength.)

Since the surroundings also radiate, the net energy exchange is

$$P = e \sigma A (T^4 - T_{env}^4), \qquad (12.11)$$

where T_{env} is the temperature of the environment. The environment may not be uniform; in such a case you would have to take into account the various sources that are present.

12.1.4   Heat Flow from a House

Consider the heat flow from the interior to the exterior of a heated one-room house. Convective effects transfer energy from the warm interior to the inside walls. Heat is conducted through the walls to the exterior surface. Convective effects transfer energy from the exterior surfaces to the air beyond. Under steady-state conditions, all of these heat flows must be equal.

• •   Load HEAT5, heat flow from a house (see Figs. 12.4 and 12.5).

The heat flow from the interior of a one-room house to the exterior is calculated. The model is based on the convective and conductive effects discussed above. Many details are not included, and so the model must be considered as a very basic one. (For example, no exchange of air is included in the model, and heat transfer through the floor is omitted.)

Heat loss from one room house through walls, ceiling, and windows.

$m \equiv 1L$ $\qquad$ $s \equiv 1T$ $\qquad$ $kg \equiv 1M$ $\qquad$ $C \equiv 1Q$ $\qquad$ $hr \equiv 3600 \cdot s$ $\qquad$ $cm \equiv 0.01 \cdot m$

$J \equiv kg \cdot m^2 \cdot s^{-2}$ $\qquad$ $F \equiv \frac{5}{9} \cdot C$ $\qquad$ $cal \equiv 4.186 \cdot J$ $\qquad$ $BTU \equiv 252 \cdot cal$ $\qquad$ $W \equiv \frac{J}{s}$

$\qquad\qquad\qquad\qquad\qquad$ $in \equiv 2.54 \cdot cm$ $\qquad$ $ft \equiv 12 \cdot in$

areas $\qquad$ $Arcl := 750 \cdot ft^2$ $\qquad$ $Arwin := 100 \cdot ft^2$ $\qquad$ $Arwll := 945 \cdot ft^2$

thickness $\qquad$ $\delta xwall := 4 \cdot in$ $\qquad$ $\delta xceil := 8 \cdot in$ $\qquad$ $\delta xwind := 0.125 \cdot in$

conductivity

$kwall := 0.0002 \cdot \dfrac{cal}{s \cdot cm \cdot C}$ $\qquad$ $kceil := 0.0002 \cdot \dfrac{cal}{s \cdot cm \cdot C}$ $\qquad$ $kwind := .002 \cdot \dfrac{cal}{s \cdot cm \cdot C}$

convection coef.

$$chvert := 0.4 \cdot 10^{-4} \cdot \dfrac{cal}{s \cdot cm^2 \cdot C^{1.25}}$$

$chhoru := 0.6 \cdot 10^{-4} \cdot cal \cdot s^{-1} \cdot cm^{-2} \cdot C^{-1.25}$ $\qquad$ $chhord := 0.5 \cdot chhoru$

Calculate the temperature of the interior and exterior wall surfaces.

$Tins := 20 \cdot C$ $\qquad$ $Tsurfin := 10 \cdot C$ $\qquad$ $Tsurfout := 0 \cdot C$ $\qquad$ $Tout := -10 \cdot C$

Given

$$chvert \cdot (Tins - Tsurfin)^{1.25} \approx kwall \cdot \dfrac{Tsurfin - Tsurfout}{\delta xwall}$$

$$2 \cdot chvert \cdot (Tsurfout - Tout)^{1.25} \approx kwall \cdot \dfrac{Tsurfin - Tsurfout}{\delta xwall}$$

$\begin{bmatrix} Tsurfin \\ Tsurfout \end{bmatrix} := Find(Tsurfin, Tsurfout)$ $\qquad$ $Tsurfin = 13.736 \cdot C$

$\qquad\qquad\qquad\qquad\qquad\qquad\qquad\qquad\qquad$ $Tsurfout = -6.402 \cdot C$

$\delta Tins := Tins - Tsurfin$ $\qquad$ $\delta Twall := Tsurfin - Tsurfout$

$\delta Tout := Tsurfout - Tout$ $\qquad$ $\delta Tins = 6.264 \cdot C$ $\qquad$ $\delta Tout = 3.598 \cdot C$

$\delta Twall = 20.138 \cdot C$ $\qquad$ $Hwll := kwall \cdot \dfrac{\delta Twall}{\delta xwall}$ $\qquad$ $Hwll = 16.594 \cdot J \cdot [m^2 \cdot s]^{-1}$

Calculate the temperature difference across the roof. $\qquad$ $Tceilin := 10 \cdot C$

$\qquad\qquad\qquad\qquad\qquad\qquad\qquad\qquad\qquad\qquad\qquad\qquad\qquad$ $Troofout := 0 \cdot C$

Figure 12.4 HEAT5, heat flow from a house. (See the next figure for the rest of the document.)

Because the document is long and should not exceed the two-page limit of the student version of MathCAD, the regions are close together; the document is still readable, however.

The house has one room with windows and a flat ceiling. A furnace maintains the interior at a constant temperature, T_{ins}. The surroundings are also at a constant temperature, T_{out}. The dimensions of the room are $25 \times 30 \times 9.5$ (feet). Each of the eight windows is 5×2.5 (feet).

```
Given
                              1.25         Tceilin - Troofout
    chhord·(Tins - Tceilin)        ≈ kceil·─────────────────
                                                 δxceil

                              1.25         Tceilin - Troofout
    chhoru·(Troofout - Tout)       ≈ kceil·─────────────────
                                                 δxceil

    ⎡Tceilin ⎤                          Tceilin = 15.091·C   Troofout = -7.181·C
    ⎢Troofout⎥ := Find(Tceilin,Troofout)
    ⎣        ⎦                          δTins' := Tins - Tceilin
    δTclrf := Tceilin - Troofout        δTout' := Troofout - Tout

    δTins' = 4.909·C          δTout' = 2.819·C          δTclrf = 22.272·C

                    δTclrf                        J
    Hclrf := kceil·──────      Hclrf = 9.176·─────
                    δxceil                        2
                                                m ·s
                                                               Twindin := 10·C
    Repeat temperature difference calculation for windows.
                                                               Twindout := 0·C
    Given
                              1.25         Twindin - Twindout
    chvert·(Tins - Twindin)        ≈ kwind·──────────────────
                                                  δxwind

                                1.25        Twindin - Twindout
    2·chvert·(Twindout - Tout)       ≈ kwind·──────────────────
                                                   δxwind

    ⎡Twindin ⎤                          Twindin = 1.103·C   Twindout = 0.853·C
    ⎢Twindout⎥ := Find(Twindin,Twindout)
    ⎣        ⎦                          δTwind := Twindin - Twindout

                    δTwind                  J          Twindout - Tout = 10.853·C
    Hwind := kwind·──────      Hwind = 65.969·─────     Twindin - Twindout = 0.25·C
                    δxwind                     2        Tins - Twindin = 18.897·C
                                             m ·s

    Hwind := Hwind·Arwin      Hwll := Hwll·Arwll       Hcl := Hclrf·Arcl

    Hwind = 612.87·W                           3          Hcl = 639.364·W
                              Hwll = 1.457·10 ·W

    Htot := Hwind + Hwll + Hcl           Artot := Arwin + Arwll + Arcl

               Hwind                  Hwll                    Hcl
    FHwin := ──────      FHwll := ──────         FHcl := ──────
                Htot                  Htot                   Htot

                    3
    Htot = 2.709·10 ·W    FHwll = 0.538        FHcl = 0.236     FHwin = 0.226
```

Figure 12.5 HEAT5 *continued.*

Check the values for the areas of the ceiling, Ar_{cl}, the windows, Ar_{win}, and the walls, Ar_{wll}.

The walls and ceiling are treated not as compound slabs but as uniform. We assign the same thermal conductivity to the walls and ceiling but permit them to have different thicknesses.

The convection coefficient for convective heat flow from a vertical plate is ch_{vert}, the coefficient for the interior vertical surface. Similarly, ch_{horu} is the convective heat flow coefficient for a horizontal plate facing up, the coefficient for the roof; ch_{hord} is the coefficient for the ceiling. Values for the exterior are double those of the interior as the air motion is generally greater on the outside.

The various heat flows are equated. For example, in the first solve block, the internal convective heat flow is equated with the conductive heat flow in the first equation. The second equation equates the convective heat flow outside and the conductive heat flow. The temperatures of the wall inside and out are determined. The temperature differences between the interior and the interior surface of the wall, between the interior and exterior surfaces of the wall, and between the exterior surface of the wall and the environment are determined.

Essentially the same calculation is repeated for the roof and for the windows. Note the difference between the temperature differences across the walls, the ceiling, and the windows. Explain the value for the temperature difference across a window.

Once the temperature differences are known, the heat flow can be computed. For example, H_{wll}, is the heat flow through the wall in energy per area per time. Determine the corresponding convective flow and verify that the conductive and convective heat flows are the same.

At the end of the document, the total heat flows are determined by including the specific areas. The total heat flow and the fraction of the heat through the walls, ceiling, and windows are determined.

If the walls had only two inches of insulation instead of four, by how much would the surface temperature of the walls change? By how much would the heat flow increase?

If the convective heat flow coefficient outside were not double the inside value but equal to it, by how much would the surface temperature of the ceiling and inside walls change? First, predict the sign of the change, then process.

If the net effect of adding storm windows were to reduce k_{wind} from 0.002 to 0.001, what would reduce heat flow more, adding two inches of insulation to the walls or installing storm windows? What results in a greater energy saving, adding two inches of insulation to the ceiling or adding storm windows?

How much heat transfer through radiation is there between the outside walls of the house and its surroundings? How does this compare with the net flow due to convection and conduction?

12.2   Ideal Gas

The experimentally observed interrelation among pressure, P, temperature, T, and volume, V, is remarkably similar in dilute gases of different composition. A summary of these observations is expressed in the equation of state for an ideal gas:

$$PV = nRT \qquad (12.12)$$

where n is the number of moles of the gas and R is the universal gas constant, $R = 8.314$ J/(mol K). The SI unit for pressure is the Pa (pascal), $Pa = N/m^2$.

• •   Load HEAT6, ideal gas (see Fig. 12.6).

The properties of an ideal gas are explored graphically. Plots of P vs. V are shown for a series of temperatures. Each curve represents pressure and volume information for a given temperature; each of the curves is referred to as an isotherm. The two different regions show the curves in linear and in log-log plots.

Some manipulations were performed to set the limits to improve plot appearance, but this is not essential. However, to prevent the appearance of lines connecting the individual curves, one limit, in this case V_m, has been reduced.

Plot P vs. T curves for a series of volumes. Plot V vs. T curves for a series of pressures.

12.2.1   The Van der Waals Equation

As the number density of gas particles increases, the behavior of the gas deviates from the ideal. Many equations of state have been suggested. The van der Waals equation of state is one familiar variation. This equation attempts to account for two phenomena. As the number density of molecules increases, the volume available for each molecule decreases. In addition, molecules are subject to intermolecular forces; on average, these forces are attractive. This attraction tends to increase the pressure the particles exert. The van der Waals equation of state attempts to take into account these two considerations. The equation is

$$\left(P + \frac{a}{V^2}\right)(V - b) = nRT, \qquad (12.13)$$

where the parameter b is a measure of the volume occupied by the gas molecules and a is a measure of the pressure increase.

• •   Load HEAT7, van der Waals equation of state (see Fig. 12.7).

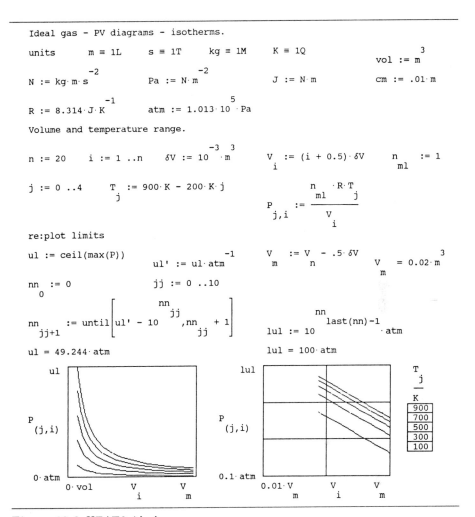

Ideal gas - PV diagrams - isotherms.

units $\quad m \equiv 1L \quad s \equiv 1T \quad kg \equiv 1M \quad K \equiv 1Q \qquad vol := m^3$

$N := kg \cdot m \cdot s^{-2} \qquad Pa := N \cdot m^{-2} \qquad J := N \cdot m \qquad cm := .01 \cdot m$

$R := 8.314 \cdot J \cdot K^{-1} \qquad atm := 1.013 \cdot 10^5 \cdot Pa$

Volume and temperature range.

$n := 20 \quad i := 1 ..n \quad \delta V := 10^{-3} \cdot m^3 \qquad V_i := (i + 0.5) \cdot \delta V \qquad n_{ml} := 1$

$j := 0 ..4 \qquad T_j := 900 \cdot K - 200 \cdot K \cdot j \qquad P_{j,i} := \dfrac{n_{ml} \cdot R \cdot T_j}{V_i}$

re:plot limits

$ul := ceil(max(P)) \qquad ul' := ul \cdot atm^{-1} \qquad V_m := V_n - .5 \cdot \delta V \qquad V_m = 0.02 \cdot m^3$

$nn_0 := 0 \qquad\qquad jj := 0 ..10$

$nn_{jj+1} := until\left[ul' - 10^{\dfrac{nn_{jj}}{}}, nn_{jj} + 1 \right] \qquad lul := 10^{last(nn)-1} \cdot atm$

$ul = 49.244 \cdot atm \qquad\qquad lul = 100 \cdot atm$

	T_j / K
	900
	700
	500
	300
	100

Figure 12.6 HEAT6, ideal gas.

A comparison is made between the van der Waals equation of state and the ideal gas equation. Substitute the following temperature values sequentially to see the deviations from the behavior described by the ideal gas law behavior associated with the van der Waals equation. Let T equal 500, 400, 360, 330, 300, 280, and 260. The a and b values are for CO_2.

Why does the volume start at $4 \cdot \delta v$? Plot $V_i - b$ vs. V_i.

In the following examples, you begin to get a sense of the role of each of the terms. Set the temperature at $500\,K$ and process. If the a term were zero, how would the curve change? Enable the $a = 0$ region and check your

Van der Waals and ideal gas equation of state are shown for a given temperature.

$m \equiv 1L$ $s \equiv 1T$ $kg \equiv 1M$ $K \equiv 1Q$

$N := kg \cdot m \cdot s^{-2}$ $J := N \cdot m$

$R := 8.314 \cdot J \cdot K^{-1}$ $Pa := N \cdot m^{-2}$ $mol := 1$ $atm := 1.013 \cdot 10^{5} \cdot Pa$

$n := 40$ $i := 1 \,..\, n$ $\delta V := 1.25 \cdot 10^{-5} \cdot m^{3}$ $V_{i} := (i + 3) \cdot \delta V$

$T := 500 \cdot K$ $a := \left[0.3637 \cdot Pa \cdot m^{6}\right] \cdot mol^{-2}$ $a := \left[0 \cdot Pa \cdot m^{6}\right] \cdot mol^{-2}$ □

$n_{mol} := 1$ $b := \left[4.27 \cdot 10^{-5} \cdot m^{3}\right] \cdot mol^{-1}$ $b := 0 \cdot m^{3} \cdot mol^{-1}$ □

$$Pi_{i} := \frac{n_{mol} \cdot R \cdot T}{V_{i}}$$ $$Pvdw_{i} := \frac{n_{mol} \cdot R \cdot T}{V_{i} - b} - \frac{a}{V_{i}^{2}}$$

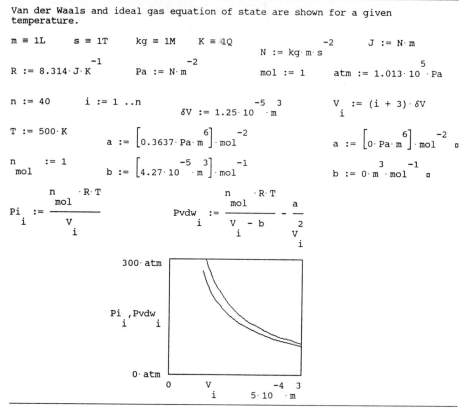

Figure 12.7 HEAT7, van der Waals equation of state.

prediction. What if b were zero and a were not? Disable the $a = 0$ region and enable the $b = 0$ region and observe. Disable the $a = 0$ and $b = 0$ regions. Plot the following two terms in the same region: $n\,RT/(V - b)$ and $-a/V^2$ vs. V. Finally, define $V_i' = i \cdot \delta V$ and plot $P_{vdwi} + a/V_i^2$ vs. $V_i - b$ and Pi_i vs. V_i' in the same region.

12.2.2   *Isothermal and Adiabatic Changes*

When discussing the P vs. V curves at constant temperature, we made no mention of how the expansion might take place. Imagine the gas to be in a cylinder with a movable piston at one end. To ensure that the temperature remains constant during an expansion or compression, the cylinder is placed in contact with a heat reservoir while the change takes place.

If the cylinder is insulated and the gas expands or is compressed, the product PV is no longer a constant. The curve associated with such a process is an adiabat. For adiabatic processes, in which there is no transfer of heat between the gas and its surroundings, we have

$$P V^\gamma = \text{const} \qquad \text{or} \qquad P_i V_i^\gamma = P_f V_f^\gamma, \qquad (12.14)$$

where γ is the ratio of specific heats at constant pressure and constant volume:

$$\gamma = \frac{c_p}{c_v}. \qquad (12.15)$$

At constant volume, all the energy is associated with the change in temperature of the substance; all the energy is associated with the change of internal energy of the substance. At constant pressure, the same change in temperature occurs and in addition the volume of the substance changes. There is work associated with the change in volume that occurs. For monatomic gases $\gamma = 5/3$; for diatomic gases $\gamma = 7/5$; and for polyatomic gases $\gamma = 4/3$.

An adiabatic process can also be expressed in terms of T and V: or P and T:

$$T_i V_i^{\gamma-1} = T_f V_f^{\gamma-1}, \qquad (12.16)$$

$$P_i^{1-\gamma} T_i^\gamma = P_f^{1-\gamma} T_f^\gamma. \qquad (12.17)$$

• • Verify the T, V and P, T relations, given the PV^γ equation.

• • Select a value for P and for V. Plot an isotherm and an adiabat through that point (omit units).

• • Plot the analogous pairs of curves in T, V and P, T diagrams.

• • How different are the curves in the last two exercises as the gas changes from monatomic to diatomic? From diatomic to polyatomic?

In addition to just trying to appreciate the functional relationships expressed in these equations, we want to connect isothermal and adiabatic processes with the operation of a heat engine, a machine that converts heat energy into work. Such machines go through a cycle of absorbing energy from a hot reservoir and performing work and dumping energy to a cold reservoir and having work performed on them. Work is performed by the machine, as the gas expands; work is performed on the machine as the gas is compressed. If the machine is to be useful, the work performed during expansion must be greater than the work done on the machine during compression. The most efficient possible engine is one that follows isothermal and adiabatic expansions or compressions. Before we discuss the cycle of a heat engine, we investigate a few more questions about adiabats and isotherms.

Any two points on a PV diagram can be connected by an adiabat and an isotherm. Let (p_1, v_1) and (p_3, v_3) represent the two points that are selected. We draw an isotherm through the point (p_1, v_1) and an adiabat through (p_3, v_3) and find the point of intersection:

$$p_1 v_1 = p_2 v_2 \qquad \text{isotherm}$$

$$p_2 v_2^\gamma = p_3 v_3^\gamma \qquad \text{adiabat.}$$

Solving the first equation for v_2, substituting in the second, and solving for p_2 gives

$$p_2 = p_3^{1/1-\gamma} \left(\frac{v_3}{p_1 v_1} \right)^{\gamma/1-\gamma}$$

The volume $v_2 = p_1 v_1 / p_2$. Using MathCAD, it is just as easy to put the original equations in a solve block and let MathCAD solve for the particular values of p_2 and v_2.

● ●   Load HEAT8, isotherms and adiabats (see Figs. 12.8 and 12.9).

When you enter the coordinates of the two points, (p_1, v_1) and (p_3, v_3), and a value for γ, the equations for the isotherm and adiabat are solved to determine the coordinates of the intersection point, (p_2, v_2).

Subscripts cannot be used in the argument of a find statement. Consequently, values for p_2 and v_2 are written without subscripts until the solutions are obtained. The values determined in a find statement can be assigned to subscripted variables. (An attempt to use subscripted results in a find statement results in the error message, "error in list.")

Just as each isotherm has associated with it a particular temperature, each adiabat has associated with it an analogous constant. K and L represent the constants associated with the points (p_1, v_1) and (p_3, v_3).

$P(V)$ and $P'(V)$ describe the isotherm and adiabat through (p_1, v_1). $Q(V)$ and $Q'(V)$ describe the isotherm and adiabat through (p_3, v_3). The other equations set the limits so that only the region of interest is plotted. Although this choice of limits presents the three points together with the corresponding isotherms and adiabats from case to case, it has the disadvantage of changing the scale and range. You may wish to add another plot region with fixed limits to see at the same time where the curves lie relative to the axes. Process. Which point corresponds to (p_1, v_1)? Which point corresponds to (p_3, v_3)? Which curves are isotherms? Which curves are adiabats? How do you know? Follow the path from (p_1, v_1) to (p_2, v_2) — the isotherm, and from (p_2, v_2) to (p_3, v_3) — the adiabat. (After processing, press [Ctrl]home to change values and [Ctrl]end and [F9] to see the results.)

Any two points in a pv diagram can be connected by an adiabat and an isotherm. The coordinates of the two points and gamma.

$$p_1 := 10 \qquad v_1 := 3 \qquad p_3 := 5.5 \qquad v_3 := 5 \qquad \tau := 1.33$$

Find p2 and v2. $\qquad p2 := 5 \qquad v2 := 5 \qquad$ guess values

given $\qquad p_1 \cdot v_1 \approx p2 \cdot v2 \qquad p2 \cdot v2^{\tau} \approx p_3 \cdot v_3^{\tau} \qquad \begin{bmatrix} p_2 \\ v_2 \end{bmatrix} := \text{find}(p2, v2)$

$$p_2 = 7.81 \qquad\qquad v_2 = 3.841$$

Characteristic values for the adiabats and isotherms.

$$K := p_1 \cdot v_1^{\tau} \qquad\qquad L := p_3 \cdot v_3^{\tau} \qquad i := 1 .. 3 \qquad T_i := p_i \cdot v_i$$

$$K \cdot v_3^{-\tau} = 5.069 \qquad T_1 \cdot v_3^{-1} = 6 \qquad\qquad \begin{array}{|c|} \hline T_i \\ \hline 30 \\ \hline 30 \\ \hline 27.5 \\ \hline \end{array}$$

$$P(V) := \frac{T_1}{V} \qquad P'(V) := \frac{K}{V^{\tau}} \qquad Q(V) := \frac{T_3}{V} \qquad Q'(V) := \frac{L}{V^{\tau}}$$

Plot limit specifications.

$$a := \text{if}\begin{bmatrix} v_3 > v_2, v_3, v_2 \end{bmatrix} \qquad \text{hul} := \text{if}\begin{bmatrix} a > v_1, a, v_1 \end{bmatrix} \cdot 1.1 \qquad \text{hul} = 5.5$$

$$b := \text{if}\begin{bmatrix} v_3 < v_2, v_3, v_2 \end{bmatrix} \qquad \text{hll} := \text{if}\begin{bmatrix} b < v_1, b, v_1 \end{bmatrix} \cdot 0.8 \qquad \text{hll} = 2.4$$

$$c := \text{if}\begin{bmatrix} p_3 > p_2, p_3, p_2 \end{bmatrix} \qquad \text{vul} := \text{if}\begin{bmatrix} c > p_1, c, p_1 \end{bmatrix} \cdot 1.1 \qquad \text{vul} = 11$$

$$d := \text{if}\begin{bmatrix} p_3 < p_2, p_3, p_2 \end{bmatrix} \qquad \text{vll} := \text{if}\begin{bmatrix} d < p_1, d, p_1 \end{bmatrix} \cdot 0.8 \qquad \text{vll} = 4.4$$

$$V := \text{hll}, \text{hll} + .2 .. \text{hul}$$

Figure 12.8 HEAT8, isotherms and adiabats. (See the next figure for the rest of the document.)

Change v_3 to 4.5 and process. Why the switch? Explore. Determine the conditions on p_3 and v_3 so that $p_1 > p_2 > p_3$ and $v_1 < v_2 < v_3$.

Change the sequence so that (p_2, v_2) is connected to (p_1, v_1) by an adiabat and to (p_3, v_3) by an isotherm. Plot some examples.

12.3 The Carnot Cycle

A Carnot cycle includes four steps; each step connects two points; the last step connects the fourth point with the first. We start the cycle at point 1:

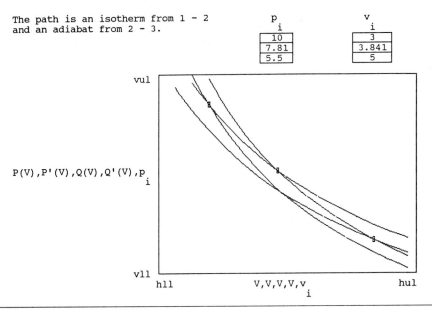

The path is an isotherm from 1 – 2 and an adiabat from 2 – 3.

p$_i$
10
7.81
5.5

v$_i$
3
3.841
5

Figure 12.9 HEAT8 *continued.*

1. isothermal expansion   (1–2)
2. adiabatic expansion   (2–3)
3. isothermal compression (3–4)
4. adiabatic compression   (4–1).

In this sequence, $p_1 > p_2 > p_3$, and $v_1 < v_2 < v_3$. Trying to specify the p, v coordinates for the first three points leads to some restrictions. Recall the previous example in which you were asked to explore these restrictions. Any pair of points may be connected by an isotherm and an adiabat but the pressure and volume sequences may not be that desired for a Carnot cycle.

Given that $v_3 > v_1$, there are two additional conditions. The temperature at (p_3, v_3) must be less than that at (p_1, v_1). The temperature at (p_3, v_3) must be greater than the final temperature reached by an adiabat from (p_1, v_1) to v_3. These conditions can be expressed as

$$p_3 < \frac{p_1 v_1}{v_3} \qquad \text{and} \qquad p_3 > \frac{p_1 v_1^\gamma}{v_3^\gamma}.$$

In other words, the point (p_3, v_3) must have $v_3 > v_1$ and must lie between the isothermal and adiabatic curves through point (p_1, v_1). In the document, these two limiting quantities are expressed as $T_1 \cdot v_3^{-1}$ and $K \cdot v_3^{-\gamma}$. Given that $v_3 > v_1$, these are the limits of p_3.

Now consider the question of work done by the gas as it expands and is compressed and as the heat is absorbed and dumped during those same processes. Recall that the definition of work is

$$W = \int F\,dx.$$

In terms of pressure, this becomes $(P = F/A)$

$$W = \int PA\,dx = \int P\,dV. \tag{12.18}$$

During expansion, the work done by the gas is positive; during compression the work done by the gas is negative.

On a PV diagram, the area under the curve representing an expansion as it goes from initial to final value is equal to the work. If the expansion is along an isotherm, the work is given by

$$W = \int_{V_1}^{V_2} \frac{nRT}{V}\,dV = nRT \int_{V_1}^{V_2} \frac{dV}{V} = nRT \ln\left(\frac{V_2}{V_1}\right). \tag{12.19}$$

If the expansion is along an adiabat, the work is given by

$$W = \int_{V_1}^{V_2} \frac{P'V'^{\gamma}}{V^{\gamma}}\,dV = P'V'^{\gamma} \int_{V_1}^{V_2} \frac{dV}{V^{\gamma}}$$

$$= P' \frac{V'^{\gamma}}{1-\gamma}\left(V_2^{1-\gamma} - V_1^{1-\gamma}\right). \tag{12.20}$$

These results can be used directly, or we can let MathCAD perform the integrations for us.

We now examine the Carnot cycle, keeping track of work performed and heat exchanged. We describe the cycle starting from the point of minimum volume and maximum pressure:

1. Isothermal expansion from (p_1, v_1) to (p_2, v_2):

$$p_1 \cdot v_1 = p_2 \cdot v_2 \qquad\qquad v_2 > v_1;$$

$$\text{work performed}: \int_{V_1}^{V_2} \frac{nRT_1}{V}\,dV.$$

Heat absorbed equals work performed.

2. Adiabatic expansion from (p_2, v_2) to (p_3, v_3):

$$p_2 \cdot v_2^\gamma = p_3 \cdot v_3^\gamma \qquad\qquad v_3 > v_2:$$

$$\text{work performed} : \int_{V_2}^{V_3} \frac{p_2 v_2^\gamma}{V^\gamma}\, dV.$$

Heat exchanged is zero.

3. Isothermal compression from (p_3, v_3) to (p_4, v_4):

$$p_3 \cdot v_3 = p_4 \cdot v_4 \qquad\qquad v_4 < v_3;$$

$$\text{work performed} : \int_{V_3}^{V_4} \frac{n\,R\,T_3}{V}\, dV.$$

Work is done on the gas; heat is given up.

4. Adiabatic compression from (p_4, v_4) to (p_1, v_1):

$$p_2 \cdot v_2^\gamma = p_4 \cdot v_4^\gamma \qquad\qquad v_1 < v_4$$

$$\text{work performed} : \int_{V_4}^{V_1} \frac{p_4 v_4^\gamma}{V^\gamma}\, dV.$$

Heat exchanged is zero.

• •   Adapt the previous document to plot a complete Carnot cycle.

• •   Load HEAT9, Carnot cycle (see Figs. 12.10 and 12.11).

For a given set of parameters, a rough plot of the Carnot cycle is shown. The work associated with each of the four steps is calculated using the explicit expressions that were shown above and using a direct integration process. The total work performed is calculated, and the efficiency is determined.

In the plot region, straight lines connect points 1, 2, and 3. (Because the document is long, we skip a functional expression here.) The line does not continue around the entire cycle as in this representation; the line from 4 to 1 may intersect either the 1 – 2 or the 2 – 3 line.

If T_1, V_1, and V_3 are fixed, by how much can T_3 vary and still have a Carnot cycle? Observe the position of (p_2, v_2) and (p_4, v_4) as T_1 is changed.

Starting with (T_1, T_3, v_1, v_3) as $(300, 100, 1, 6)$, let v_3 increase slowly and observe the changes.

Carnot cycle $N := 4$ $i := 1 .. N$

Specify, parameters, temperatures of hot and cold reservoirs, and minimum and maximum volumes.   (Values are subject to restrictions.)

$n := 1$ $R := 8.31$ $\tau := 1.67$ $\beta := \tau - 1$

$T_1 := 300$ $T_3 := 150$ $V_1 := 1$ $V_3 := 5$

Determine remaining values of p, V, and T.

$T_2 := T_1$ $T_4 := T_3$

$p_1 := \dfrac{n \cdot R \cdot T_1}{V_1}$ $p_3 := \dfrac{n \cdot R \cdot T_3}{V_3}$

$p_2 := p_3 \cdot \left[\dfrac{T_3}{T_2}\right]^{\frac{\tau}{1-\tau}}$ $p_4 := p_1 \cdot \left[\dfrac{T_1}{T_4}\right]^{\frac{\tau}{1-\tau}}$ $V_2 := \dfrac{p_1 \cdot V_1}{p_2}$ $V_4 := \dfrac{p_3 \cdot V_3}{p_4}$

$pu := p_1 \cdot 1.1$ $pl := p_3 \cdot 0.8$ $vl := V_1 \cdot 0.8$ $vu := V_3 \cdot 1.1$ $j := 1 .. 3$

p_i

$2.493 \cdot 10^3$
$1.403 \cdot 10^3$
249.3
442.99

V_i

1
1.777
5
2.814

T_i

300
300
150
150

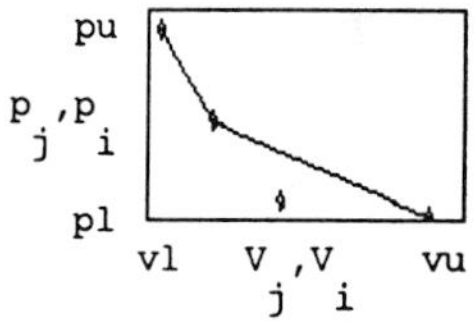

pu

p_j , P_i

pl

$vl \quad V_j , V_i \quad vu$

Determine the work performed and heat exchanged by the gas in each segment.

$W_{12} := n \cdot R \cdot T_1 \cdot \ln\left[\dfrac{V_2}{V_1}\right]$ $W_{23} := \dfrac{p_2 \cdot V_2^{\tau}}{-\beta} \left[V_3^{-\beta} - V_2^{-\beta}\right]$ $W_{12} = 1.433 \cdot 10^3$

$W_{23} = 1.86 \cdot 10^3$

$W_{34} := n \cdot R \cdot T_3 \cdot \ln\left[\dfrac{V_4}{V_3}\right]$ $W_{41} := \dfrac{p_1 \cdot V_1^{\tau}}{-\beta} \left[V_1^{-\beta} - V_4^{-\beta}\right]$ $W_{34} = -716.6$

$W_{41} = -1.86 \cdot 10^3$

$\delta Q_{12} := W_{12}$ $\delta Q_{23} := 0$ $\delta Q_{34} := W_{34}$ $\delta Q_{41} := 0$

Figure 12.10 HEAT9, Carnot cycle. (See the next figure for the rest of the document.)

By noting the shape of the Carnot cycle, you can estimate when the greatest amount of work will be performed per cycle, that is, when the area enclosed by the paths that outline the cycle is the greatest. Maximum efficiency occurs when the temperature difference between the two reservoirs is the greatest.

Does the maximum value for work per cycle occur at the same point as that of maximum efficiency? Make a rough plot of the work per cycle

Compare with values from direct integration.

$$WW_{12} := \int_{V_1}^{V_2} \frac{n \cdot R \cdot T_1}{V} dV \qquad WW_{23} := \int_{V_2}^{V_3} \frac{p_2 \cdot V_2^{\tau}}{V'^{\tau}} dV' \qquad WW_{12} = 1.433 \cdot 10^3$$

$$WW_{23} = 1.86 \cdot 10^3$$

$$WW_{34} := \int_{V_3}^{V_4} \frac{n \cdot R \cdot T_3}{V} dV \qquad WW_{41} := \int_{V_4}^{V_1} \frac{p_1 \cdot V_1^{\tau}}{V'^{\tau}} dV' \qquad WW_{34} = -716.6$$

$$WW_{41} = -1.86 \cdot 10^3$$

The total work performed.

$$W_{tot} := W_{12} + W_{23} + W_{34} + W_{41} \qquad W_{tot} = 716.6$$

$$WW_{tot} := WW_{12} + WW_{23} + WW_{34} + WW_{41} \qquad WW_{tot} = 716.6$$

$$W_{12} + W_{34} = 716.6 \qquad WW_{12} + WW_{34} = 716.6$$

Efficiency can be expressed in several equivalent ways. Work performed over heat extracted from the hot reservoir is one.

$$Eff := 1 - \frac{T_3}{T_1} \qquad Eff = 0.5 \qquad \frac{W_{tot}}{\delta Q_{12}} = 0.5$$

Figure 12.11 HEAT9 *continued.*

over the allowed range of temperatures for T_3. (For various values of T_3, observe W_{tot}.) How much work is performed when $T_3 = T_1$? How much work is performed per cycle when T_3 is at its minimum?

How does the quantity of work performed in the adiabatic expansion and compression compare? Is this a general conclusion? Can you prove this?

Do you expect me to talk?

No, Mr. Bond, I expect you to die.

Goldfinger

CHAPTER
13

Kinetic Theory

When discussing the properties of a gas, we considered such macroscopic quantities as pressure and temperature. Kinetic theory takes a different approach and provides a description of the properties of a gas from a microscopic point of view. Here we consider questions about the velocity distribution of molecules in a gas. The velocity distribution is considered as a packet of gas spreads out in time and is observed at a detector. Finally, a simple simulation of thermal equilibration is presented. A hot object is placed in a cool environment and the energy migrates according to a two-dimensional random walk.

13.1 Velocity Distribution

A molecular model for an ideal gas is that of a large number of independent, identical molecules moving randomly. Collisions are elastic, and if no energy is exchanged between the gas and its surroundings, the speed distribution remains the same and is given by

$$N(v) = 4\pi N \left(\frac{m}{2\pi kT}\right)^{3/2} v^2 e^{-mv^2/2kT}. \qquad (13.1)$$

$N(v)dv$ is the number of molecules in the gas with speeds between v and $v + dv$, k is Boltzmann's constant (the gas constant per molecule; R is the gas constant per mole), and T is the absolute temperature. N is the total number of molecules in the sample.

• •   Load KT1, Maxwell-Boltzmann distribution (see Fig. 13.1).

The distribution function is examined by creating plots of the distribution showing its dependence on temperature and mass.

New quantities in the parameter list include N_2 and O_2; these are the masses of one mole of molecular nitrogen and one mole of molecular oxygen; mN_2 and mO_2 are the corresponding molecular masses, and Av is Avogadro's number.

The sum of all particles in each velocity range should equal the total number of particles. Verify that

$$\int N(v,T)\, dv = Nm. \qquad (13.2)$$

We use Nm for the number of molecules so that there is no confusion with the distribution function itself.

Maxwell–Boltzmann velocity distribution.

Parameters

$k := 1.38 \cdot 10^{-23}$ $N2 := 0.028$ $Av := 6.022 \cdot 10^{23}$ $mN2 := \dfrac{N2}{Av}$

$O2 := 0.032$

$Nm := 10^6$ $T := 300$

The distribution equation.

$$N(v,T) := 4 \cdot \pi \cdot Nm \cdot \left[\frac{mN2}{2 \cdot \pi \cdot k \cdot T}\right]^{1.5} \cdot v^2 \cdot \exp\left[\frac{-mN2 \cdot v^2}{2 \cdot k \cdot T}\right]$$

$v := 0, 50\ ..1500$

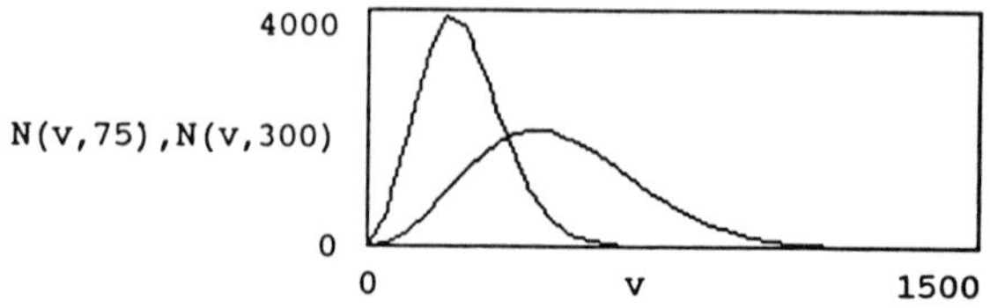

$N(v,75), N(v,300)$

Figure 13.1 KT1, Maxwell-Boltzmann distribution.

For $T = 300$, is an upper limit for v of 1500 adequate? For $T = 1200$, what fraction of the molecules have speed ranges between 0 and 1500?

The average translational kinetic energy per molecule is related to the temperature by

$$\frac{1}{2} m \overline{v^2} = \frac{3}{2} k T, \qquad (13.3)$$

where $\overline{v^2}$ is the average value of the square of the velocity. The root mean-squared velocity v_{rms} (the square root of the mean of the square of the velocity) is

$$v_{rms} = \sqrt{\overline{v^2}} = \sqrt{\frac{3 k T}{m}}. \qquad (13.4)$$

Other means of the velocity (for example, average velocity and most probable velocity) have the same dependence on temperature and mass.

The most probable speed is that associated with the peak of the velocity distribution. The root/derivative method is useful in determining this value. Keep the guess value close to the actual value to avoid spurious results; the guess values can be estimated from the plots.

Explore the relation between temperature and most probable velocity; make several comparisons. You need not plot the distributions to determine the velocities. Verify the expected dependence.

Explore the relation between mass and most probable velocity. Make several comparisons and verify the expected dependence.

At what velocity do the distribution curves for N_2 and O_2 intersect? How precisely would velocity distribution measurements have to be in order to distinguish between molecular nitrogen and molecular oxygen?

The average velocity is obtained from

$$v_{av} = \frac{\int v \cdot N(v, T) dv}{\int N(v, T) dv}. \qquad (13.5)$$

The denominator should be familiar.

The rms velocity is given by

$$v_{rms}^2 = \frac{\int v^2 \cdot N(v, T) \, dv}{\int N(v, T) dv}. \qquad (13.6)$$

For $T = 300$, determine the most probable velocity, v_{mp}, the average velocity, v_{av}, and the root-mean-squared velocity, v_{rms}. Verify that $v_{rms} > v_{av} > v_{mp}$. Superimpose plots of these individual points (plot type v, for example) on the corresponding velocity distribution.

13.2   Mean Free Path

Three quantities describing a dilute gas are the pressure, the molecular number density, and the mean free path. Pressure is a result of momentum transfer due to collisions. If the number density n_ρ — the number of particles per volume — is reduced, the pressure is reduced proportionately: $P \propto n_\rho$. Within the gas, the mean free path — the distance a molecule travels, on average, between collisions — increases as the number density decreases. Each of these quantities can be expressed in terms of either of the others.

The mean free path depends on the effective cross section (cross-sectional area) of a molecule. If the cross section is very small, the particle will undergo few collisions, and therefore it will have a large mean free path. If the molecules are approximated by spheres of the same size, a collision will occur if the center-to-center distance of adjacent molecules is less than the diameter, d. An approximation to the cross section for spherical molecules is πd^2.

A very simple way to visualize and approximate the mean free path in terms of number density follows. The number density is the number of particles per volume. The inverse of the number density is the volume per particle. If we consider this volume to be a right circular cylinder with a cross section equal to the cross section for a molecule, then the length of the cylinder is an approximate value for the mean free path. When the velocity distribution is included in the analysis, the cross section is expressed as $\sqrt{2}\,\pi d^2$. This reasoning results in a mean free path of

$$\lambda = \frac{1}{n_\rho} \frac{1}{\sqrt{2}\,\pi\,d^2} \qquad (13.7)$$

• •   Load KT2, number density, pressure, and mean free path (see Fig. 13.2).

Examine the connections between pressure, number density, and mean free path. For the calculation, one consistent set of values is necessary for the three quantities that we wish to discuss. The cross-sectional area is defined, and six functions are written. Each quantity is expressed in two separate functions, each function is expressed in terms of one of the other two variables. Paralleling this set of functions is a set of single-valued computations showing how to obtain, for example, pressure from number density or from mean free path. The units are mixed, but they can be changed to any corresponding unit that is defined. All arguments require units. Pressure readings associated with vacuum gauges are often expressed in torr, a pressure corresponding to one millimeter of mercury or $1/760$ of an atmosphere. Readings in the range of 10^{-6} to 10^{-7} torr are readily achieved and are common in many experimental settings. To reach pressures of 10^{-10} torr or less, more precautions are required.

Define $n_{\rho_i} = 10^i \cdot cm^{-3}$, where $i = 1, \ldots, 20$. Plot (all log-log scales) $P_1(n_{\rho_i})$ vs. n_{ρ_i}, $\lambda_1(n_{\rho_i})$ vs. n_{ρ_i} and $\lambda_1(n_{\rho_i})$ vs. $P_1(n_{\rho_i})$.

The distance from the earth to the sun, approximately 1.5×10^8 km, is referred to as an astronomical unit. If the number density of hydrogen

---

```
Relations between number density, pressure, and mean free path

m := 1L    s := 1T    kg := 1M      gm := .001·kg      cm := .01·m

          -2                -2                    5                    1
N := kg·m·s       Pa := N·m        atm := 1.103·10 ·Pa      torr := ——·atm
                                                                     760

                                              5
torr = 145.132·Pa      760·torr = 1.103·10 ·Pa      torr = 0.132·%·atm

Molecular hydrogen is considered.

            23       M            := 2·gm                        -2    -3
Av := 6.022·10       H2                          ρ      := 8.99·10  ·kg·m
                                                  H2

ρ is the number density at standard temperature and pressure.

              Av                        19   1             Ps := atm
ρ  := ρ   ·——            ρ = 2.707·10 ·——
       H2   M                             3
            H2                           cm

      -10
d := 2·10  ·m      molecular diameter      A := √2·π·d       collision cross
                                                  2           section

We express pressure P, mean free path λ, and number density n, in functional
form.  Each quantity can be expressed in terms of either of the others.

          Ps                         ⌈ 12    -3⌉              -8
P1(n) := ——·n                      P1⌊10  ·cm  ⌋ = 3.694·10  ·atm
          ρ

           Ps                                            -8
P2(λ) := ————                      P2(5.627·m) = 3.694·10  ·atm
         ρ·λ·A

          1 1                        ⌈ 12    -3⌉
λ1(n) := —·—                       λ1⌊10  ·cm  ⌋ = 5.627·m
          A n

           Ps                        ⌈        -8     ⌉
λ2(P) := ————                      λ2⌊3.694·10  ·atm⌋ = 5.627·m
         ρ·P·A

          ρ                          ⌈          -8     ⌉     12   -3
n1(P) := ——·P                      n1⌊3.6943·10  ·atm⌋ = 1·10  ·cm
          Ps

          1 1                                         12   -3
n2(λ) := —·—                        n2(5.627·m) = 1·10  ·cm
          A λ
```

Figure 13.2 KT2, number density, pressure, and mean free path.

molecules in interstellar space is 10 per cubic centimeter, what is the mean free path in astronomical units?

The wavelength of green light is roughly 5.5×10^{-7} m. What is the mean free path when the number density is 10^{20} molecules per cubic centimeter? Express the result in wavelengths of green light. For the same number density, what is the pressure in atmospheres? What pressure corresponds to a mean free path of one wavelength of green light?

In the previous plots of the Maxwell-Boltzmann speed distribution function, we considered the distribution at one moment. Imagine a source of gas connected to a vacuum chamber through an electronically controlled valve. (Or, we could rotate a wheel in front of the gas beam, with a tiny slot cut in it. For every rotation of the wheel, gas would briefly pass through the slot.) Gas enters the chamber in the plus x-direction.

The gas packet initially has a very tiny spread in the x-direction. The molecules have a distribution of speeds in the x-direction, but at $t = 0$ the packet has not yet spread out. As time passes, those molecules with larger x-velocities travel farther than the slower ones and the packet spreads.

• •   Load KT3, time of flight (see Figs. 13.3 and 13.4).

Observe the spread of the packet of molecules. Determine at what rate the molecules arrive at a detector some distance away.

For simplicity, treat the speed distribution as if all velocities are in the x-direction. The distribution number vs. velocity is plotted as was done earlier. Now, however, we plot the distribution vs. distance rather than velocity. In the second plot region, the packet of molecules is shown at two different times, t_1 and t_2. The plots are type s; this permits a comparison of the relative widths of steps between one distribution and the other, giving a clear sense of the spreading of the packet. To observe the packet at earlier times, let $\delta t = 10^{-5.3}$ s. To observe them at a later time, let $\delta t = 10^{-4.1}$ s. (The times are arbitrary but suit the conditions.)

Now consider the detector located at the vertical line near the right-hand side of the plot, a distance, d_{sep}, from the source. Clearly, the detector sees the fastest molecules first. At $\delta t = 10^{-5}$ s (check these as we go), particles in the highest-velocity segment (of the d_2 curve) are beginning to arrive at the detector. As δt increases from $10^{-4.5}$ s to $10^{-4.2}$ s, we can watch the progression.

Now consider events at the detector itself. The time of arrival of particles at the detector is inversely related to the velocity. The time-of-flight data constitute a nonlinear mapping of the velocity data.

In the last group of three plots, we show again the velocity distribution, followed by two examples of the number distribution in time. The upper limit of the abscissa shows the relative scales. The last plot is a close-up look at the first 0.5 ms of the distribution.

Particles with velocities greater than the most probable velocity are compressed into the leading edge of the n vs. t plot. Those with slower velocities — that is, those before the peak in the n vs. v curve — have their arrivals more spread out in time. In the last plot, we look only at the high-velocity, small-time portion of the curve.

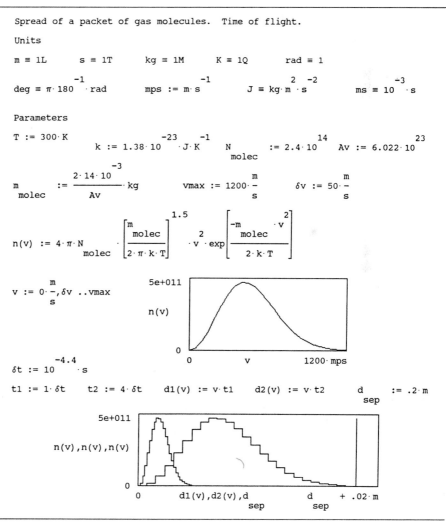

Figure 13.3 KT3, time of flight. (See the next figure for the rest of the document.)

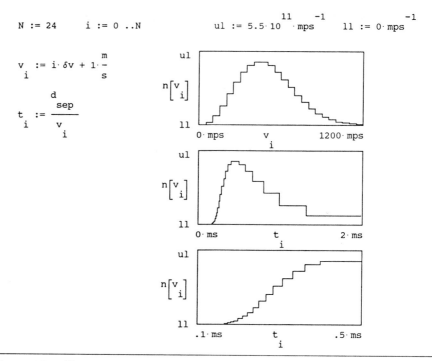

$$N := 24 \qquad i := 0 \,..\, N \qquad ul := 5.5 \cdot 10^{11} \cdot mps^{-1} \qquad ll := 0 \cdot mps^{-1}$$

$$v_i := i \cdot \delta v + 1 \cdot \frac{m}{s}$$

$$t_i := \frac{d_{sep}}{v_i}$$

Figure 13.4 KT3 *continued.*

Convince yourself that you understand how these plots correspond by restricting the range of i. For example, let $i = 5, \dots, 15$ or $14, \dots, 24$ or $0, \dots, 10$.

To distinguish between the faster velocities, we need good time discrimination. How much time separates the arrival at the detector of the molecules traveling at 1200 m/s from those traveling at 1150 m/s?

Make a semi-log plot of the time intervals between successive velocity groups. (The molecules are not really in discrete groups, but for this example we will imagine that they are.)

How would the time resolution requirements change if the distance d_{sep} was reduced to 0.1 m?

How would the number of counts per time be different if the species were N_2 and O_2? (Look at KT1.) What requirements would there be to differentiate between N_2 and O_2 if we consider only velocities less than the most probable?

13.3   **Random Walk**

Random walks can be used to simulate many different types of events. They are used here to simulate the transfer of energy units from a hot body to a cold body. In a thermal equilibration process, the hot body cools and the cool body warms. Before that problem is considered, we explore the random walk briefly, first in one dimension and then in two.

In a simple one-dimensional random walk, an object moves along a line — for example, the x-axis — in steps of equal size. The direction of the step is determined by a random process. The net displacement from the starting point is on the order of the step size times the square root of the number of steps.

• •   Plot a one-dimensional random walk. Let the number of steps be 100 and let the step size be equal to one. A very simple difference equation is all that is needed; each new value, x_{i+1}, is the previous value, x_i, plus or minus one step. To create a random sequence of ±1 values for the walk, we raise –1 to the power ceil(rnd(2)). The rnd(2) statement generates a number between 0 and 2. The ceil (rnd(2)) operation converts the number generated by the rnd process to the smallest integer larger than the number, which in this case is 1 or 2; and, of course $-1^1 = -1$; and $-1^2 = +1$. Thus repeated calls to this function generate a random sequence of ±1 values.

To be sure the concepts are familiar, let $j = 1, \dots, 10$ and let $y_j =$ rnd(2). Display y_j and ceil(y_j) in two columns side by side. Delete after you have considered the results.

Calculate the x_i as described and plot i vs. x_i.

A histogram provides a useful way to display the location of a particle for an entire random walk process. Create a histogram showing how many times the walk has led to each x-position. Let the number of intervals be the quantity $(x_{max} - x_{min})$ (use MathCAD's max and min functions). Define two indices; let j run from zero to the number of intervals, and let k run from zero to the number of intervals minus one. Define the intervals with j and plot the histogram with k. Change the probability from 50:50 to 60:40 or some other value, and observe a net drift. What rate of drift would you expect? Test your hypothesis.

With probability 50:50, let the step size depend on direction. For example, let a step in the plus x-direction be 20% greater than a step in the negative x-direction. What kind of behavior would you expect? Test your hypothesis.

• • Load KT4, two-dimensional random walk (see Figs. 13.5 and 13.6).

One to four different two-dimensional random walks can be viewed and examined. Each walk consists of n steps (j index). The index i specifies the number of independent walks. The initial positions are arranged so that the paths of the separate walks cannot lie directly on top of each other. Within one walk, however, it is possible and likely that some paths are repeated.

The set of $+1$ and -1 values for all the steps of each of the walks is represented by $f_{i,j}$. In a completely separate process, values for g_j and h_j are determined; these values specify the direction the steps will take. The

Random walk - two dimensions.

j = time seq $\quad$ i = particle index $\quad$ $n := 50$ $\quad$ $j := 0 ..n$ $\quad$ $i := 0 ..3$

Starting positions.

$x_{0,0} := 0$ $\qquad$ $y_{0,0} := 0$ $\qquad$ $x_{1,0} := 1.2$ $\qquad$ $y_{1,0} := 0.2$

$x_{2,0} := .1$ $\qquad$ $y_{2,0} := 1.1$ $\qquad$ $x_{3,0} := 1.3$ $\qquad$ $y_{3,0} := 1.3$

f supplies plus and minus signs. $\quad$ g h determine motion in x or y direction.

$$f_{i,j} := (-1)^{\text{ceil(rnd(2))}} \qquad g_j := \text{if(rnd(2)} > 1,1,0) \qquad h_j := \text{if}\left[g_j \approx 0,1,0\right]$$

$$x_{i,(j+1)} := x_{i,j} + g_j \cdot f_{i,j} \qquad y_{i,(j+1)} := y_{i,j} + h_j \cdot f_{i,j}$$

Circle representing diffusion range for N steps.

$$N := \sqrt{n} \qquad N' := 1.3 \cdot N \qquad M := 30 \qquad k := 0 ..M \qquad \theta_k := 2 \cdot \pi \cdot \frac{k}{M}$$

$$X_k := N \cdot \cos\left[\theta_k\right] \qquad Y_k := N \cdot \sin\left[\theta_k\right]$$

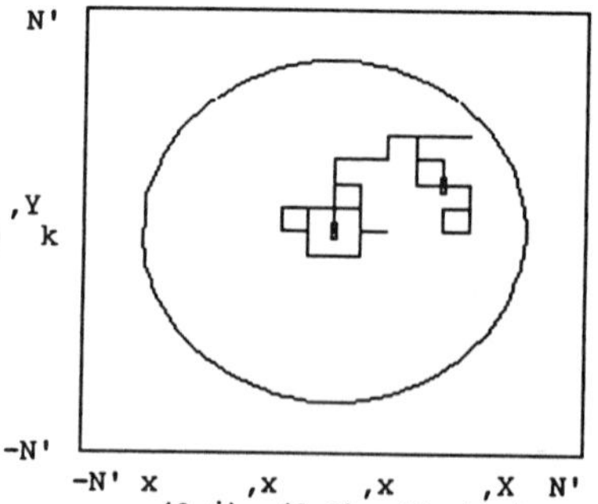

Rectangles represent starting and ending points.

$Y_{(0,j)}, Y_{(0,0)}, Y_{(0,n)}, Y_k$

$X_{(0,j)}, X_{(0,0)}, X_{(0,n)}, X_k$

Figure 13.5 KT4, two-dimensional random walk. (See the next figure for the rest of the document.)

Four walks displayed simultaneously.   N' := N'·1.2

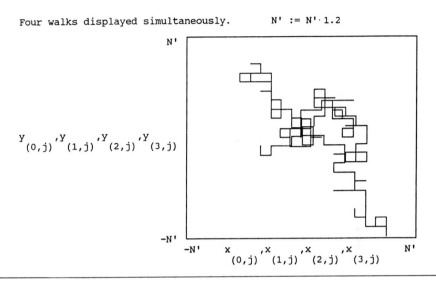

Figure 13.6 KT4 *continued.*

statement for g is a rnd statement that determines whether or not to step in the x-direction. The statement for h directs a motion in the y-direction if there is no motion in the x-direction.

A circle with radius $\sqrt{n}$ is shown together with the walk. The starting and ending points of the walk are shown as open rectangles.

To observe different random sequences, move the cursor to the $f_{i,j}$ region and press [F9]. Each time [F9] is pressed, a new set of random values are calculated. Observe a number of cases.

The second plot region is disabled. To see four different random walks simultaneously, let $i = 0, \ldots, 3$ and enable the second plot region. Four different walks are shown in the second plot region. The different walks are readily distinguished only on a color monitor. Reduce the upper limit of i to reduce the number of walks.

The g and h statements randomly select steps in the x- and y- directions. Change the g and h statements so that the probability for taking a step in the x-direction is twice that of taking a step in the y- direction. Observe a number of examples. Change the x and y step sizes and observe the net change in the pattern.

13.3.1 *Simulation of Thermal Equilibration*

A warm body placed in cold surroundings will lose heat to the exterior. As this process takes place, the warm body cools and the cool surroundings warm. After a long enough time period, the body and its surroundings come to thermal equilibrium — their temperatures become the same. We simulate such a process using a two-dimensional random walk. (A three-dimensional walk would be more realistic, but, for this software, would require too much memory and too much processing time.)

The environment is represented by a large rectangle; the object is represented by a smaller rectangle contained within the larger one. Energy units are represented as tiny open rectangles (plot type o). The number of energy units per area is a measure of the temperature. Initially, all the energy units are within the object; none are in the exterior region.

The energy units migrate according to a two-dimensional random walk. Energy is conserved: no energy units can be lost; nor can any be added during the process. An energy unit can migrate from the object to the surroundings or from the surroundings to the object without restriction. However, if an energy unit passes through the outer boundary of the environment, it reappears on the opposite side. This process is a type of periodic boundary condition. The temperature is proportional to the number of energy units per area. The temperature changes as the energy units migrate.

• •   KT5, simulation of thermal equilibration (see Figs. 13.7 and 13.8).

The document KT5 uses much memory and is computation intensive. It is advisable to exit MathCAD and reload, but do not load KT5 yet.

The statements $x_{i,0}$ and $y_{i,0}$ provide the initial positions of the $m + 1$ energy units. If $m + 1 = int^2$, where $int = 2, 3, 4$, then the units are arranged in a square array. Any value of m is, in principle, allowed. However, for $m = 3$, there are only four energy units and the results are not very interesting. Values greater than 15 require increased computation time and more memory than may be available.

Enter by hand the statements from the beginning of the document for m, i, n, j, s, $x_{i,0}$, and $y_{i,0}$. For $m = 3, 8,$ and 15, explore the x and y statements which make use of the mod and floor functions. These statements make possible an efficient way to specify the initial conditions. Print out tables of values and plot $y_{i,0}$ vs. $x_{i,0}$. What happens with intermediate values of m? When you are finished with this exploration, reload MathCAD.

Load KT5. The first plot region displays the energy units, the boundary of the object, and the boundary of the surroundings.

The f, g, and h statements have precisely the same function as in the two-dimensional walk document. The x and y statements are complicated because of the boundary conditions. Nested if statements take care of two boundary conditions at once; for example, if x is larger than the plus limit go just inside the minus limit; and if x is less than the minus limit go just inside the plus limit; otherwise, let x go to its new value.

Random walk - equilibration.

j = time seq   i = particle index

m := 15   i := 0 ..m   n := 50   j := 0 ..n

Postion m+1 particles in a rectangular array.

$$s := \sqrt{m + 1} \qquad x_{i,0} := \text{mod}(i,s) \qquad y_{i,0} := \text{floor}\left[\frac{i}{s}\right]$$

Set limits to the outer region.   Show boundary of object. Location of energy.

lim := 9   mlim := -lim + 1   plim := lim - 1   limp := lim + 1

$$k := 0 ..4 \qquad ss := s - 0.5 \qquad r_k := ss \qquad r_2 := -ss \qquad r_3 := r_2$$

$$t_k := -ss \qquad t_1 := ss \qquad t_2 := t_1 \qquad limp$$

$$y_{(i,0)}, t_k$$

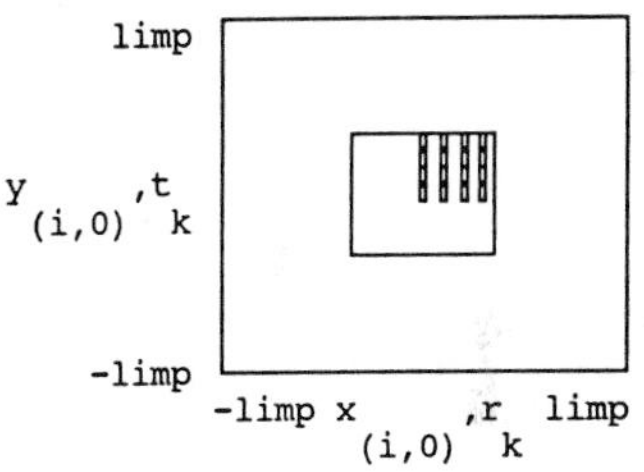

stpsz := 1

-limp

The functions for the two-dimensional walk.

-limp $x_{(i,0)}, r_k$ limp

$$f_{i,j} := (-1)^{\text{ceil}(rnd(2))} \qquad g_{i,j} := \text{if}(rnd(2) > 1,1,0) \qquad h_{i,j} := \text{if}\left[g_{i,j} \approx 0,1,0\right]$$

$$c_{i,j} := \left[\overrightarrow{\left[g_{i,j} \cdot f_{i,j}\right]}\right] \cdot stpsz \qquad d_{i,j} := \left[\overrightarrow{\left[h_{i,j} \cdot f_{i,j}\right]}\right] \cdot stpsz$$

The energy units are not to be lost.   If they wander off to the left, they reappear at the right, etc.   These are known as periodic boundary conditions.

$$\begin{bmatrix} x_{i,j+1} \\ y_{i,j+1} \end{bmatrix} := \begin{bmatrix} \text{if} \begin{bmatrix} x_{i,j} \ldots > lim, mlim, \text{if} \begin{bmatrix} x_{i,j} \ldots < -lim, plim, x_{i,j} \ldots \\ + c_{i,j} & + c_{i,j} & + c_{i,j} \end{bmatrix} \end{bmatrix} \\ \text{if} \begin{bmatrix} y_{i,j} \ldots > lim, mlim, \text{if} \begin{bmatrix} y_{i,j} \ldots < -lim, plim, y_{i,j} \ldots \\ + d_{i,j} & + d_{i,j} & + d_{i,j} \end{bmatrix} \end{bmatrix} \end{bmatrix}$$

Figure 13.7 KT5, simulation of thermal equilibration. (See the next figure for the rest of the document.)

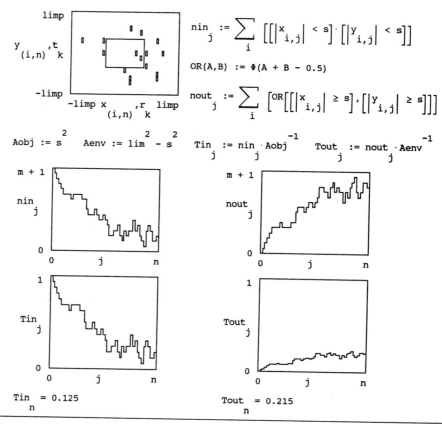

Figure 13.8 KT5 *continued.*

If the magnitude of either x or y is greater than s, then the energy unit is outside the object. A logical OR function is used to determine the status. The logical OR $(A + B)$ returns a 1 if either A or B is 1. (A and B are restricted to the values 0 and 1.) A MathCAD statement expressing logical OR is $OR(A, B) := \Phi(A + B - 0.5)$. If the argument of Φ is greater than or equal to 0, the function returns a 1; otherwise, it returns a 0. The conditions for x and y are the A and B statements.

For the count of the number of energy units inside the object, the n_{in} statement, the multiplication is equivalent to a logical AND. Both conditions must be true for the product to be 1; otherwise, it is 0.

In the final group of plot regions, the first is the arrangement of energy units after the nth step. The next two plot regions show the number of energy units inside the object and outside the object (inside the environment) as a function of the number of steps. The last two plot regions show the "temperatures," the number of energy units divided by the area. Because the number of energy units is small ($m + 1$), significant fluctuations are expected.

Place the cursor in the f-region and press [F9] for a new set of values; or just press [Esc], type pro for process, and press return.

Change the step size; use small integral values. This will permit the energy units to migrate farther without increasing the number of steps.

Reduce lim from 9 to 6 (lim must be geater than s). This change reduces the size of the environment. How should the final temperature in this case compare with that in the previous case? Explore.

Simplify the document to a one-dimensional form. Use a periodic boundary condition. Define temperature as the number of energy units per length.

> Sam, let a friend tell you, your life is going wrong.
> Records is a dead-end department,
> no Security Level worth a damn,
> it's impossible to get noticed.

> Yes, I know, fantastic, marvellous, wonderful.
> Remember me to Alison and the twins.

> Triplets.

> Triplets? God, how time flies!

> *Brazil*

C H A P T E R

14

Electric and Magnetic Fields

In most beginning texts, examples of the electric field include a few cases with discrete charges, considering specifically both the point charge and a very small collection of point charges. The development, quite logically, then moves on to continuous distributions such as a line charge or a charged ring. We examine here what might be considered transitions between discrete and continuous charge distributions.

For example, we consider the case of discrete charges uniformly spaced along a straight line, and we ask when the assemblage looks like a line charge, when like a point charge, and when like neither.

Similarly, for magnetic fields, the fields associated with a circular current loop and for a very large number of loops joined in a solenoid are considered. We explore the transition between the cases first by considering the field due to a pair of current-carrying loops and then by considering a finite solenoid constructed from a series of current loops.

Finally, in Section 14.2.2 we approach a number of very important equations from the point of view of dimensional analysis. You should not skip this section.

14.1 Electric Field

The magnitude of the electric field associated with a point charge is given by

$$E = \frac{1}{4\pi\epsilon_o}\frac{q}{r^2},\tag{14.1}$$

where E is the field, q is the charge, r is the distance from the point charge to the location where the field is evaluated, and ϵ_o is the permittivity of free space, a constant analogous to G in Newton's law of gravitation. The direction of the field is radially outward (inward) for a positive (negative) charge.

• • Load EM1, electric field due to a small number of point charges (see Fig. 14.1). (This problem was suggested in part by John Davis.)

Given two or more charges, determine the field (magnitude and direction) at any point; and if a charge is at that point, determine the force on that charge.

Enter the total number of points, N; this includes points where there are charges and other points where the field is to be determined. Then enter the locations (x_i, y_i) and the charges q_i at those locations. Enter 0 for charge, when the field is desired at a charge-free location. Separate values with commas; the values appear in tabular form. The plot shows the locations of the points as open rectangles. The origin is represented with a diamond.

At each point (x_i, y_i), the x- and y-components of the field due to all the other charges are determined (Ex_j, Ey_j). The total field is determined from the components. The force is determined from the fundamental relation $F = qE$. The field, force, and direction are presented in tabular form.

Just above the final plot regions, m $(\leq N)$ specifies the index of the point at which the field is to be plotted. In the last plot, we see the location of the point specified and the field vector. Given the other charges, consider the magnitudes and directions of the individual fields and argue for the final direction. Considering examples with the machine builds confidence in problem solving intuition.

Try some simple cases with two charges before trying more complex examples. For example, consider the field of a dipole at various points. Then consider cases with three and four charges.

If charges are distributed uniformly over a long, straight wire, the associated electric field alongside the wire no longer falls off as $1/r^2$, the

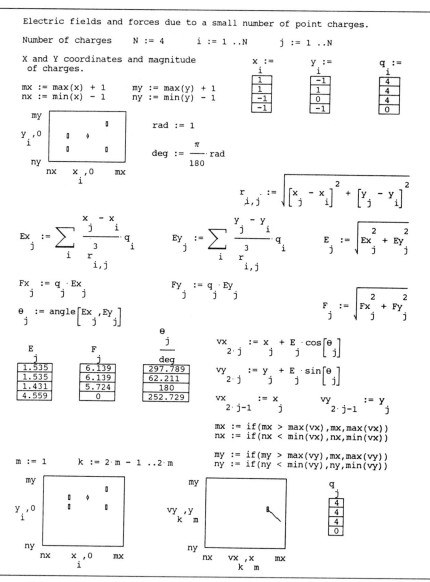

Figure 14.1 EM1, electric field due to a small number of point charges.

point charge dependence, but as $1/r$. The relation is

$$E = \frac{1}{4\pi\epsilon_o} \frac{2\lambda}{r},\qquad (14.2)$$

where λ is the linear charge density, the charge per length along the wire. This result can be obtained by integrating the contributions to the field from all the charge elements, or it can be obtained using Gauss's law.

It is a large change to go from a few point charges to a continuous charge distribution such as that of a line charge. We consider the transition between these two cases by evaluating the field for a small number of charges uniformly distributed along a straight line.

Place two positive charges q on a vertical line. Draw the perpendicular bisector to the line containing the two charges. Each charge is a distance d from the intersection of the two lines. We wish to calculate the field at a point p on the bisecting line a distance r from the line containing the charges.

The field due to each charge at the point r is (ignoring constants) $E = q/r^2$. The components parallel and perpendicular to the bisecting line are $E\cos(\theta)$ and $E\sin(\theta)$. Because the two charges are symmetrically placed relative to the bisecting line, these perpendicular components are equal and opposite; they sum to zero. The components parallel to the bisecting line add. Each parallel component is

$$E_{\text{parallel}} = E\cos(\theta) = \frac{q}{r^2 + d^2}\cos(\theta).$$

If we let $q = 1$ and note that $\tan(\theta) = d/r$, the total parallel field is given by

$$E_{\text{parallel}} = \frac{2}{r^2 + d^2}\cos\left(\text{atan}\,\frac{d}{r}\right). \tag{14.3}$$

If there are n pairs of charges, each pair symmetrically placed as was the pair just described, the total field is similarly described for each pair. In the document, $E_{ni,k}$ expresses the total field at a point due to each charge pair; sE_i is the total field at one point due to all the charges.

For purposes of comparison, we calculate several different fields: (1) the field due to n charge pairs distributed uniformly along a line, sE_i; (2) the field due to a point charge, with total charge $2n$, located at the midpoint of the line of charges, E_p; (3) the field due to a continuous infinite line charge with the same charge per length as that associated with the charge line, E_l; and (4) the field due to a continuous finite line charge with the same length and same charge per length as the charge line. The line charge density is

$$\lambda = \frac{charge}{length} = \frac{1}{2 \cdot hs},$$

where hs is half the charge separation distance between the charges spread along the line.

• •   Load EM2, electric field due to a series of point charges (see Figs. 14.2 and 14.3).

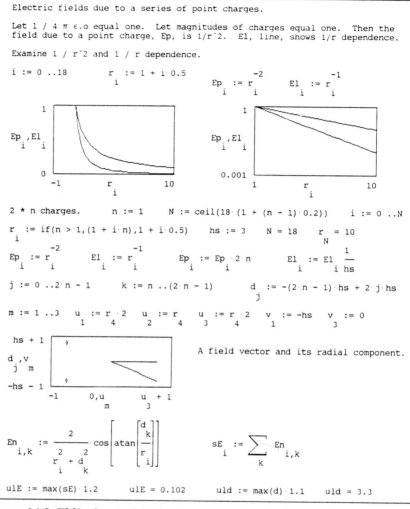

Figure 14.2 EM2, electric field due to a series of point charges. (See the next figure for the rest of the document.)

In the first two plots, a comparison is made of the field due to a point charge and due to a line charge. The plots are linear and log-log. We set all the constants equal to 1. We examine only the dependence on position. Be sure that you know which curve is which and why. It is necessary to see these dependences clearly, so we can compare our calculations with these two extreme cases.

N and r are display controls. For larger n, the number of charge pairs, the region of interest moves to larger r, so the range of r is dependent

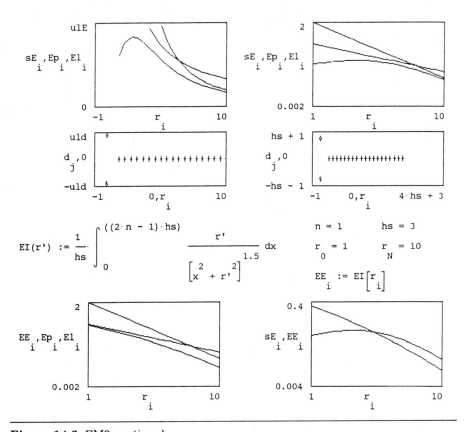

Figure 14.3 EM2 *continued.*

on n. For the curve to be reasonably smooth, the number of points is increased slightly as the range increases (the increase is not great because the computation time increases proportionately).

In the next plots, we show the fields due to the discrete distribution, the point charge, and the infinite line charge. The fields are plotted in linear and log-log plots. The location of the charges and the location of the points at which the field is calculated are also shown. There are two plots that show these points. In the first plot the abscissa corresponds to the field plot just above it. The ordinate is chosen to show all the charges. In the similar plot to the right, the x- and y-axes have the same scale so the relative spacing between charges and points at which the field is calculated is the same. This plot is not directly correlated with the log-log plot above it. Finally, we plot the field associated with a short rod, EE_i, together with that from the discrete charge distribution.

Go back to the plots of sE, E_p, and E_l and of d_j. Concentrate for now on this set of four plots. As you consider various cases, compare the total length of the charge distribution $(n-1) \cdot 2 \cdot hs$ with r_i and consider in which regions the curves sE and E_p and the curves sE and E_l have the same dependence on r.

Identify which curve is which. Why does the sE curve have a maximum at small r? Just beyond the maximum, which curve, point charge, or line charge better approximates sE? At large r, which curve better represents sE? Why?

Predict the changes that will occur if the half charge separation, hs, is reduced from 3 to 1. Test your prediction.

Let $hs = 3$ again and process the document to remind yourself of the shapes. Let $n = 2$ and process. Note how the sE field representation shifts in relation to the line charge representation.

Let $n = 4$ and process. Remember that the first of the spacing plots shows the total number of points and charges; the second shows the relative spacings. How would you characterize the sE field in comparison with E_p and E_l?

Let $n = 4$ and $hs = 1$. Look at the second spacing plot to get a sense of the spacing and range for point charges needed to give the impression of a line charge. How would you characterize the sE field in comparison with E_p and E_l?

The integral shown at the end of the document is obtained by integrating the contributions to the electric field due to individual charge elements along a uniformly charged rod. (MathCAD will do the actual integration; however, it is useful to know what it is calculating.) The field due to a charge element is given by

$$dE = \frac{1}{4 \pi \epsilon_o} \frac{dq}{R^2}. \tag{14.4}$$

The radial component of the field is given by $dE_r = dE \cdot \cos(\theta)$. Express dq in terms of the line charge; $dq = \lambda dx$. R is the total distance from a charge element dq to the point where the field is to be calculated; $R = \sqrt{r^2 + x^2}$ where r is the radial distance and x is the distance along the charged rod from its center. The angle is specified by $\cos(\theta) = r/R$.

Because the rod is symmetric, integrate from the center to one end and double the result. Set $1/4 \pi \epsilon_o = 1$:

$$dE = \frac{\lambda dx}{r^2 + x^2} \cos(\theta); \tag{14.5}$$

$$E = 2 \int_0^{\text{end}} \frac{\lambda dx}{r^2 + x^2} \frac{r}{\sqrt{r^2 + x^2}} = \frac{1}{hs} \int_0^{\text{end}} \frac{r}{(r^2 + x^2)^{3/2}} \, dx. \qquad (14.6)$$

Disable the regions for E_n, sE, and the four plot regions following them. Observe the final two plot regions for $hs = 3$ and $n = 1$, 2, 4. Both plots are log-log. Notice the degree of similarity between the discrete calculation, sE, and the continuous calculation, EE, as n goes from 1 to 4. Also compare the cases of the infinitely long line charge and the point charge. Compare the length of the charged rod with the distance to the "transition" region.

•• A third interesting case for the electric field is that due to an infinite surface. There is no dependence on r; the field is constant. (For a point charge, $E \propto r^{-2}$; for an infinite line charge, $E \propto r^{-1}$; for an infinite surface charge $E \propto r^{-0}$.) A surface could be approximated as a set of parallel line charges. Examine the field due to such an array.

Now, instead of examining the field along a line perpendicular to the line of charge (a line containing a small number of uniformly distributed discrete charges), consider the field along a line parallel to this same line of charge. A comparison is made between the discrete charge case and that of a single point charge with the same net charge.

•• Load EM3, electric field due to a series of point charges: parallel path (see Figs. 14.4 and 14.5).

First the field due to a point charge is calculated at points along a line, where the minimum distance between the line and the charge is r. The y_i' determine the position along the line where the field is calculated.

The next plot shows the arrangement of charges (diamonds) and the points at which the field is calculated (pluses). The separation between these two lines is r. The number of charge pairs is n, and the location of the charges is given by d_j. The field is calculated at the points specified by y_i. $Er_{i,j}$ is a measure of the individual contributions to the field perpendicular to the line at the y_i due to the charges at d_j; $Ep_{i,j}$ is the corresponding parallel component. Es is a measure of the total field.

Examine the parallel and perpendicular components of the field. Sketch the anticipated shape before performing the plot.

We would expect that at distances that are large compared to the length of the charge distribution, the field would be approximately that of a point charge. At very small distances, we would expect to see the local structure. At intermediate distances, we would expect to see a transition between these two cases.

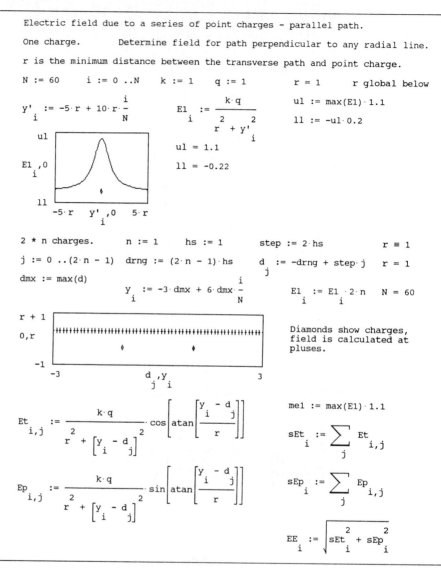

Electric field due to a series of point charges - parallel path.

One charge.   Determine field for path perpendicular to any radial line.

r is the minimum distance between the transverse path and point charge.

$N := 60$ $i := 0 \ ..N$ $k := 1$ $q := 1$ $r = 1$ r global below

$$y'_i := -5 \cdot r + 10 \cdot r \cdot \frac{i}{N}$$

$$E1_i := \frac{k \cdot q}{r^2 + {y'_i}^2}$$

$ul := \max(E1) \cdot 1.1$

$ll := -ul \cdot 0.2$

$ul = 1.1$

$ll = -0.22$

2 * n charges. $n := 1$ $hs := 1$ $step := 2 \cdot hs$ $r \equiv 1$

$j := 0 \ ..(2 \cdot n - 1)$ $drng := (2 \cdot n - 1) \cdot hs$ $d_j := -drng + step \cdot j$ $r = 1$

$dmx := \max(d)$

$$y_i := -3 \cdot dmx + 6 \cdot dmx \cdot \frac{i}{N}$$

$E1_i := E1_i \cdot 2 \cdot n$ $N = 60$

Diamonds show charges, field is calculated at pluses.

$$Et_{i,j} := \frac{k \cdot q}{r^2 + \left[y_i - d_j\right]^2} \cdot \cos\left[\operatorname{atan}\left[\frac{y_i - d_j}{r}\right]\right]$$

$$Ep_{i,j} := \frac{k \cdot q}{r^2 + \left[y_i - d_j\right]^2} \cdot \sin\left[\operatorname{atan}\left[\frac{y_i - d_j}{r}\right]\right]$$

$me1 := \max(E1) \cdot 1.1$

$$sEt_i := \sum_j Et_{i,j}$$

$$sEp_i := \sum_j Ep_{i,j}$$

$$EE_i := \sqrt{sEt_i^2 + sEp_i^2}$$

Figure 14.4 EM3, electric field due to a series of point charges: parallel path. (See the next figure for the rest of the document.)

The plot shows the magnitude of the actual field and compares it with the field that would result if all the charge were concentrated at one point at the center of the distribution. With $n = 1$, $r = 10$, and $N = 40$, is there much difference between the two field determinations? Let $r = 2$ and process. Let $r = 1$ and process. (For $r < 2$, plots are more precise if N is increased from 40 to 60.)

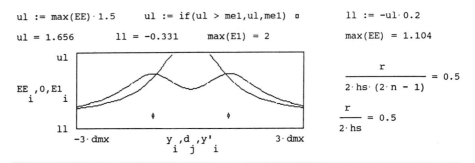

```
ul := max(EE)·1.5      ul := if(ul > me1,ul,me1) ▯      ll := -ul·0.2

ul = 1.656      ll = -0.331      max(E1) = 2      max(EE) = 1.104
```

$$\frac{r}{2 \cdot hs \cdot (2 \cdot n - 1)} = 0.5$$

$$\frac{r}{2 \cdot hs} = 0.5$$

Figure 14.5 EM3 *continued.*

Let (n, N) equal $(2, 60)$. At what distance r, compared to the charge separation distance, does the individual structure show? Is the length of the charge distribution relevant to the previous question? In terms of these lengths, at what radial distance is there a sense of an extended charge? At what distance does the group look like a point charge?

Explore the field for a variety of values of n and r.

14.1.1   Electric Flux

Gauss's law, one of the four basic laws of electricity and magnetism, relates the charge within a closed surface to the net electric flux through the surface.

$$\Phi_E = \frac{q}{\epsilon_o} \qquad \text{where} \qquad \Phi_E = \iint_{\text{closed surface}} \mathbf{E} \cdot d\mathbf{A}, \qquad (14.7)$$

where $\mathbf{E} \cdot d\mathbf{A}$ is a vector dot product. A measure of surface area, $d\mathbf{A}$ is represented by a vector normal to the surface. In documents EM4 and EM5, we consider some elementary aspects of flux through a surface.

• • Load EM4, electric flux through a surface (see Fig. 14.6).

The surface and the normal to the surface are represented in the plot region of the document. The purpose of the bulk of the code is simply to determine the coordinates of the surface and a vector normal to the surface representing $d\mathbf{A}$. To find the slope of the normal vector, recall that the slopes of perpendicular lines are related by $m_1 \cdot m_2 = -1$. The solve block is used to determine the coordinates of the endpoint of the normal vector.

The series of vertical lines (created using the subdivision parameter for plot regions) represents the electric field. If θ were zero, the angle between $d\mathbf{A}$ and $\mathbf{E}$ would be zero; the cosine of the angle (dot product) would be one, and the flux would be at its maximum. As θ increases, the flux

---

```
Flux through a surface in a uniform electric field.
```

```
The vertical subdivision lines represent the field.  The surface, at angle
θ to the horizontal, is represented by a normal vector perpendicular to it.
θ is a measure of the angle between the surface (normal) and the electric
field.
```

$$rad \equiv 1$$
$$deg \equiv \pi \cdot 180^{-1} \cdot rad$$

```
θ ≡ 30·deg     Let θ take on any value        Do not use zero degrees as
               between 1 and 90 degrees.      this gives infinite slope to
                                              the normal.  Approximate zero
cos(θ) = 0.866                                with 1*deg.
```

$$given \quad y' - y_a \approx m2 \cdot \left[x' - x_a \right] \qquad \left[x' - x_a \right]^2 + \left[y' - y_a \right]^2 \approx 0.2$$

$$\begin{bmatrix} x_2 \\ y_2 \end{bmatrix} := find(x',y')$$

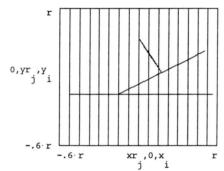

```
                                    Flux is E * dA or E dA cos(θ);
                                    the direction of dA is the normal
                                    to the surface.

                                    As you vary the angle note how
                                    many field lines pass through the
0,yr ,Y                             actual surface.
    j  i
                                    Note relation between number of
                                    field lines intersected by surface
                                    and the cos(θ).
```

```
-.6·r
       -.6·r           xr ,0,x          r
                         j    i
```

```
Let field have strength 10; area is a square, with side r.
```

$$E := 10 \qquad A := r^2 \qquad \Phi := E \cdot A \cdot cos(\theta) \qquad \Phi = 8.66$$

```
Plot information.   r ≡ 1   i ≡ 0 ..4   x ≡ 0   y ≡ 0   j ≡ 0 ..1
                                          0       0
```

$$x_1 \equiv r \cdot \frac{cos(\theta)}{2} \qquad y_1 \equiv r \cdot \frac{sin(\theta)}{2} \qquad x_3 \equiv x_1 \quad y_3 \equiv y_1 \quad x_4 \equiv 2 \cdot x_1 \quad y_4 \equiv 2 \cdot y_1$$

$$m1 \equiv \frac{y_1}{x_1} \qquad m2 \equiv \frac{-1}{m1} \qquad x_1 \equiv x_a \quad y_1 \equiv y_a \quad x' \equiv x_a - 1 \quad y' \equiv y_a + 1$$

$$xr_0 \equiv -0.5 \quad yr_0 \equiv 0 \quad xr_1 \equiv 0.95 \quad yr_1 \equiv 0.95$$

Figure 14.6 EM4, electric flux through a surface.

decreases. We can visualize this flux through the surface by "counting" the number of lines that pass through the surface.

Let θ take on different values between 1° and 90°. Relate the number of field lines through the surface with magnitude of the cosine of θ and with the value yielded by the flux calculation at the end of the document.

● ●   Load EM5, flux through a surface from a point charge (see Fig. 14.7).

Field lines radiate from a point source. The flux through two surfaces, one closed and one plane, is determined. The initial code specifies how to draw the field lines and the two surfaces.

Parameters to control are located just above the plot region. The radius of the closed surface is r; the coordinates of the center of the closed surface are (x_c, y_c); the coordinates of the center of the plane surface are (c', d'); and the length and orientation of the plane surface are L and ϕ, respectively.

Flux through a closed and a plane surface.   Press [Ctrl]PgDn to reach action.

$n := 12$ $i := 0\ ..n$ $j := 0\ ..2 \cdot n$ $z := 1.2$

$\theta_i := 2 \cdot \pi \cdot \dfrac{i}{n}$

$m := 15$ $k := 0\ ..m$ $l := 0\ ..1$

$x_{2 \cdot i} := 0$ $x_{2 \cdot i+1} := \cos\left[\theta_i\right]$ $y_{2 \cdot i} := 0 \cdot$ $y_{2 \cdot i+1} := \sin\left[\theta_i\right]$

$\alpha_k := 2 \cdot \pi \cdot \dfrac{k}{m}$ $x'_k := x_c + r \cdot \cos\left[\alpha_k\right]$ $y'_k := y_c + r \cdot \sin\left[\alpha_k\right]$ $rad \equiv 1$

$\Phi := angle(c',d')$

$deg \equiv \dfrac{\pi}{180} \cdot rad$

$c_0 := c' + \dfrac{L}{2} \cdot \sin(\Phi - \phi)$ $c_1 := c' - \dfrac{L}{2} \cdot \sin(\Phi - \phi)$

$d_0 := d' - \dfrac{L}{2} \cdot \cos(\Phi - \phi)$ $d_1 := d' + \dfrac{L}{2} \cdot \cos(\Phi - \phi)$ $dis := r - \sqrt{x_c^2 + y_c^2}$

$flux_{loop} := if(dis > 0,1,0)$

$flux_{surf} := \dfrac{(L \cdot \cos(\phi))^2}{4 \cdot \pi \cdot \left[c'^2 + d'^2\right]}$

$r \equiv 0.3$ $x_c \equiv -0.3$ $y_c \equiv 0.4$ $flux_{loop} = 0$

$c' \equiv 0.3$ $d' \equiv 0.3$ $L \equiv 0.4$ $\phi \equiv 0 \cdot deg$

$flux_{surf} = 0.071$

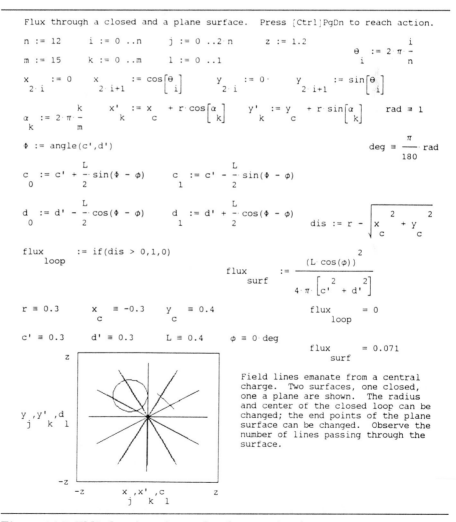

Field lines emanate from a central charge. Two surfaces, one closed, one a plane are shown. The radius and center of the closed loop can be changed; the end points of the plane surface can be changed. Observe the number of lines passing through the surface.

Figure 14.7 EM5, flux through a surface from a point charge.

The net flux through the loop and surface is shown. The flux through the surface is only approximate; the procedure fails when the surface is close to the central charge.

To get a sense of scale when specifying coordinates, note that the field lines are of length one; the plot limit, z, is 1.2. Keep the surfaces in this range.

Examine the loop first. Let (x_c, y_c, r) equal $(-0.4, 0.4, 0.4)$. Every field line that enters the surface also exits the surface, and the next flux through the loop is zero. Let $r = 0.7$. Now the central charge is enclosed, and every line passes through the surface once. There is a net flux through the surface. Try several values of x, y, and r. Observe the intersections of the field lines with the loop.

Let L equal 0.1, 0.5, and 0.9. Note the value for $flux_{surf}$. With $L = 0.5$, let ϕ equal $0°$, $30°$, $60°$, and $90°$ and observe the change in the surface orientation and the attendant change in flux. Let $d' = 0$, $c' = 1$, $L = 0.5$, and $\phi = 0$. Note the values for $flux_{surf}$ as c' decreases. Let c' equal 1, 0.8, 0.6, 0.4, 0.2, and 0.1. The flux through the surface is now greater than the flux through the loop when it contains the central charge. As $c \to 0$, the flux should go to 0.5. How is $flux_{surf}$ calculated (a ratio of what to what)? Why is this expression valid only for values of L rather less than the magnitude of the distance of the surface from the charge? Can you write an expression that will determine the flux for the plane surface whether the surface is close to or far from the central charge? Explore with various parameter values.

14.2 Magnetic Field

The magnetic field due to a long straight current carrying wire is given by

$$B = \frac{\mu_o}{4\pi} \frac{2I}{r}. \tag{14.8}$$

The magnitude of the magnetic field decreases as r^{-1} as the distance from the wire increases (just like the electric field dependence associated with a long, straight, charged wire). If the wire lies along the y-axis, then in the x-y plane, the field is in the $-z$ direction for positive x and in the $+z$ direction for negative x.

•• Plot the magnitude of the magnetic field due to a current-carrying wire as a function of distance from the wire. Let the wire lie along the

y-axis. Let x take on plus and minus values. Let $z = 0$. Avoid evaluating the field right at the wire, where $x = 0$. The sign of the calculated field should correspond to the orientation of the B-field along the z-axis.

•• Plot the magnitude of the magnetic field due to a pair of parallel current-carrying wires. Consider the cases in which currents are in the same direction and in opposite directions. Let the wires be parallel to the y-axis and placed symmetrically to either side of it in the x-y plane. Let $z = 0$. Consider values of x both between and outside the pair of wires. Avoid calculating the field at the wires. Examine the field when the currents are not equal.

The magnetic field along the axis of a current-carrying loop is given by

$$B_x = \frac{\mu_o I}{2} \frac{R^2}{(x^2 + R^2)^{3/2}}. \tag{14.9}$$

The center of the loop is at the origin; the loop lies in the y-z plane; the field is along the x-axis; and R is the radius of the loop.

•• Define a function $B(R, x)$, and plot the B-field due to a circular current loop as x goes from $-3R$ to $+3R$. Where is the B-field a maximum? Plot $B(R, x)/B_{\max}$ vs. x/R. When the field is one-half the maximum, what is the value x/R? When the field is one-quarter the maximum? One-tenth?

•• Load EM6, field due to a pair of current loops: Helmholtz coils (see Fig. 14.8).

Before examining the field due to a pair of current loops, we review the field due to one loop. The field is shown in the first plot region; the maximum is at the center of the loop. The normalized field is plotted in the next region. The given-find solve block permits us to determine the distance at which the field has fallen to any fraction, fr, of the maximum. Use of the place marker is convenient here.

Next we consider two coils of radius r oriented parallel to the y-z plane, centered on the x-axis, and situated at $x = \pm hsep$. Let currents flow in the same sense in the coils so the axial fields have the same direction and add.

The field due to each of the coils is specified, and their sum, B_t, is defined. The initial separation between the coils is $4\,r$ ($hsep = 2\,r$). The total field is plotted; vertical lines indicate the location of the coils. The individual fields are plotted just below. Compare the individual fields and the sum.

Magnetic field due to a pair of coaxial, circular, current loops.

First the field due to one loop.

$$\mu_o := 4 \cdot \pi \cdot 10^{-7} \qquad I := 1$$

$$B(r,x) := \frac{\mu_o}{2} \cdot I \cdot \frac{r^2}{\left[x^2 + r^2\right]^{1.5}}$$

Plot the field for a fixed radius r as a function of the axial distance z.

$$r \equiv 0.1 \qquad xl := 3 \cdot r \qquad x := -xl, -xl + .01 \ ..xl$$

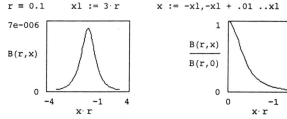

$$x' := 0.2 \cdot r \qquad fr := 0.5$$

given $\dfrac{B(r,x')}{B(r,0)} \approx fr$ $f(fr) := find(x')$ $f(0.5) = 0.766 \cdot r$

                              $d := 2 \cdot r$ $f(0.5) = 0.383 \cdot d$

A pair of loops are located along the x-axis at + hsep and - hsep, **hsep** being half the separation distance.

$$B1(r,x) := \frac{\mu_o \cdot I}{2} \cdot \frac{r^2}{\left[(x + hsep)^2 + r^2\right]^{1.5}} \qquad B2(r,x) := \frac{\mu_o \cdot I}{2} \cdot \frac{r^2}{\left[(x - hsep)^2 + r^2\right]^{1.5}}$$

$$Bt(r,x) := B1(r,x) + B2(r,x)$$

$$hsep \equiv 2 \cdot r \qquad 2 \cdot hsep = 4 \cdot r \qquad Bt(r,hsep) = 6.373 \cdot 10^{-6}$$

Figure 14.8 EM6, field due to a pair of current loops: Helmholtz coils.

Let $hsep \equiv 1\,r$ and process. (Disable the B_1, B_2 plot region for faster performance.) The overlap between the fields is greater, but the field "sags" in the center. Let $hsep \equiv 0.25\,r$. Now the separation is small enough that the total field approximates that of a single coil.

Let $hsep = 0.5\,r$. Eliminate from the plot the lines showing the coil's location from the plot (that is, let the plot region be $Bt(r, x)$ vs. x). Change the abscissa limits to $-hsep$ and $hsep$. Change the lower ordinate limit to $Bt(r, hsep)$ and the upper ordinate limit to $B(r, 0) \cdot 1.01$. Observe the shape of the field between the coils. Let $hsep = 0.51\,r$. Is the region as uniform right at the center? Does the region in which the field is almost constant change in size? If you let $hsep = 0.52\,r$, is the beginning of a "sag" in the center of the field apparent? How does the field at the center change if $hsep = 0.48\,r$? What is the "best" separation value?

Helmholtz coils, where the coil separation is nominally equal to the radius, are used as a relatively easy method for producing a region of uniform field. Sometimes the coils are used to cancel another field to create a region with a near-zero field.

14.2.1 *Solenoid*

We approach the solenoid in a similar fashion. We consider the total B-field from a set of current-carrying coils. The coils have the same radius as before and are arranged along the z-axis. The intensity of the field can be determined along the axis of the solenoid by summing the contributions due to the individual coils. In this way, the field can be determined at any point along the axis both inside and outside the solenoid. The field falls off quite abruptly near the ends of the solenoid. The field for a very long "ideal" solenoid, $B = \mu_0 n I$, is a constant; the expression gives no information about the variations in the field associated with a solenoid of finite length.

• • Load EM7, approximating a solenoid with a small number of current-carrying coils (see Fig. 14.9).

The field is created by a set of $2\,m$ uniformly spaced current-carrying coils with centers located at points d_j along the z-axis. The field is calculated at a set of locations z_i. $B_{i,j}$ is the field at z_i due to the coil at d_j; B_i' is the field at z_i due to all the coils; Bs is the field of an ideal solenoid; and sl is just a display control that can be selected to be any fraction of the maximum value for the field.

The field due to the coils, B', and the ideal solenoid field are shown in the first plot region. The locations of the coils are also displayed at the chosen height, sl.

Solenoid - on-axis magnetic field.

$\mu_o := 1$ $I := 2$ $r := 1$ $hsep := .5$ $step := 2 \cdot hsep$

$n := (2 \cdot hsep)^{-1}$ $m := 1$ m - # of pairs of coils.

n - # of coils per length

d[j - location of coils; z[i - locations where field calculated.

$j := 0 \,..\, 2 \cdot m - 1$ $drng := (2 \cdot m - 1) \cdot hsep$ $d_j := -drng + step \cdot j$

$N := 41$ $i := 0 \,..\, N$

$$z_i := -1.5 \cdot drng + 3 \cdot drng \cdot \frac{i}{N}$$

B[i,j - field at z[i due to coil at d[j; B'[i - field at z[i due to all coils
Bs - field of ideal solenoid.

$$B_{i,j} := \frac{\mu_o}{2} \cdot I \cdot \frac{r^2}{\left[r^2 + \left[z_i - d_j \right]^2 \right]^{1.5}}$$

$$B'_i := \sum_j B_{i,j}$$ $mB' := \max(B')$

$sl := mB' \cdot 0.8$

$$Bs := \mu_o \cdot n \cdot I$$

$ul := if(Bs > mB', Bs, mB') \cdot 1.05$

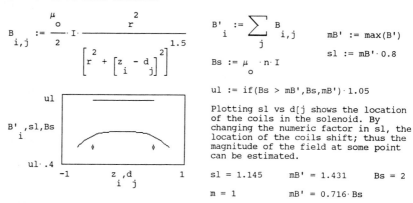

Plotting sl vs d[j shows the location
of the coils in the solenoid. By
changing the numeric factor in sl, the
location of the coils shift; thus the
magnitude of the field at some point
can be estimated.

$sl = 1.145$ $mB' = 1.431$ $Bs = 2$

$m = 1$ $mB' = 0.716 \cdot Bs$

Calculate the field at the end of the solenoid and compare with the maximum.

$$B_e := \sum_j \left[\frac{\mu_o}{2} \cdot I \cdot \frac{r^2}{\left[r^2 + \left[drng - d_j \right]^2 \right]^{1.5}} \right]$$

$mB' = 1.431$ $B_e = 1.354$

$\dfrac{B_e}{mB'} = 0.946$ $\dfrac{B_e}{Bs} = 0.677$

Figure 14.9 EM7, approximating a solenoid with a small number of
current-carrying coils.

Finally, the magnitude of the field is determined at the end of the set
of coils, and its value is compared with the maximum field and the ideal
solenoid field.

Let $m = 2, 3, 4, 5$, and 8. For $m = 8$, reduce N to 20. The "dip" is
not real; the apparent dip is due to a fortuitous choice of values of z. Let
$N = 21$ and observe the change. (Choose parameters carefully to avoid
being mislead.)

Let $m = 1$ and set up the Helmholtz coil case.

If r is reduced or increased, how will the field change? Try $N = 40$, $m = 4$, and $r = 0.5$ (or 2). How does changing $hsep$ alter the field? What is the effect of changing the magnitude of I? Summarize your observations.

14.2.2 Dimensional Analysis

Several important relationships involving electric and magnetic fields, energy, volume, and power can be obtained using dimensional analysis. The last example generates the Larmor formula, a relationship that is not easily derived but is readily delineated using dimensional analysis.

• • Determine the expressions for energy density in electric and magnetic fields. For each case there are five quantities to be considered: energy En, volume V (we are considering energy density), the constants ϵ_o and μ_o, and the fields themselves, either E or B. Four fundamental units — mass, length, time, and charge — are necessary for the analysis. Therefore, each field can be determined with one group. The two groups include the variables, $(En, V, \epsilon_o, \mu_o, E)$ and $(En, V, \epsilon_o, \mu_o, B)$.

• • Load EM8. This file contains the starting point for this problem and can be used for the rest of the problems in this section (although not every quantity is included). You must set up the solve block and determine the exponents.

Do any of these quantities have a zero exponent and drop out? In each of the two examples, solve for En/V in terms of the other variables. Check your results by determining the units of the group equal to En/V.

• • Electromagnetic radiation, including light, carries both energy and momentum. Energy flux is the energy passing through some area per unit of time; call it S. Find the dimensionless group associated with S, ϵ_o, μ_o, E, and B. Solve for S.

• • Form the dimensionless groups to determine momentum density and momentum flux.

• • Form a dimensionless group from the quantities ϵ_o, μ_o, and c, the speed of light. In this very important case, the constant is unity.

When a charge is accelerated, radiation is emitted. The Larmor equation relates the power radiated to charge and acceleration. Determine the fundamental nature of the Larmor relation through dimensional analysis.

• • Form a dimensionless group from the quantities q (charge), a (acceleration), P (power), ϵ_o, and μ_o. Use the relation from the previous example to eliminate μ_o. Form the group; then solve for P. You will discover that

$$P \propto \frac{q^2 a^2}{\epsilon_o c^3}.$$

The constant which you cannot obtain has the value $(1/4\pi) \cdot (2/3)$. Determining, from dimensional analysis, how the power radiated varies with charge and acceleration is a treat. Remember this if you should take a more advanced course in electricity and magnetism.

I can feel it, Dave; my mind is going.

2001

CHAPTER
15

Applications of the Lorentz Force

The forces on a charged particle due to electric and magnetic fields may result in a variety of motions. The possible controls over such particles have been exploited in a large variety of instruments and machines, where particles can be deflected, confined, accelerated, focused, and so on. The net force, known as the Lorentz force, is given by

$$\mathbf{F} = q\mathbf{E} + q(\mathbf{v} \times \mathbf{B}). \qquad (15.1)$$

The velocity of a charged particle under the influence of an electric field changes both in magnitude and, unless the motion is collinear with a uniform field, direction. Under the influence of a magnetic field, only the direction may be changed. In all the cases considered here, these forces are much larger than the gravitational force, and gravity is ignored. Consider first the motion of a charged particle in a constant, uniform electric field. The motion corresponds competely to that of a massive particle in a uniform gravitational field with no air resistance. For example, in two-dimensional motion, if the electric field is oriented in the y-direction, there is no acceleration in the x-direction and the x-component of the velocity is constant. If the field is directed downward, positively charged particles will follow familiar parabolic trajectories, and we use the familiar equations for trajectory and range.

15.1   Parallel-Plate Electrostatic Analyzer

The fact that charged particles move in parabolic trajectories in a constant, uniform electric field can be exploited to construct an electrostatic analyzer, a device that separates particles according to their energy. We consider first a device consisting of a pair of parallel plates separated by a distance d; a voltage V is applied across the plates. The electric field between the plates is uniform in the y-direction; the field is zero in the x-direction. The acceleration experienced by a charged particle in the region between the plates is

$$a_y = \frac{q}{m} E \qquad \text{where} \qquad E = \frac{V}{d}. \tag{15.2}$$

If the quantities q, m, and E, are constant, a_y is constant. Given the constant acceleration in one direction, the particles follow parabolic trajectories.

Particles enter the field region through a slit in the lower plate (at $y = 0$) making an angle $\theta = 45°$ with respect to the horizontal. They follow a parabolic trajectory and are detected in the $y = 0$ plane. The angle at which the particles enter the region cannot realistically be so constricted that all particles enter only at angle θ. The acceptance angle α is a measure of the deviation ($\pm$) from the central angle θ. If the particles deviate from ideal entry in the z-direction as well and enter at angle β, the initial values of the x- and y-components of the velocity are reduced. When both these possibilities are taken into consideration, the initial values for v_{ox} and v_{oy} are

$$v_{ox} = v_o \cos(\theta + \alpha) \cdot \cos(\beta) \qquad v_{oy} = v_o \sin(\theta + \alpha) \cdot \cos(\beta). \tag{15.3}$$

• •   Verify the relations for v_{ox} and v_{oy}. By what fraction are v_{ox} and v_{oy} reduced if β is 5°?

• •   Load EML1, the parallel-plate electrostatic analyzer (see Fig. 15.1).

For reasons that will become clear momentarily, choose $\theta = 45°$. Initially, let conditions be ideal: $\alpha = 0$, $\beta = 0$. The separation between the plates is d; the applied voltage is V.

For the modest field of 15 V, across 10 cm, what is the acceleration of an electron in g's? (A description of acceleration in g's may be useful for describing accelerations of massive objects on earth, but not necessarily for charged particles.)

The initial velocity in the x-y plane is $v_{oxy}(En)$; the velocity is specified in terms of the initial energy, En. $R(En)$ is the range of the particle

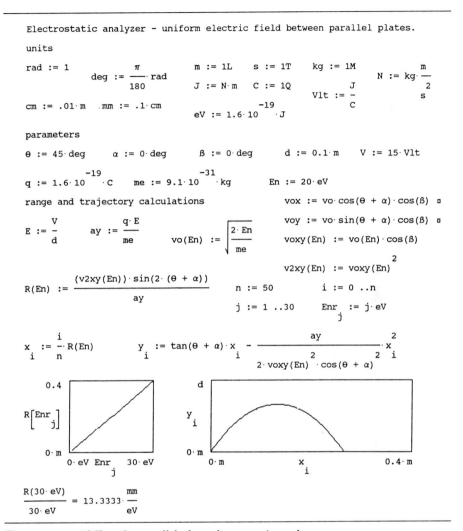

Figure 15.1 EML1, the parallel-plate electrostatic analyzer.

(check the kinematic equations). The trajectory equation expresses y_i in terms of x_i.

The range is plotted as a function of energy. From this we can determine a very important property of the analyzer, the dispersion. Dispersion, in this case, is a measure of the spatial separation of the particles per energy difference. In this particular case, the range vs. energy curve is linear and the slope gives the dispersion. It is expressed here in mm/eV. Dispersion can be expressed in many forms. In optics, the dispersion can be expressed as angular separation per difference in wavelength.

Dispersion is a separation divided by a difference in the property being measured.

Resolution is often discussed along with dispersion. This property is a measure of the ability to distinguish the difference between adjacent groups of a similar character. For example, in this device if the acceptance angle is not zero, all the particles of a given energy will not focus at one point. They will be distributed; there will be a characteristic line shape associated with the detection of many particles. Resolution, a dimensionless number, is a measure of the total energy, for example, divided by the energy width of the line. The narrower the width of the line for a given energy, the higher is the resolution. The higher the resolution, the smaller the real differences in energy that can be distinguished from each other. When we consider the magnetic spectrometer, we will determine a line shape by means of a simulation.

Change the range function to $R(En, \alpha)$. Let α_k range from $-5°$ to $5°$ in steps of $1°$. Plot $R(En, \alpha_k)$, $R(En, 0)$ vs. α_k. How much does the range vary? What energy does this correspond to? (This can be handled easily in a solve block or by using the dispersion information.)

If this energy difference is a measure of the line width, what is the resolution?

If the initial angle were $30°$, what would the dispersion and resolution be? At $60°$? What is the difference in range between the extreme trajectories for a central angle of $30°$ and deviation angles of $\pm 5°$ and the case where the central angle is $45°$? Plot the trajectories.

While it was already known that the maximum range occurs at $45°$ and that larger or smaller angles result in a smaller range, we see now that for trajectories starting with a range of angles this phenomena results in what we refer to as first-order focusing (there is no focusing at $30°$). The folding over of the ranges for initial angles symmetrically spaced about $45°$ reduces the spread of the ranges and increases resolution.

θ and α specify the angles in the x-y plane. β is a measure of the initial velocity in the z-direction. Let $\alpha = 0$. Change v_{oxy} to be a function of β. Change the range to be a function of α and β. Plot $R(0, \beta)$ vs. β as β goes from $-5°$ to $5°$. (Just change enough to get the plot; don't worry about unessential regions.) Finally, plot $R(5°, \beta_k)$, $R(0, \beta_k)$, $R(0,0)$ vs. β_k. If the limits of α and β are $5°$, over what distance range might the electrons appear? To what range of energies does this spread correspond?

Since all deviations from the ideal entry angle of $\theta = 45°$, $\alpha = 0$, and $\beta = 0$, result in a reduced range, the greatest range is the truest measure of the particle's energy.

Would this system work for protons? Test your ideas. Plot the electric field strength necessary to focus protons at a distance of 0.5 m as a function of energy of the proton. Are these realistic field strengths?

15.2   Magnetic Focusing Spectrometer

A 180° magnetic focusing spectrometer has focusing properties somewhat similar to those of the electrostatic analyzer described in the previous section. The spectrometer has a uniform B-field in the z-direction. Particles enter the field at the origin, in the x-y plane. The initial velocity is in the y-direction. Particles are detected along the x-axis after having completed approximately one-half of a circular orbit.

Unlike the electrostatic analyzer, the orbits of the particles in a magnetic focusing spectrometer are circular. Because B is vertical and velocity is in the x-y plane, $v \times B = v\,B$. Thus the force due to magnetic field $F = q(v \times B) = q\,v\,B$, which is constant in the x-y plane and perpendicular to v. These are the necessary conditions for uniform circular motion, a constant force perpendicular to the velocity.

Applying Newton's second law, we can determine the radius of the orbit:

$$q\,v\,B = m\frac{v^2}{r} \qquad \text{or} \qquad r = \frac{m\,v}{q\,B}. \tag{15.4}$$

A magnet is sometimes referred to as a momentum selector because the radius r is proportional to the momentum, mv. The angular frequency of a charged particle moving in a circular orbit is

$$\omega = \frac{v}{r} = \frac{q\,B}{m}. \tag{15.5}$$

• •   For B-fields of $0.1\,T$ and $1\,T$, create log-log plots of radius vs. energy for both electrons and protons. Let the kinetic energy take on values of 10^n eV where $n = 1, \ldots, 5$ for the electrons and $n = 3, \ldots, 7$ for the protons. This calculation ignores relativistic effects, which arise as particle velocities approach the speed of light.

• •   Protons are to follow an orbit with $r = 0.5\,m$. Plot B-field strength vs. proton kinetic energy.

• •   In the electrostatic analyzer that we considered in the previous section, we found that the range was proportional to the energy. Is the same true for the magnetic analyzer that we have just described? Plot range (diameter of circular orbit) as a function of energy. What B-field would result in the same range for 20 eV electrons as we found for the electrostatic analyzer of the previous section?

• • If $B = 1.31 \cdot 10^{-4} T$, plot the diameter of the electron orbits vs. E as the energy ranges from $1\,\text{eV}$ to $20\,\text{eV}$. Which analyzer, electrostatic or magnetic, has the better dispersion at $20\,\text{eV}$? (How does the strength of this B-field compare with that of the earth's magnetic field?)

Line shape. If a particle originates at the origin and remains in the x-y plane but its initial angle deviates from the y-axis by the angle α, the range is reduced. The effect is similar to the reduction in range associated with α in the electrostatic analyzer. There is first-order focusing.

The focusing can be visualized by imagining a circular orbit with one point of the circumference fixed at the origin and the diameter on the x-axis. This corresponds to an orbit with entry along the y-axis. Now rotate the circle about the fixed point, the origin, $\pm\alpha$. Convince yourself that these circular orbits intersect the x-axis at $2\,r\cos(\alpha)$ for both $\pm\alpha$. So just as in the case of the trajectories in the electrostatic analyzer, trajectories of the same energy but deviating by $\pm\alpha$ from the ideal entry angle focus at the same point, a little short of the ideal entry case.

Given a magnetic focusing spectrometer with an acceptance range of $\pm\alpha$, if electrons are emitted from a source with equal probability into any angle, what will the line shape look like? We know that the case $\alpha = 0$ has the greatest range. We know there is first-order focusing. This gives us an inkling of the shape; we can, however, get a much better sense of the shape by running a simulation.

To determine the line shape for a source of finite width $\pm s$ and a spectrometer with an acceptance angle $\pm\alpha$, define the function

$$x(x_o, \theta) := 2 \cdot r \cdot \cos(\theta) + x_o, \tag{15.6}$$

where x_o is the initial x-value (which is limited by the width of the source) and θ is the angle at which the particle is emitted. We can calculate the value of the function many times since the initial values for x_o and θ can take on random values within the allowed range ($\pm s$, $\pm\alpha$). After a sufficient number of range values have been determined, the data are plotted in a histogram; the result is a simulation of a line shape displaying the number of trajectories within a given range as a function of range.

• • Load EML2, line shape (see Fig. 15.2).

Note that s is half the source width, and α is the magnitude of the maximum deviation angle. The values for the various ranges are z_i. The operation

$$\text{rnd}(2 \cdot C) - C \tag{15.7}$$

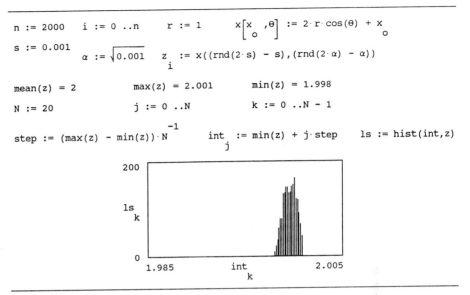

Figure 15.2 EML2, line shape.

generates random numbers in the range $\pm C$. The number of electron orbits, for which the range is calculated, is n. (Start with a smaller n if you have a slow machine.) The shape that you see plotted in EML2 is characteristic of the best compromise between source width and acceptance angle ($\alpha = \sqrt{s}$). The line shape can be examined in more detail by changing the plot limits. Process a few times to obtain some sense of the statistical fluctuations. (Repeat by moving the cursor to the first rnd region and processing [F9].) In each case make an estimate of the resolution.

Examine cases where one source of error is zero. Let $s := 0$ and $\alpha := 0.1$. Let $s := 0.001$ and $\alpha = 0$. Observe a number of cases.

Let $s := 0.003$ and $\alpha = \sqrt{0.001}$. The result is a line with a central region of nearly uniform height. Observe a number of cases.

Let $s := 0.001$ and $\alpha = 0.1$; this gives a longer trail-off at lower energies. Observe a number of cases.

15.3 Cylindrical Electrostatic Analyzer

Our final analyzer example is that of a cylindrical focusing electrostatic analyzer. The analyzer consists of two plates that lie along circles concentric with the z-axis. Both plates start at the x-axis and extend through an angle of approximately 130° into the second quadrant. Particles enter the space between the plates from the positive x-axis in the plus y-direction. A detector is placed at that location where first-order focusing occurs.

When a potential is applied across the plates, the electric field between them varies as $1/r$, the field associated with a long charged wire or cylinder. The field strength is selected to permit particles of a given energy to pass along a circular arc. Substituting in the equation of motion the quantity qE for the force, we have

$$q|E| = \frac{mv^2}{r} \qquad \text{so} \qquad |E| = \frac{mv^2}{qr}. \qquad (15.8)$$

The field is directed radially inward for positively charged particles. θ is the angle through which the particle has passed in its transit from the entry point to the detector. The x- and y-components of the acceleration can be expressed as

$$ax = \frac{qE}{m}\cos(\theta) \qquad \text{and} \qquad ay = \frac{qE}{m}\sin(\theta). \qquad (15.9)$$

• •   Load EML3, cylindrical electrostatic analyzer (see Figs. 15.3 and 15.4).

Calculations are performed for three trajectories. Values for the energies and angles of deviation must be specified for each. The field is adjusted to pass the zeroth ray along a circular trajectory between the plates.

This calculation requires lots of memory. If a number of calculations have been performed since you last loaded MathCAD, you should quit MathCAD and reload. Units are not included so as not to use additional memory.

The velocity form of the Verlet algorithm is used to compute the particle trajectories. (The Euler-Cromer algorithm is fine for more rapid exploration but in this case is slightly less precise; it also requires less memory. If you experience memory difficulties, simplify the algorithm to the Euler-Cromer form. All the features examined are essentially the same. The angle at which the trajectories focus is less precisely located but is very close, and the specific value of the angle is not the central issue.)

The parameters and initial conditions are specified. The energy, entered in eV, is converted to Joules by multiplying by the charge of the proton, q. The range of the time scale in which successive steps of the iterative process are considered is in nanoseconds.

Electrostatic cylindrical analyzer.

parameters $q := 1.6 \cdot 10^{-19}$ $m := 9.1 \cdot 10^{-31}$ $rad := 1$ $deg := \dfrac{\pi}{180} \cdot rad$

three trajectories $j := 0 \, .. \, 2$ $En_j :=$

15
13
17

$\alpha_j :=$

0 · deg
0 · deg
0 · deg

$En_j := En_j \cdot q$ $v_j := \sqrt{2 \cdot En_j \cdot m^{-1}}$

Iterated solution. Initial conditions, parameters user defined functions to implement the velocity form of the Verlet algorithm.

$n := 80$ $i := 0 \, .. \, n$ $x_{0,j} := 1$ $y_{0,j} := 0$ $\theta_{0,j} := 0$

$vx_{0,j} := v_j \cdot \sin[\alpha_j]$ $vy_{0,j} := v_j \cdot \cos[\alpha_j]$

$r(x,y) := \sqrt{x^2 + y^2}$

$ax_{0,j} := -v_0^2 \cdot r[x_{0,j}, y_{0,j}]^{-1} \cdot \cos[\theta_{0,j}]$ $ay_{0,j} := -v_0^2 \cdot r[x_{0,j}, y_{0,j}]^{-1} \cdot \sin[\theta_{0,j}]$

$a(x,y) := -v_0^2 \cdot r(x,y)^{-1}$ $ns := 10^{-9}$ $\delta t := 14 \cdot ns$

$D(x,vx,ax) := x + vx \cdot \delta t + 0.5 \cdot ax \cdot \delta t^2$

$c(x,vx,ax,y,vy,ay) := \cos(angle(D(x,vx,ax), D(y,vy,ay)))$

$s(x,vx,ax,y,vy,ay) := \sin(angle(D(x,vx,ax), D(y,vy,ay)))$

$VX(x,vx,ax,y,vy,ay) := vx + 0.5 \cdot (ax + a(x,y) \cdot c(x,vx,ax,y,vy,ay)) \cdot \delta t$

$VY(x,vx,ax,y,vy,ay) := vy + 0.5 \cdot (ay + a(x,y) \cdot s(x,vx,ax,y,vy,ay)) \cdot \delta t$

$$
\begin{bmatrix} ax_{(i+1),j} \\ ay_{(i+1),j} \\ vx_{(i+1),j} \\ vy_{(i+1),j} \\ x_{(i+1),j} \\ y_{(i+1),j} \end{bmatrix} := \begin{bmatrix} a[x_{i,j}, y_{i,j}] \cdot c[x_{i,j}, vx_{i,j}, ax_{i,j}, y_{i,j}, vy_{i,j}, ay_{i,j}] \\ a[x_{i,j}, y_{i,j}] \cdot s[x_{i,j}, vx_{i,j}, ax_{i,j}, y_{i,j}, vy_{i,j}, ay_{i,j}] \\ VX[x_{i,j}, vx_{i,j}, ax_{i,j}, y_{i,j}, vy_{i,j}, ay_{i,j}] \\ VY[x_{i,j}, vx_{i,j}, ax_{i,j}, y_{i,j}, vy_{i,j}, ay_{i,j}] \\ x_{i,j} + vx_{i,j} \cdot \delta t + \frac{1}{2} ax_{i,j} \cdot \delta t^2 \\ y_{i,j} + vy_{i,j} \cdot \delta t + \frac{1}{2} ay_{i,j} \cdot \delta t^2 \end{bmatrix}
$$

Figure 15.3 EML3, cylindrical electrostatic analyzer. (See the next figure for the rest of the document.)

Several functional forms are defined to simplify the equation block; these include the position, acceleration, cosine, and sine. Note that *VX* and *VY* are expressions of the velocity form of the Verlet algorithm.

In the equation block, acceleration, velocity, and position are iterated for each component. Finally, the trajectories are plotted both in

Draw a radial marker at the focusing angle.

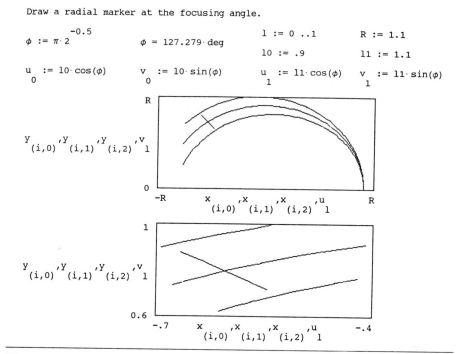

$$\phi := \pi \cdot 2^{-0.5}$$

$$\phi = 127.279 \cdot \deg$$

$$l := 0 \,..\, 1$$

$$R := 1.1$$

$$l0 := .9$$

$$l1 := 1.1$$

$$u_0 := l0 \cdot \cos(\phi)$$

$$v_0 := l0 \cdot \sin(\phi)$$

$$u_1 := l1 \cdot \cos(\phi)$$

$$v_1 := l1 \cdot \sin(\phi)$$

Figure 15.4 EML3 *continued.*

entirety and in close up at the focus region. A marker, a short radial line, is drawn at the angle $\phi = \pi/\sqrt{2}$, which is the theoretical focus angle.

First determine a rough measure of the dispersion. The initial conditions are for energies of 15, 13, and 17 eV with a 0° deviation angle. The quantities l_0 and l_1, just above the plot regions, specify the distance from the origin of the starting and ending points of the radial marker. Change these values until the radial line just touches the outer and inner trajectories. This provides a measure of the spread which occurs in this case for different values of energy. Express the dispersion in mm/eV. Is the dispersion a constant? Is there an asymmetry? If yes, on which side is the dispersion greater?

Now examine the focusing. Let all the energies be 15. Let the α's be 0°, 5°, −5°. Does the focusing occur where predicted? Adjust the values for l_0 and l_1 to get a sense of the line width. Estimate the resolution.

Find the kinetic energy, E', for a particle with deviation angle $\alpha = 0°$, whose trajectory intersects the 15 eV, $\alpha = -5°$ trajectory at the focusing angle. Use your value for the dispersion to select an energy. Then show the two trajectories.

Find the kinetic energy, E', for a particle with deviation angle $\alpha = -5°$, whose trajectory intersects the 15 eV, $\alpha = 0°$ trajectory at the focusing angle.

If, instead of having the expected $1/r$ field dependence, the field was radial but the magnitude was independent of the radial position (that is, constant in magnitude), what would the new focusing angle be?

15.4   Trajectories in Crossed E- and B-Fields

A problem commonly found in elementary texts is that of a particle passing through a region of crossed electric and magnetic fields (for example, E_y, B_z, and velocity in the x-direction). The magnitudes of the fields are adjusted so that particles of a particular energy pass through this velocity filter undeflected :

$$F_{Ey} + F_{By} = 0 \quad \text{or} \quad q\,E_y = q\,v\,B_z \quad \text{and} \quad v = \frac{E_y}{B_z}. \quad (15.10)$$

Leaving the problem at this stage may leave some false impressions.

• •   Load EML4, Wien filter (see Fig. 15.5).

Let K be the electron energy for which there should be no deflection. The initial parameters are specified and the field strengths are adjusted to pass an electron with energy K. Let K' be the kinetic energy of the particle sent through the filter. The deviation angle from the x-axis is θ.

Let $K' = 10\,\text{eV}$ and process. Then let $K' = 9.9\,\text{eV}$ and $10.1\,\text{eV}$. Determine the dispersion? Do trajectories diverge from the central path as expected?

Looks can be deceiving. Change the upper limit of the abscissa from $4\,\text{cm}$ to d ($15\,\text{cm}$) and process with $K' = 10\,\text{eV}$. Then, let $K' = 2\,\text{eV}$ and $K' = 12\,\text{eV}$. The motion is probably more complex than might be surmised from the case of zero deflection.

The problem of crossed E- and B-fields is a rich one. We explore this problem further. The configuration remains the same: the electric field is in the y-direction, and the magnetic field is in the z-direction.

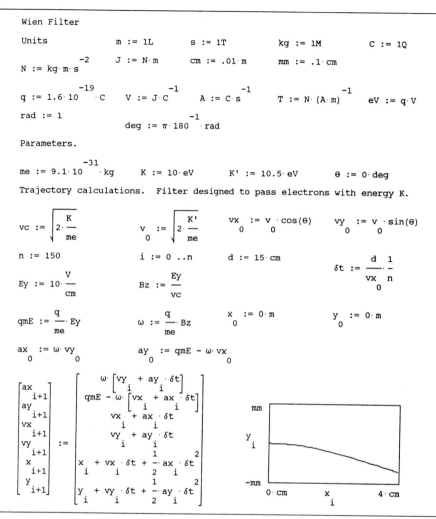

```
Wien Filter

Units              m := 1L        s := 1T         kg := 1M          C := 1Q

          -2       J := N·m       cm := .01·m     mm := .1·cm
N := kg·m·s

        -19                   -1              -1                -1
q := 1.6·10   ·C    V := J·C       A := C·s         T := N·(A·m)       eV := q·V

rad := 1                        -1
                   deg := π·180   ·rad

Parameters.

        -31
me := 9.1·10    ·kg      K := 10·eV       K' := 10.5·eV       θ := 0·deg

Trajectory calculations.  Filter designed to pass electrons with energy K.

        ___                  ___
       / K                  / K'       vx  := v ·cos(θ)     vy  := v ·sin(θ)
vc := / 2·——        v  :=  / 2·——        0    0               0    0
     √    me          0  √    me

n := 150                 i := 0 ..n       d := 15·cm

                                                              d   1
             V                     Ey                  δt := —— · -
Ey := 10· ——          Bz := ——                               vx  n
             cm                    vc                           0

       q                     q              x  := 0·m         y  := 0·m
qmE := —— · Ey        ω := —— · Bz           0                 0
       me                    me

ax  := ω· vy            ay  := qmE - ω· vx
  0        0              0
```

• •   Load EML5, motion of a charged particle in crossed E- and B-fields (see Fig. 15.6).

This program requires lots of memory. It may be prudent to exit MathCAD and reload.

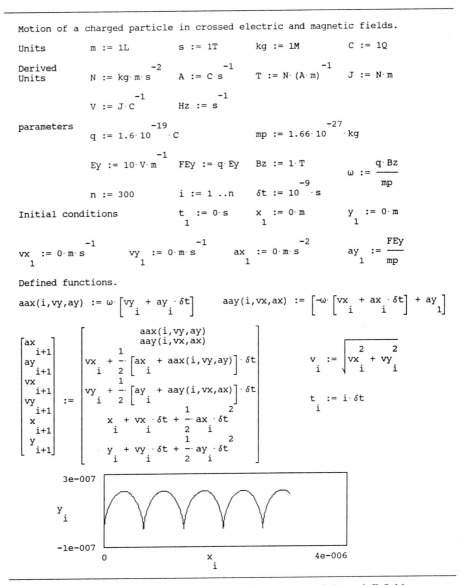

Motion of a charged particle in crossed electric and magnetic fields.

Units $m := 1L$ $s := 1T$ $kg := 1M$ $C := 1Q$

Derived $N := kg \cdot m \cdot s^{-2}$ $A := C \cdot s^{-1}$ $T := N \cdot (A \cdot m)^{-1}$ $J := N \cdot m$
Units

               $V := J \cdot C^{-1}$ $Hz := s^{-1}$

parameters $q := 1.6 \cdot 10^{-19} \cdot C$ $mp := 1.66 \cdot 10^{-27} \cdot kg$

               $Ey := 10 \cdot V \cdot m^{-1}$ $FEy := q \cdot Ey$ $Bz := 1 \cdot T$ $\omega := \dfrac{q \cdot Bz}{mp}$

               $n := 300$ $i := 1 .. n$ $\delta t := 10^{-9} \cdot s$

Initial conditions $t_1 := 0 \cdot s$ $x_1 := 0 \cdot m$ $y_1 := 0 \cdot m$

$vx_1 := 0 \cdot m \cdot s^{-1}$ $vy_1 := 0 \cdot m \cdot s^{-1}$ $ax_1 := 0 \cdot m \cdot s^{-2}$ $ay_1 := \dfrac{FEy}{mp}$

Defined functions.

$aax(i, vy, ay) := \omega \cdot \left[vy_i + ay_i \cdot \delta t \right]$ $aay(i, vx, ax) := \left[-\omega \cdot \left[vx_i + ax_i \cdot \delta t \right] + ay_1 \right]$

$$\begin{bmatrix} ax_{i+1} \\ ay_{i+1} \\ vx_{i+1} \\ vy_{i+1} \\ x_{i+1} \\ y_{i+1} \end{bmatrix} := \begin{bmatrix} aax(i, vy, ay) \\ aay(i, vx, ax) \\ vx_i + \frac{1}{2} \cdot \left[ax_i + aax(i, vy, ay) \right] \cdot \delta t \\ vy_i + \frac{1}{2} \cdot \left[ay_i + aay(i, vx, ax) \right] \cdot \delta t \\ x_i + vx_i \cdot \delta t + \frac{1}{2} \cdot ax_i \cdot \delta t^2 \\ y_i + vy_i \cdot \delta t + \frac{1}{2} \cdot ay_i \cdot \delta t^2 \end{bmatrix}$$

$v_i := \sqrt{vx_i^2 + vy_i^2}$

$t_i := i \cdot \delta t$

Figure 15.6 EML5, motion of a charged particle in crossed E- and B-fields.

The initial conditions have the particle start at the origin with zero velocity. The Verlet algorithm is used.

Position the y vs. x plot near the bottom of the screen and process. Observe the motion that was described. Go through the argument again so that it becomes clear to you.

What is an approximate value for the drift velocity?

Based on your knowledge of the motion and your reading of the y vs. x graph, make rough plots by hand of eight situations (the cases are given below). Try to determine the general shape; don't worry about numerical values. Draw two and let MathCAD plot the same two, and so on. These are tough, but give them your best shot. For each one whose general shape you predict, count one. A score of two is not bad; three is very good; four is excellent; and eight is amazing. Take your time; think carefully about each one before having MathCAD spill the beans. Try to ask questions that will nail down a few points.

The eight cases are (1) vx_i vs. x_i, (2) vy_i vs. x_i, (3) ax_i vs. x_i, (4) ay_i vs. x_i, (5) vx_i vs. y_i, (6) vy_i vs. y_i, (7) ax_i vs. y_i, and (8) ay_i vs. y_i. There is a great deal of information here worth pondering.

Try some other initial conditions; let (vx_1, vy_1) equal $(-1, 0)$ and $(-2, -5)$.

Delete the plot regions that were just created. Now add a weak, constant electric field in the x-direction. Let $E_x = 1 \text{V}/\text{m}$. Define FEx and adjust ax_1 and aax appropriately. Predict the effect on the motion of this added field and verify.

If you're not going to kill me, I have things to do.

Darkman

C H A P T E R

16

Circuits

We approach dc circuits through two common examples, the voltage divider and the Wheatstone bridge. In these, examples we find application of Kirchhoff's laws and a general approach to dc circuit problems. We then consider briefly the charging and discharging of capacitors, and the use of diodes. Combining a resistor, a capacitor, and a diode in one unit, we construct a simple power supply. Finally, we consider how to construct a simple amplifier using a tunnel diode in a voltage divider circuit.

16.1 DC Circuits

Voltage and current are quantities that are basic to all circuits. The voltage, V, (in volts) between two points in a circuit indicates the work done on a charge as it moves from one point to the other in a circuit. A battery is a voltage source; there is a potential difference between its terminals. Voltage must be measured between two points. Frequently voltage measurements are indicated at one point with an implicit reference to a ground point at V = 0. (Take a piece of copper pipe and drive it well into the ground; the voltage of the pipe is ground potential.) If the voltage is specified at a point, the implication is that the number actually specifies the potential difference between that point and ground.

Current is a measure of charge per time flowing through a cross section of a wire or other circuit element. The current unit is the ampere. The direction of current is from more positive potential to less positive potential. Current flows through a circuit element.

Two laws useful in dc circuit analysis are (1) Kirchhoff's current law and (2) Kirchhoff's voltage law. These two laws can be stated briefly as: (1) at any point in a circuit, the sum of all currents flowing *in* equals the sum of all currents flowing *out*, and (2) the algebraic sum of the voltage changes across all the elements around any closed loop of a circuit is zero.

We consider the circuit elements: the resistor, the capacitor, and the diode. The voltage-current relationship in each of these devices is different. The resistor is a device in which the current through the element is proportional to the voltage across it. This statement is known as Ohm's law:

$$i = \frac{v_2 - v_1}{R}, \tag{16.1}$$

where v_2 and v_1 are the voltages at either end of the resistor. The unit of resistance is the ohm (Ω). The voltage change across a resistor, moving in the direction of current flow, is $-iR$; this is called a voltage drop. Going across the resistor against the current results in a voltage increase of iR.

In the circuits considered here, the connecting wire between elements is assumed to have zero resistance. Consequently, there is no voltage drop across these wires. All points on a length of wire are at the same voltage.

Since resistors in series have the same current passing through them, the net resistance for resistors in series (see Fig. 16.1) is the sum of the individual resistances:

$$R_{\text{tot}} = R_1 + R_2 + \cdots = \sum_i R_i. \tag{16.2}$$

Since resistors in parallel have the same voltage across them, resistors in parallel (see Fig. 16.2) add in terms of their reciprocals:

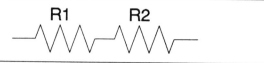

Figure 16.1 Two resistors in series.

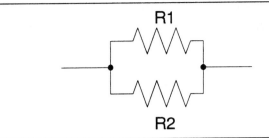

Figure 16.2 Two resistors in parallel.

$$\frac{1}{R_{\text{tot}}} = \frac{1}{R_1} + \frac{1}{R_2} + \cdots = \sum_i \frac{1}{R_i} = \sum_i R_i^{-1}$$

$$R_{\text{tot}} = \left(\sum_i R_i^{-1} \right)^{-1}. \qquad (16.3)$$

16.1.1   *The Voltage Divider*

The voltage divider is a very common circuit element. The divider consists, in this case, of two resistors in series (see Fig. 16.3). A dc voltage source is included, so there is a voltage to divide. We want to know the voltage at the point (node) between the two resistors R_1 and R_2; we refer to this voltage as v_{out}. (The common point between the negative terminal of the battery and resistor R_2 is at ground.)

Current flows from the voltage source through R_1, through R_2, and back through the battery. Because R_1 and R_2 are in series, the current

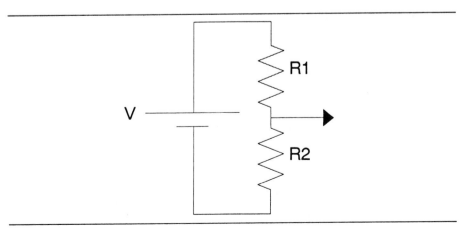

Figure 16.3 Voltage divider circuit.

through them is the same. Apply Kirchhoff's voltage law, starting at the minus terminal of the battery. The path takes us across the battery from the minus terminal to the plus terminal (so the voltage change is $+v$), across R_1 in the direction of the current $(-i\,R_1)$, across R_2 in the direction of the current $(-i\,R_2)$, and back to the starting point, so the sum is zero:

$$v - i\,R_1 - i\,R_2 = 0 \qquad \text{or} \qquad i = \frac{v}{R_1 + R_2}.$$

Starting from ground, we can get to the v_{out} node by going across R_2 against the current. The output voltage from the divider is then

$$v_{\text{out}} = +i\,R_2 = \frac{R_2}{R_1 + R_2}\,v. \tag{16.4}$$

• • Verify that the same expression for v_{out} is obtained by the path that includes the battery and R_1.

• • Load CIRC1, voltage divider circuit (see Fig. 16.4).

A voltage divider consisting of a voltage source and two resistors in series is analyzed for the case that the sum of the two resistors is a constant. Although the results are easily obtained, they should be well understood; the voltage divider will be revisited in several following applications.

A solve block, with the find statement in functional form, returns the values for the current and R_1 when R_2 is specified. One equation in the solve block is an expression of Kirchhoff's voltage law for the circuit. The other assures that the total resistance remains a constant. Try several values for R_2 and observe the results.

A sequence of values is obtained by defining R_{2i} and substituting it in calls on f, the find statement of the solve block. In the first plot region, the magnitudes of both the resistances and the total current are plotted against the R_1 values. In the second plot region, the voltages across R_1 and R_2 and their sum are plotted against the R_1 values. Be careful to determine which curve is associated with which resistor. Which curve corresponds to v_{out}? The abscissa for these plots is R_{1i}. How would the curves appear if plotted against R_{2i}?

Let $R_1 = 100\ \Omega$. Let R_2 go from 20 to 980 Ω (the sum $R_1 + R_2$ is no longer a constant). Plot v_{out} vs. R_2. What is the voltage range for v_{out}? How is this case different from the previous one?

Summarize the properties of a voltage divider.

• • Load CIRC2, voltage divider with load (see Fig. 16.5).

The voltage divider circuit.

The common potentiometer used in the laboratory is a resistor with three connection points, three terminals. Two of the connections are at either end of the resistor and do not change. The third contact can slide along the length of the resistor. At any given contact point the total resistance is divided into two resistances R1 and R2. The sum is, of course, constant.

$V := 5$ $R1 := 1$ $R2 := 4$

$R := R1 + R2$ $I := 1$

given $V - I \cdot R1 - I \cdot R2 \approx 0$ $R1 + R2 \approx R$

$f(R2) := find(I, R1)$

$R2 := 3$

$f(R2) = \begin{bmatrix} 1 \\ 2 \end{bmatrix}$ I is the upper value, R1 the lower.

The values for I and R1 can be written using subscripts on the function f.

$I := f(R2)_0$ $R1 := f(R2)_1$ $I = 1$ $R1 = 2$

We can look at an entire sequence by letting R2 take on a sequence of values.

$n := 5$ $i := 0 .. n$ $R2_i := \begin{bmatrix} i \\ - \\ n \end{bmatrix} \cdot R$ $I_i := f\begin{bmatrix} R2_i \end{bmatrix}_0$ $R1_i := f\begin{bmatrix} R2_i \end{bmatrix}_1$

$V1_i := I_i \cdot R1_i$ $V2_i := I_i \cdot R2_i$ $VV_i := V1_i + V2_i$

$R1_i, R2_i, I_i$

$V1_i, V2_i, VV_i$

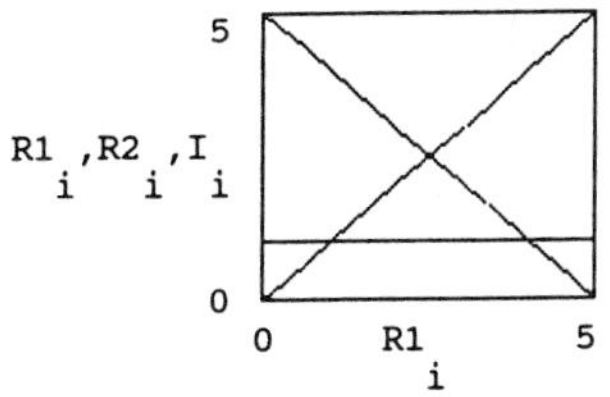

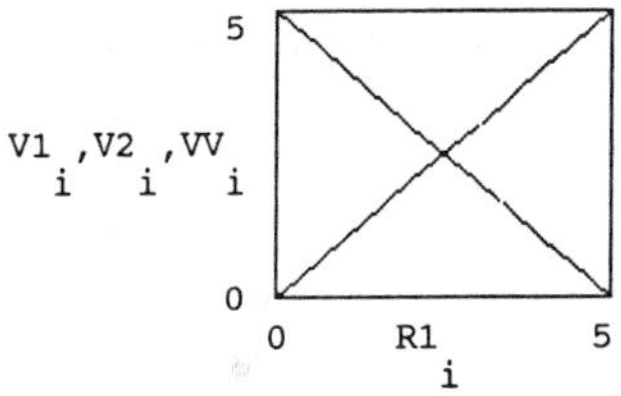

Figure 16.4 CIRC1, voltage divider circuit.

The behavior of a loaded voltage divider deviates somewhat from the behavior we studied in the previous example. This circuit is sufficiently common that you should be very familiar with it. A load resistor R_3 is added to the voltage divider we discussed in the previous example. The resistor is in parallel with R_2. A solve block is again set up in functional form. In the solve block are three equations: a statement of the voltages for each of the two loops in this circuit and a current statement at the branching node. (There are three simple loops that could be drawn for this circuit; only two are independent; any two will do.) For any value of R_3, f returns the currents I_1, I_2, and I_3; v_{out} is calculated using the value of I_3.

Let R_3 take on values much larger than, equal to, and much smaller than R_2. Predict the values of the output voltages before calculating $f(R_3)$.

The voltage divider circuit.

Connect a third resistor from the central tap point to the lower edge of the potentiometer and examine the new circuit.

I2 is the current through R2; I3 is the current through R3. Their sum is the current through R1.

```
------------        V := 5       R1 := 3        R2 := 2        R3 := 10
|           |
V          R1  Vout              I1 := 1        I2 := 1        I3 := 1
            |----->
            |      |    given    V - I1·R1 - I2·R2 ≈ 0
           R2     R3
            |      |             V - I1·R1 - I3·R3 ≈ 0
------------------
                                 I1 ≈ I2 + I3
```

$$F(R3) := find(I1, I2, I3)$$

$$F(10) = \begin{bmatrix} 1.071 \\ 0.893 \\ 0.179 \end{bmatrix}$$

We look at a series of values. $n := 5$ $i := 0 ..n$

$i := 0 ..10$ $R3_i := (10 - i) \cdot 1$ $I2_i := F\left[R3_i\right]_1$ $I3_i := F\left[R3_i\right]_2$

$Vout_i := I3_i \cdot R3_i$

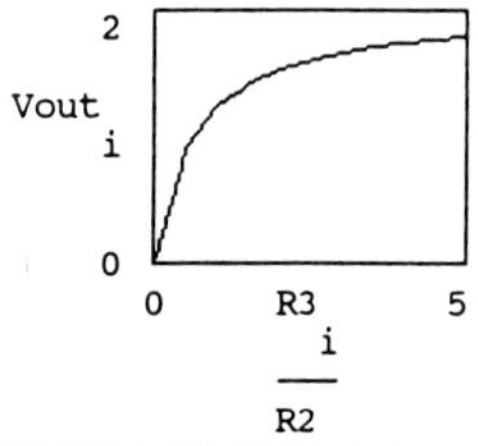

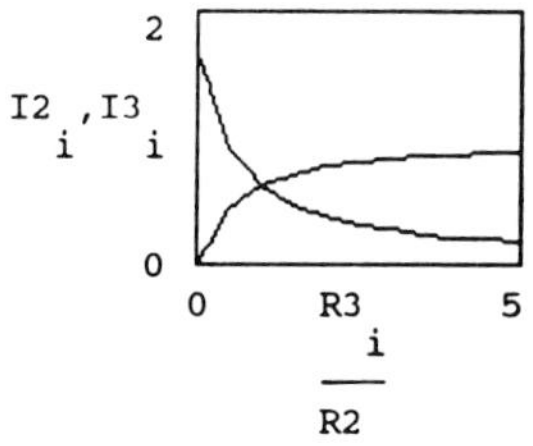

Figure 16.5 CIRC2, voltage divider with load.

If $R_2 = R_1$ and $R_3 = \infty$, what is v_{out}?

If $R_2 = R_1$ and $R_3 = 0$, what is v_{out}?

In the second part of the document, the values of I_2, I_3, and v_{out} are determined as a function of R_3. Is the total current constant? Plot the total current supplied by the voltage source as a function of R_3 and explain the curve, including the value at the intercept.

If $R_1 = 1\Omega$, how do the curves change in terms of qualitative shape and overall magnitude?

Note that v_{out} is defined as $I_3 \cdot R_3$. Why not $I_2 \cdot R_2$?

What effect will there be on the curves if the multiplier of $(10 - i)$ in R_3 is changed to 2 or 5?

Make a summary statement of the effect of R_3 on the divider circuit.

16.1.2 Wheatstone Bridge

The Wheatstone bridge has many applications. Here we want to find the value of an unknown resistance by a comparison/balancing process with known resistances. The unknown is placed in a circuit with known resistances. The known values are adjusted until a null — a zero for a particular current — is achieved; then the unknown matches the known value.

The circuit is two two-resistor voltage dividers connected to the same voltage source (see Fig. 16.6). The outputs of the two dividers are connected to each other through a low-resistance galvanometer, a device able to indicate the presence of small currents. Moving-coil galvanometers typically have resistances in the range of 25 to 500 Ω and sensitivities (a kind of inverse dispersion specification) range from 0.2 to 0.0001 $\mu A/$ mm.

• •   Load CIRC3, Wheatstone bridge (see Fig. 16.7).

The four resistors in the voltage dividers of the bridge circuit, R_1 through R_4, have as their base value R. The galvanometer resistance is represented by R_5. For the time being, it is irrelevant which resistor is the unknown. What we want to gain is some sense of the current in R_5 as the bridge is unbalanced, and also, given a particular galvanometer sensitivity, how accurately an unknown can be determined.

A solve block is set up with six equations and six unknowns: three voltage loop equations and three current node equations. The find statement is in functional form. A specification of the resistance, R_3, returns the currents in each of the resistors and the total current. The specific R in the $f(R)$ expression can be changed to any one of the five resistors.

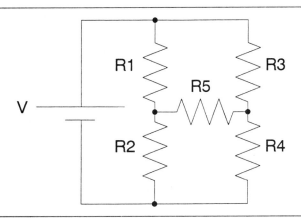

Figure 16.6 Wheatstone bridge circuit.

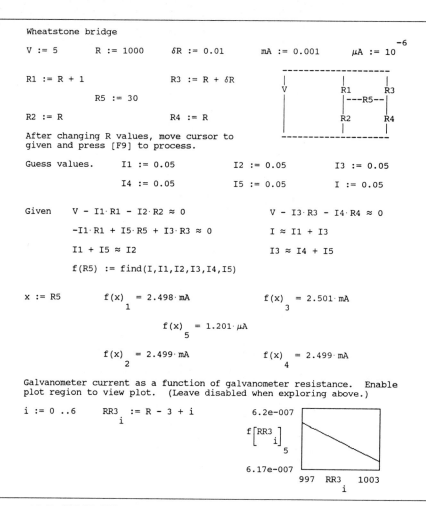

Figure 16.7 CIRC3, Wheatstone bridge.

RRx_i represents a series of values to be substituted in f. Leave the plot region disabled when you are using the first section of the document.

In the plot region, R_3 goes from $R - 3$ to $R + 3$ in steps of one. The plot is of I_5 vs. R_3. What is the slope, the ratio of galvanometer current per change in R_3, in $\mu A/\Omega$?

How much does the galvanometer's resistance affect the ability to balance the circuit? Change the find statement to $f(R_5)$. Let RR take on the values from 20 to 420 in steps of 40. Plot the current in the galvanometer vs. the RR resistance values. For small changes in R_3, is the response symmetric around the balance value?

Let $R_3 = R - \delta R$ and repeat. What happens to the current values?

Let $R_5 = 0\,\Omega$. Keep the ratio $R/\delta R = 10^5$. How does the current I_5, $f(0)_5$, change as R takes on the values 10^n where $n = 0, 1, \ldots, 4$? (Remember, keep the ratio constant.) Repeat with $R_5 = 100\,\Omega$.

A standard way of balancing the bridge is with $R_1 = R_3$ and $R_2 = R_4$, but it is not necessary that $R_1 = R_2$. How does the sensitivity change if $R_2 = 5 \cdot R_1$ and $R_3 = R + \delta R$?

How does the sensitivity change if $R_1 : R_2 = R_3 : R_4$ and $R_3 = 10\,R_1$? Examine for the case where R_1 varies from its balance value and for the case where R_3 varies from its equilibrium value. Let the deviations cover the same percent change.

16.2   Capacitors

A capacitor is a device capable of storing separated charge. The relation

$$Q = CV \tag{16.5}$$

indicates that the stored charge, $\pm Q$, stored is proportional to the applied voltage; the capacitance, C, is the constant of proportionality. The unit of capacitance is the farad, F.

The net capacitance of capacitors in parallel is the sum of the individual capacitances:

$$C_P = \sum C_i. \tag{16.6}$$

The net capacitance of capacitors in series is the reciprocal of the sum of reciprocals (as for resistors in parallel):

$$C_S = \left(\sum C_i^{-1} \right)^{-1}. \tag{16.7}$$

• •   Write functions for the combinations: $C_P(C_1, C_2)$ and $C_S(C_1, C_2)$. Test them.

When voltage is applied to a dc circuit containing a capacitor, current flows, charging the capacitor. After some time, the voltage across the capacitor approaches the applied voltage and no further current flows. Capacitors block constant current; transients such as charging currents are not considered to be dc, even though they may come from a dc source.

In circuits with resistors and capacitors, we frequently need to know the time that it takes for a capacitor to charge or discharge. Consider the following circuit. An uncharged capacitor is in series with a voltage source, a resistor, and an open switch. When the switch is closed, the circuit is complete and current begins to flow. The current is given by

$$i = \frac{V}{R} e^{-t/RC}. \tag{16.8}$$

As the capacitor charges, the current decreases exponentially. The charge on the capacitor, initially zero, builds in time:

$$q = C V (1 - e^{-t/RC}). \tag{16.9}$$

When $t = 0$, $i = V/R = i_o$; when $t = RC$, $i = i_o \cdot 1/e$. The product RC is known as the time constant. Verify that the units of RC are time.

After one time constant, the current is reduced to $1/e$ of its maximum value at $t = 0$.

When $t = 0$, the charge $q = 0$. After one time constant,

$$q = C V \left(1 - \frac{1}{e}\right) = 0.632 \, C V.$$

• •   Let $V = 10$, $R = 5 \cdot 10^5$, and $C = 10 \cdot 10^{-6}$. At $t = 0$ a switch is closed, completing the circuit composed of a battery, capacitor, and resistor in series. The capacitor is initially uncharged. Plot the current as a function of time, $i(t)$. Define $i'(t) = d/dt \, i(t)$. How does $i'(0) \cdot RC$ compare with v/R? Interpret this. (Remember, RC has units of time.)

Plot the charge on the capacitor as a function of time. Define $q'(t) = d/dt \, q(t)$. How does $q'(0)$ compare with $i(0)$? How does $q' RC$ compare with CV? Interpret.

When a capacitor discharges, both the charge and current decrease:

$$q(t) = C V e^{-t/RC} \qquad i(t) = \frac{V}{R} e^{-t/RC}. \tag{16.10}$$

• •   Plot these curves on a semilog plot. How many time constants elapse as q falls from its maximum value at $t = 0$ to $1/2$, $1/4$, $1/10$ of the maximum?

16.3 Diodes

When a diode is on, its dynamic resistance, $\Delta V/\Delta I$, is, roughly, constant and relatively small. When a diode is off, its resistance is comparatively very large. The diode is a nonlinear device, in contrast to the resistor. The solid-state diodes, to which we refer, have two terminals, which are called the cathode and anode, for historical reasons. The diode is on if the anode is approximately 0.7 V more positive than the cathode. We consider our ideal diode to be either on or off. Because of the change in resistance with the change in direction of the applied voltage, the diode is frequently used as a rectifier, permitting current to flow in one direction but not the other.

Consider a circuit comprised of voltage source, diode, and resistor in series (see Fig. 16.8). The anode of the diode is connected to the positive terminal of the voltage source. If the source voltage, v_{in}, is greater than 0.7 V, the diode is on and current flows. If v_{in} is less than 0.7 V, the diode is off and no current flows. When there is no current, $iR = 0$ and $v_{\text{out}} = 0$. When $v_{\text{in}} > 0.7$ V, the voltage across the resistor is $v_{\text{in}} - 0.7$. So we can write

$$v_{\text{out}i} = \text{if}(v_{\text{in}i} > 0.7, v_{\text{in}i} - 0.7, 0).$$

The voltage across the diode itself, $v_{\text{dio}i} = v_{\text{in}i} - v_{\text{out}i}$. If $v_{\text{out}} = 0$, $v_{\text{dio}} = v_{\text{in}}$; that is, the entire voltage appears across the diode.

• •   For $v_{\text{in}i} = 3\sin(\omega t)$, sketch two cycles of the input and output voltages. Sketch two cycles of $v_{\text{out}i}$ and $v_{\text{dio}i}$ together.

• •   Load CIRC4, simple diode circuit: half-wave rectifier (see Fig. 16.9).

An input signal is specified. An if statement defines the output, the voltage across the resistor, of this diode circuit. The voltage across the diode is specified in terms of the input voltage and the voltage across the resistor.

In the first plot region, the input and output voltages are shown together. Why is the amplitude of the v_{out} curve always less than that of v_{in}?

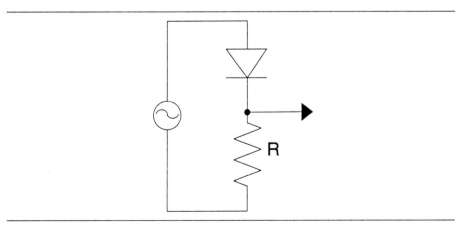

Figure 16.8 Half-wave rectifier circuit.

Circuit consists of ac source, diode, and resistor in series.   The cathode
of the diode is connected to the resistor.

```
------------              vout is the output voltage relative to ground.
|          |
Vac        diode          vdio is the voltage across the diode
|          |
|          |-----vout
|          |
|          R
|          |
----grnd----
```

$n := 60 \qquad i := 0 \,..n \qquad f := 2 \qquad T := \dfrac{1}{f} \qquad \delta t := \left[\dfrac{2 \cdot T}{n}\right] \qquad t_i := i \cdot \delta t$

$vin_i := 3 \cdot \sin\left[2 \cdot \pi \cdot f \cdot t_i\right]$

$vout_i := if\left[vin_i > 0.7, vin_i - 0.7, 0\right] \qquad\qquad vdio_i := vin_i - vout_i$

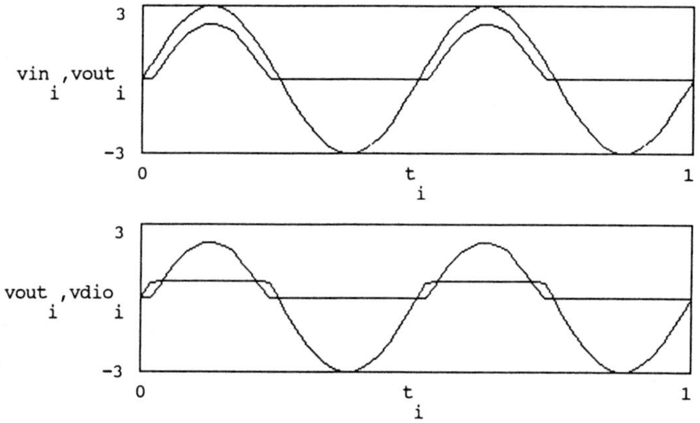

Figure 16.9 CIRC4, simple diode circuit: half-wave rectifier.

Why is the v_{out} curve positive or zero, but never negative?

Explain the shape of the curve showing the voltage across the diode. Is there a portion of the curve for which v_{dio} and v_{out} are identical? Explain.

Examine the shapes of the v_{out_i} and v_{dio_i} curves. What will the sum of $v_{out_i} + v_{dio_i}$ look like? Verify your answer.

Predict how the plot will change if $v_{in_i} = 1 \cdot \sin(2\,\pi\,f\,t_i)$. Change the amplitude and verify your answer. What if the amplitude is 15?

This circuit is known as a half-wave rectifier; the output follows the input during half of the voltage cycle.

•• A full-wave rectifier can be constructed with four diodes (see Fig. 16.10). When $v_{in} > 1.4\,\text{V}$, both diodes D_1 and D_2 will turn on and current

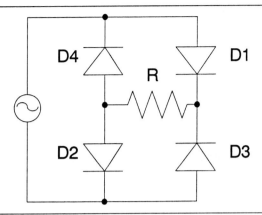

Figure 16.10 Full-wave rectifier: diode bridge.

will flow through R. When $v_{\text{in}} < -1.4\,\text{V}$, both diodes D_3 and D_4 will turn on and current will flow through R in the *same* direction as it did when $v_{\text{in}} > 1.4\,\text{V}$.

Change the v_{out} statement in CIRC4 so that it describes a full-wave rectifier. Two conditions are to be specified; therefore, two if statements (or a nested pair of if statements) are needed. The statement

$$v_{\text{out}i} = \text{if}(v_{\text{in}i} > 1.4, v_{\text{in}i} - 1.4, 0)$$

specifies one condition. Fill in the missing values for the second if statement:

$$v_{\text{out}i} = \text{if}(v_{\text{in}i} \quad , \quad , v_{\text{out}i}).$$

The second condition overrules part, but not all, of the specification of the first statement. Combine these two if statements into a single nested pair of if statements. Write these conditions with one if statement using the magnitude of v_{in}, $|v_{\text{in}i}|$.

For small signals, the loss in signal amplitude resulting from the voltage drop across the diode(s) may be unacceptable. In this case, the use of additional circuitry is required. An operational amplifier may be included to create a precision rectifier. This circuit compensates for the diode drop. Rectification by the diode still occurs, but the apparent voltage drop due to the diode becomes approximately zero instead of the 0.7 V that we see here.

If we add a capacitor to the original one-diode half-wave rectifier circuit, we can radically change the output of the circuit. Let the capacitor be in parallel with the resistor (see Fig. 16.11).

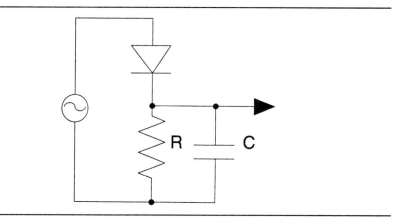

Figure 16.11 Half-wave rectifier with capacitor.

When the diode is on, the capacitor voltage rapidly follows the voltage across the resistor/capacitor pair, $(v_{\text{in}i} - v_{\text{dio}i})$. The capacitor voltage changes promptly because the resistance through which it must charge is small. However, when the diode is off, the capacitor can discharge through R. (It cannot discharge through the diode, because when the diode is off its resistance is very large.)

For an input voltage source that varies sinusoidally with time, an ac source, the capacitor charges during one half (approximately) the cycle and discharges in the other half. If the time constant RC is short compared to the period of oscillation of the voltage source, the capacitor discharges almost completely in each cycle. However, if the time constant is on the order of or several times greater than the period of oscillation of the input voltage source, then the capacitor discharges only a small fraction of its total charge before it is charged again.

The if statement for such a condition is slightly more complex than the previous case. The voltages at the two terminals of the diode are $v_{\text{in}j}$ and $v_{\text{c}j}$ (v_{c} and v_{out} are identical). If $v_{\text{in}j} - v_{\text{c}j} > 0.7$ V, the diode is on, and the voltage across the capacitor is the voltage applied. Otherwise, the capacitor discharges through the resistor R and the voltage decreases as

$$V = V_o\, e^{-t/RC}. \tag{16.11}$$

However, because we are iterating this procedure in small steps it is convenient to approximate the exponential with

$$e^{-t/RC} \approx 1 - \frac{\delta t}{RC}. \tag{16.12}$$

• •   Load CIRC5, half-wave rectifier plus capacitor; a simple dc power supply (see Fig. 16.12).

The first if statement reflects the discussion above. Either the diode is on and the voltage across the capacitor, which is the output voltage, follows the applied voltage or the diode is off and the capacitor discharges through the resistor.

Circuit consists of ac source, diode, and a parallel resistor and capacitor in series with the diode.

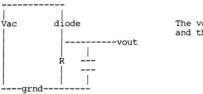

The voltage across the capacitor, vc, and the output voltage, vout, are identical.

$n := 100$ $i := 0 ..n$ $f := 60$ $T := \dfrac{1}{f}$ $\delta t := \dfrac{2 \cdot T}{n}$ $t_i := i \cdot \delta t$

$vin_i := 3 \cdot \sin\left[2 \cdot \pi \cdot f \cdot t_i\right]$ $j := 0 ..n - 1$ $vc_0 := 0$

$R := 2 \cdot 10^3$ $C := 10^{-5}$ $R \cdot C = 0.02$ $T = 0.017$ $R \cdot C = 1.2 \cdot T$

$$vc_{j+1} := if\left[vin_j - .7 > vc_j, vin_j - .7, vc_j \cdot \left[1 - \frac{\delta t}{R \cdot C}\right]\right]$$

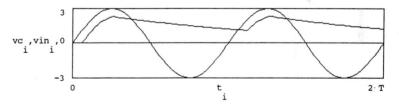

The output of a four diode bridge circuit with a capacitor in parallel with the resistor. Charging occurs twice as frequently.

$$vc_{j+1} := if\left[\left|vin_j\right| - 1.4 > vc_j, \left|vin_j\right| - 1.4, vc_j \cdot \left[1 - \frac{\delta t}{R \cdot C}\right]\right]$$

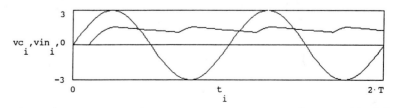

Figure 16.12 CIRC5, half-wave rectifier plus capacitor; a simple dc power supply.

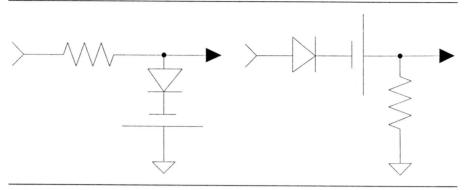

Figure 16.13 Diode clamps.

The initial values for R ($2\,\mathrm{k}\Omega$) and C ($10\,\mu\mathrm{F}$) result in a time constant slightly longer than the period of oscillation associated with the input signal. (The place marker is useful in expressing RC in terms of T.)

If R is increased from $2\,\mathrm{k}\Omega$ to $8\,\mathrm{k}\Omega$, how will the output curve change? Observe the output curve for a variety of R and C values and input voltages.

If $RC \approx$ several T, the output voltage is nearly constant. What is the approximate value of the dc output voltage when $R = 2\,\mathrm{k}\Omega$? When $R = 8\,\mathrm{k}\Omega$? How much current is drawn through the $8\,\mathrm{k}\Omega$ load resistor?

The second if statement specifies the output voltage for a four-diode full-wave bridge rectifier. In this case, charging occurs twice as frequently and the decay period is cut in half; thus the output is somewhat smoother.

For the case of the load resistor, R, having the value $2\,\mathrm{k}\Omega$, compare the "ripple" — that is, the difference between the maximum and minimum output voltages — for the two circuits?

• •   Write appropriate if statements and show outputs for the diode circuits in Fig. 16.13. For each of these circuits, there are four permutations — two orientations for the diodes and two orientations for the battery.

16.4   Tunnel Diode Amplifier

The tunnel diode has the unusual property of local negative resistance over part of its operating range. Making use of this feature, an amplifier circuit can be created by assembling a voltage divider made up of a tunnel diode and a resistor (see Fig. 16.14).

Negative resistance means that for some range of voltages for an increase in voltage there is a corresponding decrease in current. This condition occurs for the tunnel diode over a small part of its operating range. To achieve amplification, the diode must be biased, made to operate, at least in part, in the negative resistance range of the tunnel diode (the bias sets the voltage point at which the diode operates when the signal voltage is zero).

● ●   Load CIRC6, tunnel diode amplifier (see Figs. 16.15 and 16.16).

First recall the behavior of a normal resistor ($R = V/I = \delta V/\delta I$). In a plot of current vs. voltage, the slope of the curve ($\delta I/\delta V$) is the inverse of the resistance. If we were to specify $I(V)$, then the derivative of that function would be the inverse of the resistance, $dI/dV = \mathrm{inv}R(V)$, and $R(V) = \mathrm{inv}R(V)^{-1}$.

These operations are performed in the first section of the document. The current, the inverse of the resistance, and the resistance are plotted as a function of voltage. When the resistance of a resistor is evaluated in this way, it is a constant, identical to the originally specified value even though it is obtained here through the differentiation process.

If the slope of the curve increases, does this indicate an increase or decrease in resistance?

The same procedure is then followed for the tunnel diode. A prototypical characteristic curve, $i(v)$, a cubic expression, is shown. The derivative

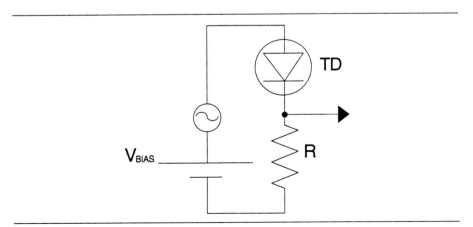

Figure 16.14 Tunnel diode amplifier.

Tunnel Diode - characteristics.

Typical ohmic resistor - current voltage characteristics. Resistance
determined from the inverse of the slope.

R := 2   I := 1   V := 0,0.2 ..5

$$I(V) := \frac{V}{R} \qquad invR(V) := \frac{d}{dV} I(V)$$

$$R(V) := invR(V)^{-1}$$

invR(1) = 0.5   R(1) = 2

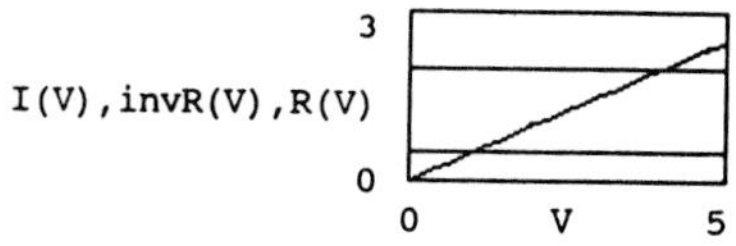

I(V),invR(V),R(V)

v := 0,0.1 ..3

Tunnel diode - current voltage characteristics and effective resistance

$$i(v) := v^3 - 4 \cdot v^2 + 4.2 \cdot v \qquad invr(v) := \frac{d}{dv} i(v) \qquad r(v) := invr(v)^{-1}$$

V := 1   a := root(invr(V),V)   a = 0.719

V := 2   b := root(invr(V),V)   b = 1.948

V := 1.5 $c := root\left[\frac{d}{dV} invr(V),V\right]$ c = 1.333

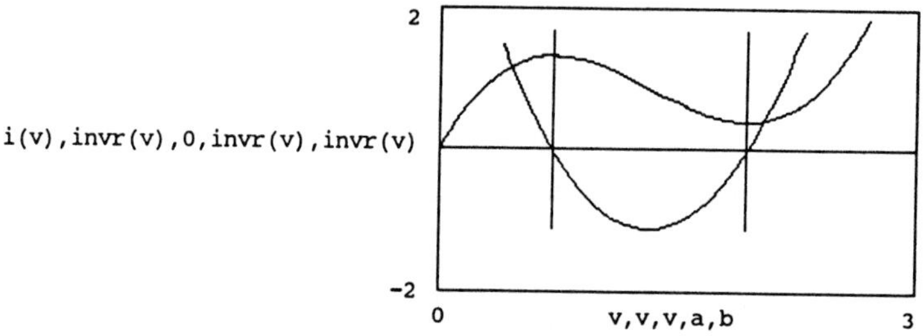

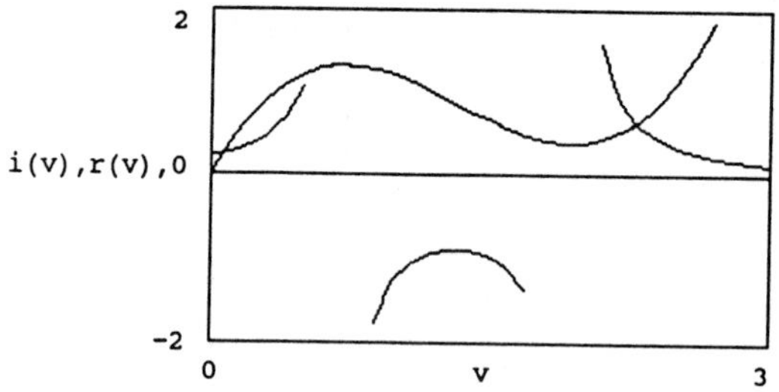

i(v),invr(v),0,invr(v),invr(v)

v,v,v,a,b

i(v),r(v),0

v

Figure 16.15 CIRC6, tunnel diode amplifier. (See the next figure for the rest of the document.)

yields the inverse of the dynamic resistance, invr(v); the dynamic resistance is determined by taking the inverse.

To determine the range over which the resistance is negative, we determine those voltage values where the slope of the characteristic curve, $i(v)$,

Tunnel diode amplifier

Input signal - dc component $v_{dc} := 1$

ac component $\omega := 5000$ $T := 2 \cdot \pi \cdot \omega^{-1}$ $\delta t := T \cdot .05$ $t := 0, \delta t \ .. 2 \cdot T$

$$v_{sig}(t) := 0.1 \cdot \sin(\omega \cdot t)$$

Input signal. $vin(t) := v_{dc} + v_{sig}(t)$ $R := 0.8$

Output signal.   Form is the same as that of a two resistance voltage divider.

$$vout(t) := \frac{R}{R + r(vin(t))} \cdot vin(t)$$

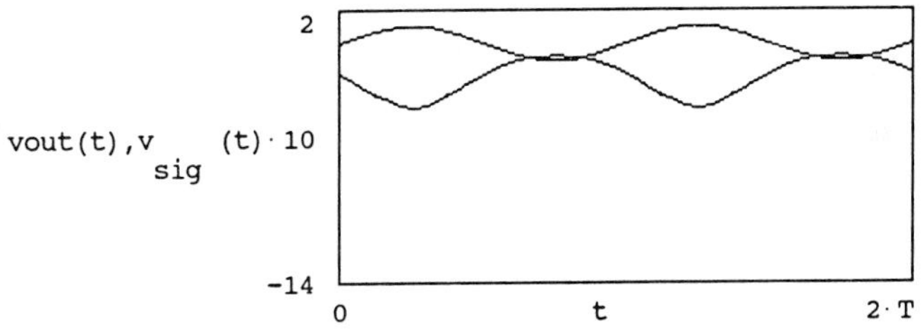

$vout(t), v_{sig}(t) \cdot 10$

Figure 16.16 CIRC6 *continued*

is zero. These points mark the transition between positive and negative resistance. We also calculate the voltage at which the negative resistance is a maximum. Keep these three values — a, b, and c — in mind when changing the bias voltage. How does the resistance vary as the voltage increases from a and b? From b and c?

We plot the characteristic curve, and the derivative, $invr(v)$. The edges of the negative resistance region are marked. Examine the slope of the characteristic curve and compare it with the plot of the derivative. Note where the $invr(v)$ curve is positive and where it is negative; these also indicate the same areas of positive and negative resistance (taking an inverse does not change the sign).

In the second plot region, the $i(v)$ curve is plotted again, together with $r(v)$. Examine the resistance curve in detail. Explain the shape of the curve, qualitatively.

So far, this is all characteristics (just the properties of the tunnel diode), no amplifier circuit. Now construct a voltage divider circuit with the tunnel diode in place of one of the resistors. Include as the voltage source both an adjustable dc voltage supply (the bias voltage) and an ac signal source.

According to our study of voltage dividers, the output from this voltage divider circuit is, as we have seen,

$$v_{out} = \frac{R}{R + r_{td}} \cdot v_{in}, \tag{16.13}$$

where the tunnel diode resistance has been substituted for one of the resistances. The input signal is the sum of the dc bias voltage and the ac signal voltage. If

$$v_{dc} + |v_{sig}| < a \qquad \text{or if} \qquad v_{dc} - |v_{sig}| > b,$$

then the negative resistance region is avoided and there is no amplification. However, if

$$v_{dc} + |v_{sig}| < b \qquad \text{and} \qquad v_{dc} - |v_{sig}| > a,$$

then the circuit is operating entirely within the negative resistance region and amplification will occur. The operating range need only be partly within the negative resistance range for some amplification to occur. As the resistance is dependent on voltage, so is the amplification. Consequently, signals at different voltages will be amplified to different degrees and distortion will occur.

Let the signal voltage be relatively small, with an amplitude of 0.1 V, for example. Let the bias voltage take on a sequence of values. The voltage applied to the diode ranges between the values $v_{dc} \pm |v_{sig}|$. Observe the change in character of the output voltage as seen in the sequence of plots as v_{dc} starts at 0.5 V and increases in steps of 0.1 V. Keep in mind the values a, b, and c and the tunnel diode resistance as these changes occur. Cover the voltage range from a to c. Adjust the multiplier of v_{sig}, in the plot region, so that the plotted amplitudes of the input and output signals are roughly the same.

At what point is the gain of the amplifier (ratio of output and input signals) the greatest? When the voltage b is within the extremes of the input signal, why does the output appear as it does? Compare output signals when the maximum of the input is just less than b, and when the minimum of the input is just greater than b. Explain the difference.

Unfortunately, tunnel diodes are difficult to use and consequently are not common circuit elements. However, the concepts of characteristic curve, bias voltage, amplification, and distortion are general.

> Back off man, I'm a scientist.
>
> *Ghostbusters*

C H A P T E R

17

Optics

In this chapter, we take a close look at Snell's law of refraction, consider refraction from the point of view of Fermat's principle of least time, and consider the case where the index of refraction varies continuously within a medium. Because the law of refraction informs us of angles but does not give us any information about the relative intensities of the refracted and reflected beams, we take a quick look at the Fresnel equations.

We describe refraction in terms of light rays that travel in straight lines within a uniform medium. This description is legitimate if the smallest dimension of an object presented to the incident radiation is very much greater than the wavelength of that radiation. If this condition is fulfilled, diffraction effects can be ignored.

We then examine the interference of electromagnetic radiation in Young's two-slit experiment. The summing of phasors is considered, as is the problem of interference from a series of slits. The problem of single-slit diffraction is treated, and finally the ability to resolve source is considered. The Rayleigh criterion is explored. Radiation patterns from several sources close together are presented in exercises dealing with subtracting lines from a composite.

17.1   Refraction

In section 2.4, on curve fitting, we examined some data associated with incident and refracted light passing from air to water. Comparing the data with Snell's law, we obtained a best fit value for the index of refraction. Snell's law can be stated as

$$n_1 \sin(\theta_1) = n_2 \sin(\theta_2), \tag{17.1}$$

where the angles θ_1 and θ_2 are measured from the normal to the interface between the two media characterized by their indices of refraction, n_1 and n_2. Expressed in terms of velocities, the index of refraction is

$$n = \frac{c}{v}, \tag{17.2}$$

where c is the speed of light in a vacuum and v is the velocity of light in the medium with index n. The law is sometimes specified with one index, for example, n_{21}, which is the index of refraction of medium 2 relative to medium 1.

$$n_{21} = \frac{n_2}{n_1} = \frac{v_1}{v_2}. \tag{17.3}$$

Typical values of n range from 1 to 1.7.

• •   Plot the velocity of light in a medium vs. n, as n ranges over the values suggested. By what percent does the velocity change for the values of n considered?

• •   Load OPT1, Snell's law of refraction (see Figs. 17.1 and 17.2).

Snell's law is straightforward, yet important enough that it should be explored. In the document, the x-axis is the interface between two different optical media. The index of refraction is n_1 when $y > 0$ and n_2 when $y < 0$. The angle of incidence is θ_1 and the angle of refraction is θ_2. Both angles are measured from the normal.

Enter values for n_1 and n_2 . Observe the value for θ_{1MAX}. Enter a value for θ_1 where $|\theta_1| < |\theta_{1MAX}|$. This restriction depends on the critical angle, which we will discuss later.

Of all the code after the assignment of n_1, n_2, and θ_1, the calculation of θ_2 is pertinent; θ_2 is the refracted angle. The purpose of the rest of the code is to set up the details of the figure — the incident and refracted rays and the arcs indicating the incident and refracted angles. Some scaling is set up so that the rays will have appropriate relative lengths. Limits for the plot region are also specified so that the ratio of height to width remains constant.

Snell's law.

Enter values for n1, n2, and θ1, the angle of incidence.
The magnitude of θ1 cannot be greater than the magnitude of θ1_MAX.

n1 ≡ 1   n2 ≡ 1.6

$$\theta1\_MAX \equiv if\left[n1 > n2, asin\left[\frac{n2}{n1}\right], \frac{\pi}{2}\right]$$ rad ≡ 1

θ1_MAX = 90· deg $deg \equiv \frac{\pi}{180} \cdot rad$

θ1 ≡ 45· deg |θ1| < |θ1_MAX| $x_1 \equiv tan(\theta1)$ $y_1 \equiv 1$ $y_3 \equiv -y_1$

$$\theta2 \equiv asin\left[\frac{n1}{n2} \cdot sin(\theta1)\right]$$ $x_3 \equiv \left|y_3\right| \cdot tan(|\theta2|) \cdot \frac{-x_1}{|x_1|}$ $f \equiv 1$

θ2 = 26.228· deg $i \equiv 1 \ldots 3$

given $\left[x_3^2 + y_3^2\right]^{0.5} \cdot f \approx \left[x_1^2 + y_1^2\right]^{0.5}$ f := find(f)   f = 1.269

                                                                          $x_3 := f \cdot x_3$ $y_3 := f \cdot y_3$

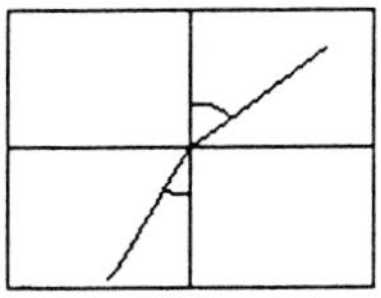

mx

$y_i, s_j, u_j, 0$

−mx

          −mx x_i, r_j, t_j, x_i mx

More information needed to draw the figure.

$$mxx \equiv if\left[\left[\left|x_1\right| > \left|x_3\right|\right], \left|x_1\right|, \left|x_3\right|\right]$$ $mx \equiv if\left[mxx > y_1, mxx, y_1\right] \cdot 1.35$

m ≡ 5   j ≡ 0 ..m

          $step1 \equiv \frac{\theta1}{m}$ $step2 \equiv \frac{\theta2}{m}$

                                                          step2 = 0.092

$$\phi_j \equiv \left[angle\left[x_1, y_1\right] + step1 \cdot j\right]$$ $\tau_j \equiv \frac{3 \cdot \pi}{2} - step2 \cdot j$ $R \equiv \frac{mx}{3}$

$r_j \equiv R \cdot cos\left[\phi_j\right]$ $s_j \equiv R \cdot sin\left[\phi_j\right]$ $t_j \equiv R \cdot cos\left[\tau_j\right]$ $u_j \equiv R \cdot sin\left[\tau_j\right]$

Figure 17.1 OPT1, Snell's law of refraction. (See the next figure for the rest of the document.)

Sometimes, it is possible to bring only the most pertinent regions to the beginning of a document and place all the computation at the end. To do this, you must make use of the global equality. However, you cannot make a solve block global.

It may be convenient to move the θ_1 region near the plot region; the global equality permits this. It may also be desirable to move the n_1 and n_2 regions. Change the assignment to global if this is done.

Try a series of angles, large and small; consider different refraction indices, including the cases of $n_1 > n_2$ and $n_2 > n_1$.

The refracted angle in terms of the incident angle.   The derivative points out how rapidly the refracted angle changes compared to the incident angle.

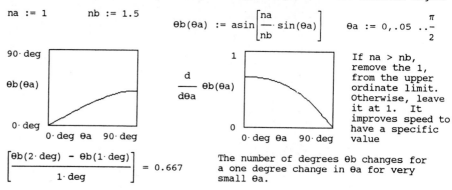

$$\theta b(\theta a) := asin\left[\frac{na}{nb} \cdot sin(\theta a)\right]$$

na := 1   nb := 1.5

$\theta a := 0, .05 \ .. \dfrac{\pi}{2}$

If na > nb, remove the 1, from the upper ordinate limit. Otherwise, leave it at 1.   It improves speed to have a specific value

$$\left[\frac{\theta b(2 \cdot deg) - \theta b(1 \cdot deg)}{1 \cdot deg}\right] = 0.667$$

The number of degrees θb changes for a one degree change in θa for very small θa.

Figure 17.2 OPT1 *continued.*

If medium 2 is glass with $n_2 = 1.5$ and if medium 1 is air with $n_1 = 1$, observe the case of $\theta_1 = 45°$. Now pour water, $n = 1.33$, on top of the glass so that the air is replaced with water. With θ_1 the same, how does θ_2 change?

With $n_1 = 1.48$ and $n_2 = 1.5$, examine the case for $\theta_1 = 10°$, $40°$, $70°$, and $89°$. (We will want to recall this result later in the chapter.)

A change in θ_1, $\delta\theta_1$, results in a change in θ_2 of $\delta\theta_2$. $\delta\theta_2/\delta\theta_1$ depends on the value of θ_1. For the case $n_1 < n_2$, does $\delta\theta_2/\delta\theta_1$ increase or decrease as θ_1 increases from smaller to larger angles?

Now move to the remainder of the document. We refer to media 1 and 2 as a and b just to keep the names distinct. (This is not strictly necessary unless, after this portion of the document, you wished to refer to the quantities in the first part of the document.)

Snell's law is written in functional form. Indices and a range of incident angles are specified. The refracted angle is plotted as a function of the incident angle. The derivative of this function, the change in the refracted angle relative to the incident angle, is also plotted vs. the incident angle.

As θ_a goes from $0°$ to $90°$, we see at once the range of angles spanned by θ_b. It is also interesting that the slope of this curve is greatest for small angles.

Let $n_a = 1.49$ and $n_b = 1.5$. Examine the same curves.

For a ray originating in the medium of larger index of refraction, the angle of refraction is larger than the angle of incidence. If the angle of refraction is $90°$, then the incident angle is the critical angle.

Define the critical angle $\theta_c(n_a, n_b)$. Let $n_a = 1$ and let n_b take on a range of values from 1.1 to 1.9 in steps of 0.1. Plot the critical angle as a function of the index of refraction n_b. Let $n_a = 1.33$ and $n_b = 1.4, \ldots, 1.9$. Plot $\theta_c(n_a, n_b)$ vs. n_b.

17.1.1   Fermat's Principle of Least Time

Fermat's principle of least time states that the path taken by a beam of light between any pair of points is that path which takes the least time.

• •   Load OPT2, Fermat: least time (see Fig. 17.3).

To test this conjecture, two points, (x_1, y_1) and (x_2, y_2), are selected on either side of an interface. A series of x-values intermediate between the values of x_1 and x_2 are selected along the interface, and the time is determined from point 1 to the particular point on the interface, $t_1(x)$, and from that point to point 2, $t_2(x)$. The two times are summed, $t(x)$. Plots of these times vs. x are shown. The position for which the minimum time occurs is determined using the root/derivative procedure. Finally, the time corresponding to that x-location for the minimum is determined.

To verify that the location is in correspondence with Snell's law, we determine the angles of incidence and refraction, θ_1 and θ_2, in terms of the x- and y-values of the three points involved. Values of $n_1 \cdot \sin(\theta_1)$ and $n_2 \cdot \sin(\theta_2)$ are then compared. (Given the number of different numerical processes, the agreement is quite good.)

Explain how the angles θ_1 and θ_2 are determined.

Finally, we plot the path for the minimum time.

Try a series of values for x_1 and x_2 ($x_1 \neq x_2$), and for n_1 and n_2. (As the problem is set up, $y_1 > 0$ and $y_2 < 0$.)

If n_1 is fixed and n_2 is increased, how will the point on the interface move? Explain why in terms of velocities.

• •   Perform a similar calculation to show that the reflection process is also one of least time. Show that the least time is consistent with the law of reflection, where the angle of incidence equals the angle of reflection. Within one medium, is minimum distance an equally good rule?

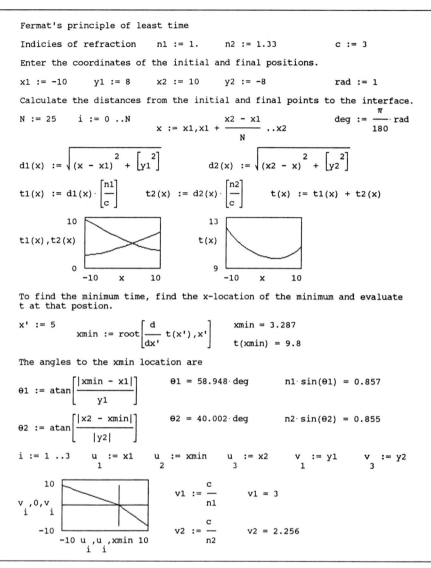

Fermat's principle of least time

Indicies of refraction   n1 := 1.   n2 := 1.33   c := 3

Enter the coordinates of the initial and final positions.

x1 := -10   y1 := 8   x2 := 10   y2 := -8   rad := 1

Calculate the distances from the initial and final points to the interface.

N := 25   i := 0 ..N

$$x := x1, x1 + \frac{x2 - x1}{N} \ ..x2 \qquad deg := \frac{\pi}{180} \cdot rad$$

$$d1(x) := \sqrt{(x - x1)^2 + \left[y1\right]^2} \qquad d2(x) := \sqrt{(x2 - x)^2 + \left[y2\right]^2}$$

$$t1(x) := d1(x) \cdot \left[\frac{n1}{c}\right] \qquad t2(x) := d2(x) \cdot \left[\frac{n2}{c}\right] \qquad t(x) := t1(x) + t2(x)$$

t1(x),t2(x)   (graph, vertical axis 0 to 10, horizontal axis x from -10 to 10)

t(x)   (graph, vertical axis 9 to 13, horizontal axis x from -10 to 10)

To find the minimum time, find the x-location of the minimum and evaluate t at that postion.

$$x' := 5 \qquad xmin := root\left[\frac{d}{dx'} \ t(x'), x'\right] \qquad xmin = 3.287$$
$$t(xmin) = 9.8$$

The angles to the xmin location are

$$\theta1 := atan\left[\frac{|xmin - x1|}{y1}\right] \qquad \theta1 = 58.948 \cdot deg \qquad n1 \cdot sin(\theta1) = 0.857$$

$$\theta2 := atan\left[\frac{|x2 - xmin|}{|y2|}\right] \qquad \theta2 = 40.002 \cdot deg \qquad n2 \cdot sin(\theta2) = 0.855$$

i := 1 ..3 u_1 := x1 u_2 := xmin u_3 := x2 v_1 := y1 v_3 := y2

$v_i, 0, v_i$ (graph, vertical axis -10 to 10, horizontal axis from -10 to 10, labeled u_i, u_i, xmin)

$$v1 := \frac{c}{n1} \qquad v1 = 3$$

$$v2 := \frac{c}{n2} \qquad v2 = 2.256$$

Figure 17.3 OPT2, Fermat: least time.

As stated, Fermat's principle is not strictly correct. The path is not necessarily a minimum. It is an extremum and has a stationary value. It could be a maximum, a minimum, or even a point of inflection where the tangent is horizontal.

For example, locate at the foci of an elliptical mirror the two points between which the rays travel; all paths are equal in distance and time. Adjacent points contribute similarly to the reflection. It is also possible to construct a mirror with such curvature that the point of reflection is a maximum time.

Fermat's principle is similar to the concept of least action in mechanics. Fermat's principle can be written

$$\delta \int \frac{ds}{v}. \tag{17.4}$$

This is an extremum. The extremum of the integral of kinetic energy minus potential energy can be reduced to

$$\delta \int v \, ds. \tag{17.5}$$

This, too, is an extremum.

17.1.2   *Refraction in an Inhomogeneous Medium*

In inhomogeneous media, the index of refraction of a substance may change gradually with position. Optical examples where such a change can be observed include the "wet road" effect and rays from the sun at sunset. Earthquake waves are subject to similar behavior.

- • Load OPT3, refraction in an inhomogeneous medium (see Fig. 17.4).

Consider a medium in which the index of refraction decreases with depth. We approximate this medium with a series of slabs: The index of refraction is constant within one slab; the index varies from one slab to the next. The index, in this case, is described by

$$n(y) = n_0 + \delta n \cdot \frac{y}{d}, \tag{17.6}$$

where y is incremented in steps of d, the thickness of the slabs of constant index.

A ray is directed downward into the medium; the incident angle is θ_1. Snell's law is applied at the interface between successive slabs. Recall from the OPT1 study of Snell's law that when the indices are close (for example, 1.48 and 1.5), the bending at most angles is slight. As the index decreases with depth, the refraction angle increases. The refracted angle at one interface is the incident angle at the next. Thus, as the ray descends, the incident angle continue to increase.

This computation process stops at the critical angle. Snell's law states that $\sin(\theta_2) = (n_1/n_2) \cdot \sin(\theta_1)$. When $(n_1/n_2) \cdot \sin(\theta_1) \geq 1$, or, for the until statement, when $1 - n_1/n_2 \sin(\theta_1)$ goes negative, the critical angle has been reached and the process stops.

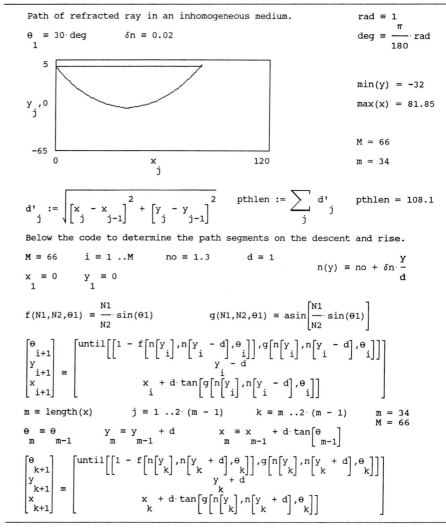

Path of refracted ray in an inhomogeneous medium.

$rad \equiv 1$

$\theta_1 \equiv 30 \cdot deg$ $\delta n \equiv 0.02$

$deg \equiv \dfrac{\pi}{180} \cdot rad$

$min(y) = -32$

$max(x) = 81.85$

$M = 66$

$m = 34$

$$d'_j := \sqrt{\left[x_j - x_{j-1}\right]^2 + \left[y_j - y_{j-1}\right]^2}$$ $pthlen := \sum_j d'_j$ $pthlen = 108.1$

Below the code to determine the path segments on the descent and rise.

$M \equiv 66$ $i \equiv 1 .. M$ $no \equiv 1.3$ $d \equiv 1$

$x_1 \equiv 0$ $y_1 \equiv 0$

$n(y) \equiv no + \delta n \cdot \dfrac{y}{d}$

$f(N1,N2,\theta 1) \equiv \dfrac{N1}{N2} \cdot sin(\theta 1)$ $g(N1,N2,\theta 1) \equiv asin\left[\dfrac{N1}{N2} \cdot sin(\theta 1)\right]$

$$\begin{bmatrix} \theta_{i+1} \\ y_{i+1} \\ x_{i+1} \end{bmatrix} \equiv \begin{bmatrix} until\left[\left[1 - f\left[n\left[y_i\right],n\left[y_i - d\right],\theta_i\right]\right],g\left[n\left[y_i\right],n\left[y_i - d\right],\theta_i\right]\right] \\ y_i - d \\ x_i + d \cdot tan\left[g\left[n\left[y_i\right],n\left[y_i - d\right],\theta_i\right]\right] \end{bmatrix}$$

$m \equiv length(x)$ $j \equiv 1 .. 2 \cdot (m - 1)$ $k \equiv m .. 2 \cdot (m - 1)$ $m = 34$
$M = 66$

$\theta_m \equiv \theta_{m-1}$ $y_m \equiv y_{m-1} + d$ $x_m \equiv x_{m-1} + d \cdot tan\left[\theta_{m-1}\right]$

$$\begin{bmatrix} \theta_{k+1} \\ y_{k+1} \\ x_{k+1} \end{bmatrix} \equiv \begin{bmatrix} until\left[\left[1 - f\left[n\left[y_k\right],n\left[y_k + d\right],\theta_k\right]\right],g\left[n\left[y_k\right],n\left[y_k + d\right],\theta_k\right]\right] \\ y_k + d \\ x_k + d \cdot tan\left[g\left[n\left[y_k\right],n\left[y_k + d\right],\theta_k\right]\right] \end{bmatrix}$$

Figure 17.4 OPT3, refraction in an inhomogeneous medium.

At this point the ray is reflected, the angle ($\theta_m = \theta_{m-1}$) and then the process reverses, and the ray returns to the surface. At each step x is incremented by

$$\delta x = d \cdot \tan(refraction\ angle).$$

The incident angle, θ_1, can take on values between 1° and 75°. The iteration range M should be greater than or equal to the number of steps m to reach the critical angle, otherwise reflection will occur before the critical angle is reached.

Process the document.

Plot $n(y_i)$ vs. y_i. Use plot type s. Is the radius of curvature of the path smaller in regions of larger n or smaller n?

At what angle does the ray penetrate the deepest?

At what angle does the ray traverse the largest distance in the x-direction? Explore this region carefully (use steps of $1°$); there is a surprise here.

At what angle is the path length the ray follows the greatest?

17.2   Fresnel Equations: Intensity of Reflected and Refracted Rays

A ray incident on an interface will be partly reflected and partly transmitted. The angle of the transmitted ray is given by Snell's law. The amplitudes of the reflected and transmitted rays are given by the Fresnel equations. The intensities are related to the squares of the amplitudes. These equations can be derived from Maxwell's equations. The equations themselves are not difficult, but they are sufficiently complicated that their form is not immediately obvious.

The equations for the reflection and transmission coefficient are

$$R_p = \left(\frac{(n_1 \cos(\theta_1) - n_2 \cos(\theta_2))}{(n_1 \cos(\theta_1) + n_2 \cos(\theta_2))} \right)^2 \tag{17.7}$$

$$T_p = \frac{n_2 \cos(\theta_2)}{n_1 \cos(\theta_1)} \left(\frac{2 n_1 \cos(\theta_1)}{(n_1 \cos(\theta_1) + n_2 \cos(\theta_2))} \right)^2 \tag{17.8}$$

$$R_l = \left(\frac{(n_2 \cos(\theta_1) - n_1 \cos(\theta_2))}{(n_1 \cos(\theta_2) + n_2 \cos(\theta_2))} \right)^2 \tag{17.9}$$

$$T_l = \frac{n_2 \cos(\theta_2)}{n_1 \cos(\theta_1)} \left(\frac{2 n_1 \cos(\theta_1)}{(n_1 \cos(\theta_2) + n_2 \cos(\theta_2))} \right)^2 , \tag{17.10}$$

where p and l refer to the perpendicular and parallel polarizations, respectively, of the electric field vector; vectors that represent the incident, reflected, and transmitted rays lie in a plane — parallel polarization, for example, means that the electric field vector is parallel to this plane.

• •   Load OPT4, Fresnel equations (see Fig. 17.5).

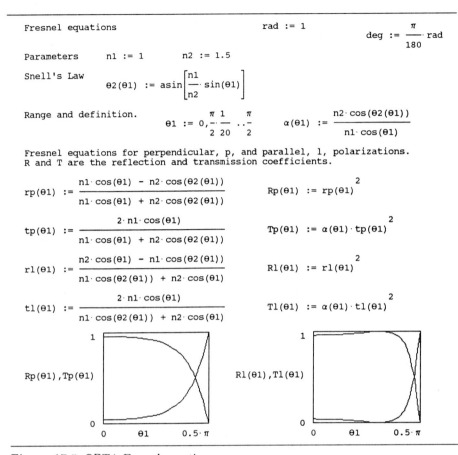

Fresnel equations   rad := 1

$$deg := \frac{\pi}{180} \cdot rad$$

Parameters   n1 := 1   n2 := 1.5

Snell's Law
$$\theta2(\theta1) := asin\left[\frac{n1}{n2} \cdot sin(\theta1)\right]$$

Range and definition.
$$\theta1 := 0, \frac{\pi}{2} \cdot \frac{1}{20} \ .. \frac{\pi}{2} \qquad \alpha(\theta1) := \frac{n2 \cdot cos(\theta2(\theta1))}{n1 \cdot cos(\theta1)}$$

Fresnel equations for perpendicular, p, and parallel, l, polarizations.
R and T are the reflection and transmission coefficients.

$$rp(\theta1) := \frac{n1 \cdot cos(\theta1) - n2 \cdot cos(\theta2(\theta1))}{n1 \cdot cos(\theta1) + n2 \cdot cos(\theta2(\theta1))} \qquad Rp(\theta1) := rp(\theta1)^2$$

$$tp(\theta1) := \frac{2 \cdot n1 \cdot cos(\theta1)}{n1 \cdot cos(\theta1) + n2 \cdot cos(\theta2(\theta1))} \qquad Tp(\theta1) := \alpha(\theta1) \cdot tp(\theta1)^2$$

$$rl(\theta1) := \frac{n2 \cdot cos(\theta1) - n1 \cdot cos(\theta2(\theta1))}{n1 \cdot cos(\theta2(\theta1)) + n2 \cdot cos(\theta1)} \qquad Rl(\theta1) := rl(\theta1)^2$$

$$tl(\theta1) := \frac{2 \cdot n1 \cdot cos(\theta1)}{n1 \cdot cos(\theta2(\theta1)) + n2 \cdot cos(\theta1)} \qquad Tl(\theta1) := \alpha(\theta1) \cdot tl(\theta1)^2$$

Figure 17.5 OPT4, Fresnel equations.

Here, θ_2 is the refracted angle as determined by Snell's law, in terms of the incident angle. The reflection, R, and transmission, T, coefficients indicate the fraction of the incident energy that is either reflected or transmitted.

Snell's law permits the expression of θ_2 in terms of θ_1. The equations are written as functions of θ_1. The perpendicular and parallel cases are plotted separately. Examine the curves for various values of n.

In the case of R_l, the reflection coefficient is zero at one point. This angle is Brewster's angle. Determine Brewster's angle using the root/derivative method.

When θ_1 equals Brewster's angle, what is θ_2? At this angle, what is the sum of θ_1 and θ_2? (Polarized sunglasses are especially useful near this angle, because they are designed to absorb the perpendicular polarization.)

Find the angle at which the reflected and transmitted rays carry the same energy. Do this for both the parallel and perpendicular polarizations. Examine the dependence on n.

For unpolarized incident radiation, the total reflection coefficient, R, and the total transmission coefficient, T, are the averages of the perpendicular and parallel coefficients. Define and plot these coefficients R and T.

How do the R and T curves change as n_2 takes on smaller and larger values (for example, $n = 1.2$ or $n = 1.9$)?

Plot the amplitudes r_p, t_p vs. θ_1 and r_l, t_l vs. θ_1.

For an air-glass ($n = 1.5$) interface and unpolarized incident radiation, for what range of incident angles in air does the refracted ray have 80% or more of the incident energy? 70%? 50%?

17.3 Interference

Radiation from a coherent source (for example, laser light) is incident upon a pair of rectangular slits. The radiation is diffracted as it passes through each slit. In effect, each slit acts as a source of coherent radiation. Radiation from the slits that arrives at a particular point interferes. The electric field vectors of the two waves superpose. That is, the net electric field at some point, P, is the sum of the individual electric fields, which in general have a nonzero phase difference between them. At point P, the electric fields from the two slits can be written

$$E_1 = E_0 \sin(\omega t + \phi_1) \qquad \text{and} \qquad E_2 = E_0 \sin(\omega t + \phi_2).$$

The phase difference between the two waves depends on the path difference between the slits and the point P. Let the distance between slit 1 and point P be S_{1P} and the distance between slit 2 and point P be S_{2P}. The path difference, $\delta p = S_{1P} - S_{2P}$, is related to the phase difference $\delta\phi$ between the two waves as

$$\frac{\delta\phi}{2\pi} = \frac{\delta p}{\lambda}. \tag{17.11}$$

If $\delta p / \lambda$ has an integral value, then there is constructive interference. If $\delta p / \lambda$ has an odd half integral value, then there is destructive interference.

The standard arrangement for Young's two-slit experiment are: (1) The planes containing the slits and the viewing screen are parallel; (2) The slit separation is d; the distance between the planes containing the slits and the screen is D; $D \gg d \gg \lambda$; (3) In the viewing plane, distances are measured along a line perpendicular to the orientation of the slits; (4) The observation distance is y; the observation angle θ is equal to $\tan^{-1}(y/D)$.

Given these conditions, the path difference, δp, can be written

$$\delta p = d\sin(\theta),$$

and the interference conditions for constructive interference can be written

$$\frac{\delta p}{\lambda} = \frac{d\sin(\theta)}{\lambda} = m \qquad (17.12)$$

where $m = 0, 1, 2, \ldots$. For destructive interference we have

$$\frac{\delta p}{\lambda} = \frac{d\sin(\theta)}{\lambda} = m + \frac{1}{2} \qquad (17.13)$$

where $m = 0, 1, 2, \ldots$.

•• In a carefully laid out Young's two-slit experiment, a peak can be located to the nearest 0.3 mm. In this case, $\lambda = 5 \cdot 10^{-7}$ m, $d = 10^{-4}$ m, and $D = 4$ m. (If you wish to use units for this example, define the meter to be mt; m is the traditional symbol for the order of interference, and n is tied up as the index of refraction.)

Define $y(m)$ to be the y-location on the screen of the maxima, associated with constructive interference, of order m. Take into account both the sine and tangent functions as needed; do not use the small-angle approximation. Define a second function $y'(m)$ locating the maxima as a function of m, only in this case use the small-angle approximation.

Plot $y(m)$ vs. m and plot $y(m) - y'(m)$ vs. m as $m = 1, \ldots, 30$. At what value of m is the difference between $y(m)$ and $y'(m)$ equal to 0.3 mm? What is the spacing between successive orders for small m? How does this spacing depend on d? If d were reduced to $5 \cdot 10^{-5}$ m (in MathCAD use mt), would the difference between $y(m)$ and $y'(m)$ be apparent at lower, higher, or unchanged values of m?

•• Using the equations for E_1 and E_2 at the beginning of this section, define the terms and plot a few cycles of $E_1 + E_2$. Let $\phi_1 = 0$. Let ϕ_2 successively take on the values 0, π, 2π. Which phase angles correspond to constructive and destructive interference?

Knowing that

$$\sin(\alpha) + \sin(\beta) = 2\sin\left(\frac{\alpha + \beta}{2}\right)\cos\left(\frac{\alpha - \beta}{2}\right), \qquad (17.14)$$

show that $E_1 + E_2$ can be written

$$2\,E_0 \cos\left(\frac{\phi_1 - \phi_2}{2}\right)\sin\left(\omega t + \frac{\phi_1 + \phi_2}{2}\right). \qquad (17.15)$$

What are the amplitude and phase angle of this sum? Plot this function and verify that this form and the form plotted by adding E_1 and E_2 directly are equivalent.

With a larger number of similar vectors, it is still possible to write the sum as the product of a phase-related term and a time-dependent term.

To calculate the intensities, it is necessary to know whether the radiation is coherent or incoherent. For coherent radiation, the vector sum of the individual amplitudes is determined and this resultant amplitude is squared. This value is proportional to the intensity. Because of phase differences, the intensity at a particular point can be less than that produced by any of the individual sources.

For incoherent radiation, the individual amplitudes are squared. These squared amplitudes are then added; this sum is proportional to the intensity. In this case, the intensity at a point cannot be less than that produced by any individual source.

• • Plot the intensity associated with E_1, with E_2, and with the sum $E_1 + E_2$. For the case $\phi_1 = 0$ and $\phi_2 = \pi$, how does the intensity of E_1 or E_2 compare with the intensity of the sum? Where is the energy?

When we observe the interference pattern from the two-slit experiment, for example, we observe the intensity not as a function of time but as a function of position. Our eyes, and most other light-sensitive devices, are incapable of observing the time variation of rapidly fluctuating signals — they detect an average.

• • Verify that the average of $\sin^2(\theta)$ over one cycle is $1/2$. Let MathCAD perform the integration for you.

The intensity is proportional to the square of the amplitude, so we have

$$I \propto (2\,E_0)^2 \cos^2\left(\frac{\phi_1 - \phi_2}{2}\right)\sin^2\left(\omega t + \frac{\phi_1 + \phi_2}{2}\right)$$

$$I_{av} = I_0 \cos^2\left(\frac{\phi_1 - \phi_2}{2}\right) \qquad \text{and} \qquad I_0 \propto (2\,E_0)^2.$$

The phase difference is related to the path difference by

$$\delta\phi = \frac{2\pi}{\lambda} d \sin(\theta), \tag{17.16}$$

so we can rewrite

$$I_{av} = I_0 \cos^2\left(\frac{\pi d \sin(\theta)}{\lambda}\right). \tag{17.17}$$

It is frequently convenient to define an angle α:

$$\alpha = \frac{\pi d \sin(\theta)}{\lambda} = \frac{kd}{2} \sin(\theta)$$

and express the average intensity as

$$I_{av} = I_0 \cos^2(\alpha). \tag{17.18}$$

• •   Plot I_{av} vs. $d\sin(\theta)$ as $d\sin(\theta)$ goes from $-2.5\,\lambda$ to $2.5\,\lambda$. (Remember this example when we approach this problem in the next chapter using the fast Fourier transform.)

17.3.1   *Phasors*

Visualize E_1 in terms of a rotating vector. A vector of magnitude E_0 rotates counterclockwise about the origin. The angle between the positive x-axis and the vector is $\theta_p = \omega t + \phi$. The projection of this phasor, this rotating vector, on the y-axis (the vector's y-component) represents E_1. This is consistent with our description $E_1 = E_0 \sin(\omega t + \phi_1)$. We can represent E_2 similarly. The phase difference between the two vectors, $\phi_2 - \phi_1$, is independent of time. Consequently, as the two vectors rotate, they are always separated by the same angle, their phase difference. If we form the vector sum of E_1 and E_2 at $t = 0$ and allow the resultant to rotate, the y-projection is the same as the sum of the projections of the two individual vectors.

Associated with the two-slit experiment would be two phasors. For a series of n slits, there would be n phasors. If the slits are uniformly spaced, the phase difference between successive phasors would be the same. We consider the case of uniformly spaced slits (constant phase difference).

• •   Load OPT5, phasor sums (see Fig. 17.6).

The number of phasors, N, is, for example, equal to the number of rectangular slits illuminated by the source. We assume that the phasors are all of the same magnitude. The phase difference between successive phasors is ϕ.

Calculate the amplitude for a given number of phasors at a specified phase angle.

(N and θ are entered near the plot.)

The number of phasors to be summed. $n := 1 \, ..N$

Enter the phase angle. $rad \equiv 1$

$$deg \equiv \frac{\pi}{180} \cdot rad$$

$$X_0 := 0 \qquad Y_0 := 0$$

Calculate the components. $r := 1$

$$x_n := r \cdot \cos((n-1) \cdot \phi) \qquad y_n := r \cdot \sin((n-1) \cdot \phi)$$

Calculate the location of the ends of each successive phasor.

$$X_n := X_{n-1} + x_n \qquad Y_n := Y_{n-1} + y_n \qquad m := 0 \, ..N$$

To draw the resultant, draw from the beginning to the end.

$$k := 0 \, ..1 \qquad RX_1 := X_N \qquad RY_1 := Y_N$$

$$RM := \left[RX_1^2 + RY_1^2 \right]^{0.5} \qquad mx := if(\max(X) > \max(Y), \max(X), \max(Y)) \cdot 1.1$$

Make the plot square.

$N \equiv 3 \qquad \phi \equiv 40 \cdot deg \qquad RM = 2.532$

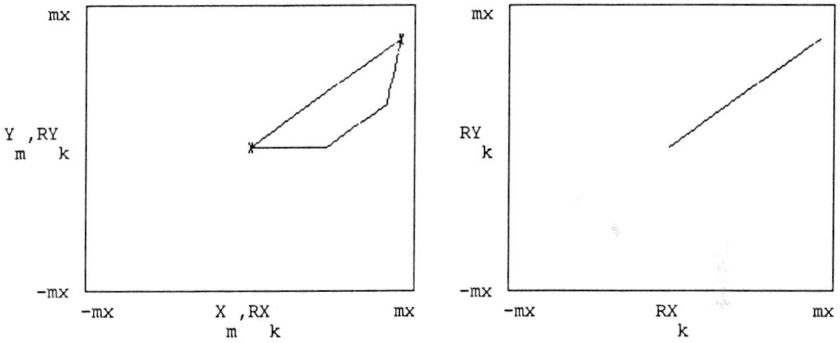

Figure 17.6 OPT5, phasor sums.

Let $N = 2$. Let ϕ take on the values 0°, 45°, 90°, 135°, 180°. Notice that the magnitude of the resultant RM changes as the phase angle changes. Where is RM a maximum? A minimum? Let ϕ equal 225° and 360°. Make a rough plot, by hand, of the amplitude RM vs. ϕ (save this for comparison when looking at OPT6). The intensity is proportional to the square of the resultant.

Let $N = 3$. What is the maximum amplitude? At what phase angle does the first minimum occur? Demonstrate. At what phase angle will the next maximum occur? What is the amplitude of this maximum as compared to the maximum at $\phi = 0$? Record this value.

Find the next minimum and next maximum. What are the phase angles?

Let $N = 10$. Let $\phi = 5°$. What is the maximum possible amplitude with $N = 10$? At what phase angle will the first minimum occur? Demonstrate. Where will the next maximum occur? What is its amplitude? Record the value. Verify that you have the correct angle by increasing and decreasing the phase angle in steps of 1°; watch RM for the value of the magnitude. Is the maximum exactly where you thought it would be? Now find the last maximum before $\phi = 360°$. Record the magnitude. Let $\phi = 180°$. Compare the magnitudes of the maxima.

• •   Load OPT6, amplitude vs. angle for an N-phasor system (see Fig. 17.7).

In OPT6, phasors are summed at a series of angles. The phasors are not shown, only the sums. This document is computation intensive, so be patient with large values of N.

Let $N = 2$ (near plot region with global equality). Observe the plot structure and compare it with the hand plot you made for $N = 2$ in OPT5. The peaks are of equal amplitude and of magnitude N.

Let $N = 3$. At what angles do the maxima occur? What is the magnitude of the maximum occurring at $\phi = 180°$? Refer to your notes from OPT5 for comparison.

Where do the minima occur? Express the phase difference for destructive interference between adjacent slits in terms of π and N, the total number of slits.

Let $N = 4$. How many subsidiary maxima are there? How many minima are there between successive principal maxima?

Let $N = 10$. (This takes a minute but it's worth it.) Notice how the subsidiary maxima lean to the "outside." (Recall that in the previous exercise you were asked if the maxima were exactly where you thought they would be.) The maxima are not precisely centered between the minima. Also note that the maxima are not of the same magnitude. Compare with your values from the previous exercise. Examine the plot with the ordinate as a log scale. Why does the plot appear in separate segments?

We note that as N increases, the peaks with phase angle $2\pi n$ dominate. The maximum amplitude of the peaks increases (proportional to N), and the peaks become narrower.

Calculate the amplitude for multi-slit interference.

The number of phasors to be summed. Insert any value
between 2 and 20.   (Too large values will run subscripts
out of bounds.)

$M := 6 \cdot N$ The number of angles at which the sum is to be evaluated.
(If the calculation is too slow, reduce the numerical factor
to 5 or 4.)

$$i := 1 ..N \qquad j := 0 ..M$$

$$\theta_j := \frac{j}{M} \cdot 2 \cdot \pi$$

Calculate the x and y components of each phasor. $r := 1$

$$x_{i,j} := r \cdot \cos\left[(i - 1) \cdot \theta_j \right] \qquad y_{i,j} := r \cdot \sin\left[(i - 1) \cdot \theta_j \right]$$

$$X_{0,0} := 0 \qquad Y_{0,0} := 0$$ For the plot, choose a reference
point; here, start at the origin.

Determine the x and y components of the sums.

$$X_{i,j} := X_{(i-1),j} + x_{i,j} \qquad Y_{i,j} := Y_{(i-1),j} + y_{i,j}$$

Calculate the amplitude at each angle.

$$R_j := \left[\left[X_{i,j} - X_{0,0} \right]^2 + \left[Y_{i,j} - Y_{0,0} \right]^2 \right]^{0.5}$$

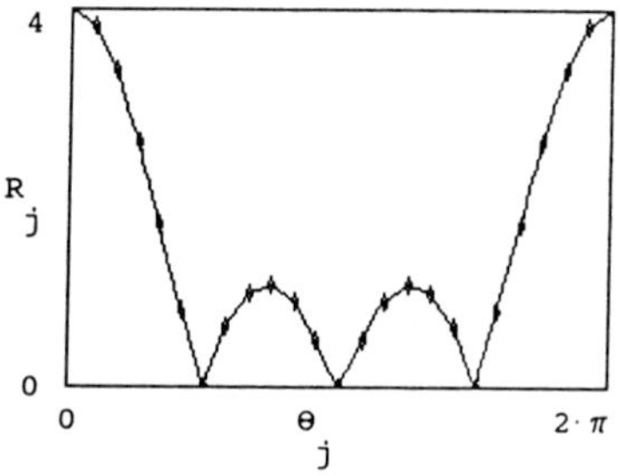

$N \equiv 4$

Figure 17.7 OPT6, amplitude vs. angle for an N-phasor system.

For each of the cases just considered, for $N = 2, 3, 4$, and 10, at what observation angle does the principal maximum where $\phi = 2\pi$ occur? For the same cases, what is the angular width of each of these maxima?

• •   Go back to OPT5. What if the intensities from the sources are not equal? Make the necessary changes in the document to assign different amplitudes (r) to the different phasors.

For $N = 2$, try the ratio 2:1. How are the maxima and minima affected?

Try a number of examples.

What physical conditions might be associated with unequal intensities?

• • Make similar adjustments to OPT6. Let $N = 2$. Consider the ratio 2:1. Compare with the previous example using phasors.

Order does make a difference. Let $N = 3$. Let the intensities take on the values (0.6, 1.2, 1.4), (0.6, 1.4, 1.2), and (1.2, 0.6, 1.4).

Let $N = 4$ and consider the four cases (0.6, 1.2, 1.4, 0.6), (1.2, 0.6, 0.6, 1.4), (1.2, 0.6, 1.4, 0.6), and (1.2, 1.4, 0.6, 0.6).

17.4   Diffraction

In considering the process of diffraction from a single rectangular slit, imagine that the slit is subdivided into a series of even narrower rectangles. Let each rectangle, each section of the slit opening, act as a source of waves (Huygens wavelets). For this configuration of sources, minima occur when

$$a \sin(\theta) = m\,\lambda \qquad m = 1,\, 2,\, 3 \ldots, \qquad (17.19)$$

where a is the slit width.

We can approach this relation as a limiting case of interference. For multislit interference, $\delta p = d \sin(\theta)$, where δp is the path difference between adjacent slits and $\delta \phi$ is the corresponding phase difference. If we consider this set of slits to be points within our diffraction slit, then we have

$$d = \frac{a}{N} \qquad \text{and} \qquad \delta p = \frac{a}{N} \sin(\theta)$$

where a is the slit width, d is the separation between adjacent source points within the slit, and N is the number of source points.

For destructive interference, the phase difference is π radians for every $N/(2\,m)$ source points. Between adjacent source points the phase difference is

$$\delta \phi = \frac{\pi}{N/(2\,m)}. \qquad (17.20)$$

Substituting in our relation for path difference and phase difference, we find

$$\frac{a}{N} \sin(\theta) \frac{1}{\lambda} = \frac{\pi}{(N/2\,m)} \frac{1}{2\,\pi}$$

so that the condition for single-slit diffraction minima is given by

$$a \sin(\theta) = m\,\lambda \qquad m = 1,\, 2,\, 3 \ldots.$$

In terms of wavelength, the phase difference between adjacent "points" within a single slit can be written as

$$\delta\phi = \frac{2\,\pi}{\lambda}\,\delta p = k\,d\sin(\theta).$$

In expressing the intensity, one-half the total phase difference from one end of the slit to the other, β, is a useful quantity:

$$\beta = \frac{N\,\delta\phi}{2} = \frac{N\,k\,d\sin(\theta)}{2} = \frac{k\,a}{2}\sin(\theta). \tag{17.21}$$

It can be shown that the total electric field, E, is proportional to $\sin(\beta)/\,\beta$. The intensity as a function of the angle θ is given by

$$I(\theta) = I_0\left(\frac{\sin(\beta)}{\beta}\right)^2 \qquad \text{where} \qquad \beta = \beta(\theta). \tag{17.22}$$

• •   Plot $I(\theta)/\,I_0$ vs. β as β goes from $-\,3\,\pi$ to $3\,\pi$. (Remember this when we approach the same problem in the next chapter using the fast Fourier transform.)

The minima occur when $\sin(\beta) = 0$ or

$$\beta = m\,\pi \qquad \text{where} \qquad m = 1, 2, 3, \ldots.$$

These minima are the same as those expressed by equation 17.19. The maxima occur where

$$\frac{dI}{d\beta} = 0.$$

• •   Show that

$$\frac{dI}{d\beta} = I_0\,\frac{2\sin(\beta)\,(\beta\cos(\beta) - \sin(\beta))}{\beta^3}.$$

If $dI/\,d\beta = 0$, then $(\beta\cos(\beta) - \sin(\beta)) = 0$ or $\tan(\beta) = \beta$. Values of β that solve this equation give the locations of the maxima.

Determine the location of the maxima of the E-field. Is the relationship for these maxima the same as the location of maxima as determined from intensity expressions?

• •   Plot $\tan(\beta)$, β vs. β as β goes from 0 to $5\pi/\,2$. In plotting $\tan(\beta)$, select β values judiciously so as to avoid calculations at $\beta = (\pi/\,2)\cdot(2\,m+1)$. Also, it may be useful to limit the ordinate range. The intersection points of the curves $\tan(\beta)$ and β are the solutions. Use a given-find procedure to find the first two nontrivial solutions to the equation $\tan(\beta) = \beta$. Find the corresponding θ.

In Young's two-slit experiment, the intensity at any point depends on both diffraction and interference effects:

$$I_{\text{diff}} \propto \frac{\sin^2(\beta)}{\beta^2} \qquad I_{\text{int}} \propto \cos^2(\alpha).$$

Thus

$$I \propto \frac{\sin^2(\beta)}{\beta^2} \cdot \cos^2(\alpha),$$

where

$$\alpha = \frac{k\,d}{2}\sin(\theta) \qquad \text{and} \qquad \beta = \frac{k\,a}{2}\sin(\theta).$$

• • Plot I vs. θ as θ goes from $\sin^{-1}(-3\lambda/a)$ to $\sin^{-1}(3\lambda/a)$. Let the constant of proportionality be one. (See Chapter 18.)

• • Examine the intensity pattern for the four-slit case. Include both interference and diffraction effects. Plot the total intensity as a function of angle.

• • For a circular aperture of radius r (as opposed to a rectangular slit or aperture) the intensity is given by

$$I \propto \left(\frac{J_1(k\,r\sin(\theta))}{k\,r\sin(\theta)} \right)^2.$$

Plot I vs. $k\,r\sin(\theta)$ as $k\,r\sin(\theta)$ takes on values from -10 to 10. Verify that

$$\theta_{\min} = \tan^{-1}\left(1.22\,\lambda/2\,r\right) \simeq 1.22\frac{\lambda}{2\,r}. \qquad (17.23)$$

The Bessel functions are standard functions inMathCAD, just as are sine and cosine.

If sources are close together, the diffraction patterns resulting from the passage of light through an aperture may overlap. A recording device — for example, the eye — sees the sum of the intensities. (There is no coherence between the radiation from the separate sources; thus we sum the intensities.) In observing a light pattern, it may not be clear how many sources are involved and what their relative strengths may be. One criterion for being able to distinguish between two images, two central maxima, is known as Rayleigh's criterion. The concept is that two sources of equal intensity are resolved if the central maximum of one pattern falls at the first minimum of the other. Another way to express this criterion is to give the intensity amplitude at the dip between the two central maxima.

The first minimum of a slit diffraction pattern occurs at the angle

$$\theta = \frac{\lambda}{a} \qquad \text{(in the small-angle approximation)}.$$

The diffraction patterns from the two sources separated as specified by the Rayleigh criterion have equal intensities at half this angle. Substituting in the equation for β this angle for θ we obtain

$$\beta = \frac{k\,a}{2}\sin(\theta) = \frac{2\pi\,a}{\lambda}\frac{\lambda}{2}\frac{}{2\,a} = \frac{\pi}{2}.$$

The intensity of each pattern at $\theta = \lambda/2\,a$ is

$$I \propto \frac{\sin^2(\beta)}{\beta^2} = \frac{1}{(\pi/2)^2} = \frac{4}{\pi^2}.$$

The total intensity is double this value because there are equal contributions from the two sources,

$$\frac{I}{I_0} = \frac{8}{\pi^2}. \qquad (17.24)$$

Thus the Rayleigh criterion indicates a dip between maxima of approximately 20% to say that the peaks are resolved. Clearly this choice is somewhat arbitrary. Other criteria can be defined.

• •   Suggest a different set of criteria for the resolution of two diffraction patterns of equal intensity. Test your suggestion.

• •   Load OPT7, single-slit diffraction (see Figs. 17.8 and 17.9).

In OPT7, we examine the sums of diffraction patterns and consider how well they are resolved. The amplitude is plotted instead of intensity in some cases because the magnitudes of the subsidiary intensity maxima are too small in linear plots. Recall that the maxima and minima are in the same location in both amplitude and intensity plots. We plot I vs. θ three times: linear, linear with the ordinate range reduced, and semi-log. We also show y vs. I.

The intensity patterns from two single-slit sources of width a, separated by a distance d, are $I_1(y)$ and $I_2(y)$. (The sources are at $\pm d/2$. Note that y, not θ, is the argument.) There is no coherence between the radiation from the two sources.)

When d is of the order of the slit separation in the two-slit interference experiment (10^{-4}) the central maxima are not resolved. We could not easily distinguish between the case where there are two sources of amplitude one very close together and the case where there is one source of amplitude two. Test this. Try several different values of slit separation.

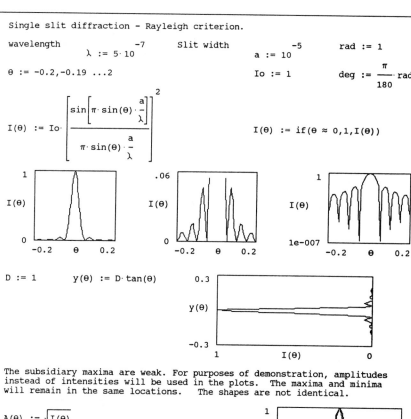

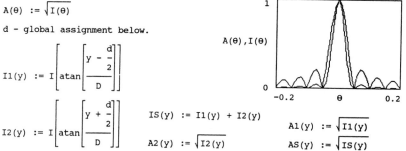

Figure 17.8 OPT7, single-slit diffraction. (See the next figure for the rest of the document.)

In the final set of three plots, we see the individual amplitudes, the total intensity, and the intensity of one source in both linear and semi-log plots.

Process with $d \equiv 10^{-2} \cdot (1, 3, 4, 5, 6,$ and perhaps $10)$. Observe the location of the first minimum in relation to the central maximum and the size, if any, of the dip between the maxima.

When d is of the order of the slit separation in the two-slit interference
experiment (10^-4), the peaks are not distinguishable. From the graph, one
could not distinguish, for example, between two sources of amplitude one very
close to each other and one source of amplitude two. Change the d value.

y := -0.15,-0.145 ..0.15   d ≡ 10^{-2} ·1

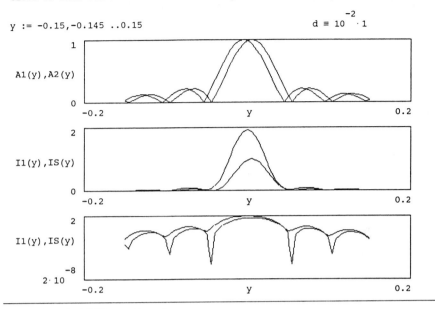

Figure 17.9 OPT7 *continued.*

As you go through the sequence observe how I_1 fits into the sum in
both the linear and log-log plots. At what point does it seem reasonable
to think that there is more than one source? At what point are the two
sources resolved according to the Rayleigh criterion?

With $d = 4 \cdot 10^{-2}$, observe the sum when $I_2 = I_2 \cdot 1.2$ and $I_2 = I_2 \cdot 5$.
The specification of resolution becomes more problematic as the difference
in intensities is taken into account.

• •   Three intensity functions UI_1, UI_2, and UI_3 are given in OPT8.
Your task is to determine the intensity and location of the sources that
make up the composite line. Two of the composites contain two lines; one
contains three. Slit widths are equal; intensities need not be. I_0 and c are
the parameters to determine.

It is suggested that you define a function with specific values of I_0
and c, for example, $I_a(y) = I$(your value for amplitude, your value for
shift $- c$, y). Plot the unknown and your fitted line in the same region.
Guided by the experience you gained from OPT7, adjust your I_0 and c
values to correspond to one component of a composite line. When satisfied
with the adjustment of the parameters, plot $UI(y) - I_a(y)$ vs. y. If the

unknown is composed of two lines and if $UI - I_a$ looks like a single line shape then the parameters of the first line were well chosen. Next define $I_b(y) = UI(y) - I_a(y)$ and then find the parameters associated with I_b by fitting another line shape to it. If there are three lines, the procedure will have to be repeated. Amplitudes, in this example, can range between 0 and 2; c can be plus or minus; I_0 and c values are all even tenths.

I've seen things that you people wouldn't believe.
Attack ships on fire off the shoulder of Orion.
I've watched c-beams glitter in the dark near the Tannhäuser gate.
All those moments will be lost in time ... like tears in rain.

Blade Runner

C H A P T E R

18

Fast Fourier Transform

When a signal, such as the representation of a portion of a song on an oscilloscope screen, is expressed in terms of a Fourier series, the components are still expressed in the time domain. A general expression of a Fourier series is

$$f(t) \sim \frac{1}{2}a_0 + \sum_n (a_n\cos(nt) + b_n\sin(nt)). \qquad (18.1)$$

The coefficients give us information about the amplitudes of the various components. The description is one of time and amplitude.

However, it is also possible to display information about the signal in the frequency domain, where the information displayed would show the frequency-amplitude composition. The time and frequency representations of a signal are different, but equivalent, representations of the same function. The fast Fourier transform (FFT) is a means of converting the signal from the time domain to the frequency domain. The inverse transform will convert from the frequency domain to the time domain.

Sampling Rate

The FFT is a discrete transform. It operates not on a continuous function but on a set of values of the function taken at uniform time intervals. Recall how a function is plotted in MathCAD, for example, a sine curve:

$$n := 10 \qquad i := 0 \ldots n \qquad f := 1 \qquad \omega := 2 \cdot \pi \cdot f$$

$$t_i := 2 \cdot \pi \cdot \frac{i}{N} \qquad y_i := \sin(\omega \cdot t_i). \qquad (18.2)$$

It is helpful to think of these plots as being constructed from a sampled set of values. The t_i are uniformly spaced in time. A line may be drawn through the points to provide a continuous curve, but the basis for the plot is a finite number of uniformly spaced points.

From previous plotting examples using MathCAD, a loose rule of thumb was developed as to the number of points needed per cycle to represent a function adequately. That number, however, was determined more from a sense of aesthetics than of necessity. From an information standpoint, what is the necessity? How frequently do we need to sample a signal in order to represent it adequately. Before we answer that directly, think back to the last movie that you saw with a clear shot of the wheels of a vehicle accelerating from rest. First the wheels rotate as would be expected, then the direction of rotation reverses. (A film of an airplane propeller would behave similarly.) Why?

Movie cameras take 24 still photographs each second. The film is held in register in the film plane, and the shutter is opened briefly, permitting the film to be exposed; while the shutter is closed, the hold on the film is relaxed and the film is advanced one frame, and the process is repeated. Projectors operate similarly. The film is held in register, and a shutter (albeit of a different type from that in the camera) permits light to pass through the film and illuminate the screen; while the shutter blocks the light, the film is advanced one frame, and the process is repeated.

The motion of the wheel (or propeller) is thus sampled 24 times per second. Imagine the camera moving in the frame of the wheel so that translational motion is removed. Assume that there is one point on the wheel, not at the center, that we can easily observe.

At a slow rate of rotation, each successive frame (sample) shows the observation point advanced from the previous frame. We perceive a simple rotation. As the rotation rate increases, the angle rotated between successive views increases. When the angle reaches 180°, or two samples per rotation, the perceived rotation rate is at its maximum. Further increases in the rotation rate make the wheel appear to be rotating in the opposite

$$\delta := 170 \qquad \delta' := 190 \qquad\qquad n := 10 \qquad\qquad i := 0 \,..n$$

$$a_i := i \cdot \delta \qquad a_i := \mathrm{mod}\big[a_i, 360\big] \qquad a'_i := i \cdot \delta' \qquad a'_i := \mathrm{mod}\big[a'_i, 360\big]$$

$$a''_i := 360 - a'_i \qquad a''_i := \mathrm{mod}\big[a''_i, 360\big]$$

a_i	a'_i	a''_i
0	0	0
170	190	170
340	20	340
150	210	150
320	40	320
130	230	130
300	60	300
110	250	110
280	80	280
90	270	90
260	100	260

Figure 18.1 Comparison of 170° and 190° rotation steps: a represents motion in the clockwise direction for the 170° case; a' represents motion in the clockwise direction for the 190° case; and a'' represents the 190° motion as measured in the counterclockwise direction. The counterclockwise motion described by a'' is equivalent to 170° motion in the clockwise direction.

direction. As the rotation rate continues to increase, the apparent angular velocity will decrease until it reaches zero (see Fig. 18.1).

For example, consider two motions. In one, a uniform wheel with one viewing spot turns through the angle $\delta = 170°$ between successive views of the wheel; the rotation angle is represented as $a_i = i \cdot \delta$. Because we cannot distinguish different revolutions but only the location of the spot, we write $a_i = \mathrm{mod}(a_i, 360)$. This provides only the locations and not the net angle through which the wheel has turned. If the turning angle between successive views were $\delta' = 190°$, then the location would be given by a'_i in Fig. 18.1.

If we imagined that the wheel were turning in the opposite direction, then the a'_i data would be that as shown as a''_i. These data are the same as those for a_i. That is, motion through successive angles of 190° appears the same as motion through successive angles of 170° in the opposite direction. We see the smaller angle and the backward motion.

When the rotation rate and the sampling rate are equal (one sample per cycle), the wheel appears to be stationary. (This effect is similar to that used in a common laboratory technique for measuring rotation rates using a strobe light.) Further increases in the angular velocity result in a repetition of the previous observations: increasing angular velocity up to a maximum, reversal of apparent direction, and decreasing angular velocity. Consequently, several different rotation rates yield the same apparent motion, and if the film sequence is our only data, we can no longer distinguish the different rates. This folding process, the representation of high frequencies masquerading as low frequencies, is known as *aliasing*. We will come back to this later in the chapter.

The only way to be sure of seeing the correct rotation rate is to put on the wheel a governor, which will limit the maximum rotation frequency to half the sampling frequency. Stated slightly differently, no less than two samples per cycle of the highest frequency are required.

This maximum frequency, known as the Nyquist critical frequency, is given by

$$f_{Nc} = \frac{1}{2} \cdot f_{\text{sampl}} \tag{18.3}$$

or

$$\frac{1}{\delta t_{Nc}} = \frac{1}{2 \cdot \delta t_{\text{sampl}}} \tag{18.4}$$

and

$$\delta t_{\text{sampl}} = \frac{\delta t_{Nc}}{2} = \frac{1}{2 f_{Nc}}, \tag{18.5}$$

where δt_{sampl} is the maximum allowable time interval between samples. (Note: The name δt_{sampl} will be written without the subscript, δt, in all future use.) Given a sampling rate, f_{Nc} is the maximum frequency that can legitimately be observed. (Some information, possibly all, is lost at the limit.) Or given a frequency, f_{Nc}, that must be observed, f_{sampl} is the minimum sampling frequency. Similarly, δt_{sampl} in the above expression is the maximum allowable time interval between samples if f_{Nc} is to be observed. (A smaller sampling time would mean a greater frequency and would be suitable.)

The role of the governor above is essentially that of a mechanical lowpass filter; a lowpass filter passes frequencies below a level determined by the filter and blocks higher frequencies. Electrical signals are similarly prepared by being passed first through a lowpass filter. When this is done, we know the maximum possible frequency that can be passed and can then select a sampling rate that assures that no information is lost and that aliasing will not occur.

This description is not just an analogy. It describes the problem of sampling signals in general. Earlier, we noted the relation between uniform circular motion and simple harmonic motion. From the Fourier series examples, we know that any signal may be constructed from a summation of individual sinusoidal waves. The sampling problem for a rotating wheel is the same as that for a signal, where one rotation rate of the wheel corresponds to one Fourier component of our signal.

A practical rule about MathCAD's FFT is that the number of samples should be an integral power of two (and the power should not be less than three). If the data do not include that number, pad the data with zeros so that the vector in which the data are stored has the requisite number of elements.

In the time domain, the data are representations of the value of some signal or function, taken at uniform time intervals, δt. In the frequency domain, the data are representations of the value of amplitude and phase of the Fourier components of the original function. The data are presented at uniform frequency intervals, δf.

When data are plotted in the time and frequency domains, the traditional procedure is to represent the data in the time domain with a continuous curve, while the frequency domain is often represented as amplitudes at specific frequencies. That is, the frequency information is presented at discrete frequencies. Instead of a continuous curve, the graph is of uniformly spaced free-standing vertical lines.

18.2   Nuts and Bolts of MathCAD's FFT

A single frequency wave, a sinusoidal wave, is not an inappropriate place to start. The waveform is familiar, and other examples are all variations on the basic theme. In particular, take the FFT of $y(t) = \sin(\omega t)$.

However, before we treat a specific example in detail, an outline of a general approach for implementing a transform is presented. A detailed MathCAD solution is then presented.

- Specify the maximum frequency value of the various discrete frequencies of the function being sampled. Determine the time interval δt between samples (see eq. 18.5). (This is the largest allowed time interval. Smaller time intervals are fine; they would result in more samples.)

- Specify the number of samples and the time interval between samples. These choices may be constrained. A specific time period may need to be spanned. The number of samples conveniently handled is limited; the FFT requires that the number of samples be a power of two. Or it might be desirable to have the spacing of elements in the frequency domain take on a particular value. The values δt, δf, and N are interrelated.

- Create a vector — for example, y_i — that contains the sampled values of the desired function. The procedure is very much the same as preparing data to be plotted.

- The FFT is executed with one command — for example, $q = \text{fft}(y)$ — subscripts are not included in the statement. A plot of the results is the most immediate way to assess the transform. It is convenient to use the error bar plot type (e). If both the magnitude of the transform and zero are plotted along the ordinate, the transform is presented in the conventional form of amplitudes at specific frequencies. The actual calculation performed by MathCAD's operator fft is

$$q_j = \frac{1}{\sqrt{N}} \sum_{k=1}^{N} v_k e^{2\pi i(j/N)k}. \tag{18.6}$$

- The inverse transform is also accomplished with one statement — for example, $iq = \text{ifft}(q)$. In some cases, it is useful to start in the frequency domain and construct a particular wave using the inverse FFT. The actual calculation performed is similar to that for the FFT except for the minus in the exponential:

$$iq_j = \frac{1}{\sqrt{N}} \sum_{k=1}^{N} w_k e^{-2\pi i(j/N)k}. \tag{18.7}$$

It is often useful to determine several other quantities that can aid in keeping track of the data in the transform process. The quantities are specified in terms of a particular frequency, f, the period ($T = 1/f$), the sampling time interval δt, and the number of samples, N (the number of intervals is $(N-1)$).

The total time sampled: $t_{\text{tot}} = (N-1) \cdot \delta t$.

The number of cycles in this time: $n_c = t_{\text{tot}}/T = t_{\text{tot}} \cdot f$.

The number of samples per cycle: $s_c = (N-1)/n_c = T/\delta t$.

The spacing between elements in the frequency domain:

$$\delta f = f \cdot \frac{s_c}{N} = \frac{1}{\delta t \cdot N}. \tag{18.8}$$

Specific steps for implementing MathCAD's FFT of a single sine wave are:

1. Specify the frequency, f. Define T and ω. Determine the maximum possible $\delta t_{\max}$.

2. Specify the δt to be used.

3. Specify the number of data samples, N. The number should be some power of two, $N = 2^m$, where m is some integer greater than two. Values for N could be 8, 16, 32, 64, ... In selecting this value, it may be useful also to define and evaluate the number of samples per cycle, s_c.

4. Calculate the interval δf in the frequency domain.

5. Specify an index i to range from 0 to $N - 1$. (It takes on N values.)

6. Calculate the times associated with the given sampling rate, $t_i := i \cdot \delta t$.

7. Calculate the values for the function at those times, $y_i := \sin(\omega \cdot t_i)$.

8. Take the FFT, $q := \mathrm{fft}(y)$. Just use the vector name; no subscripts are needed.

9. The variable q has fewer elements than y; one more than half as many. Because the first element is in the 0-position, the index of the last element of q is $N / 2$; call it M. Set up an index j, for q, that runs from 0 to M. Evaluate q_j. Ignore values with magnitudes like 10^{-13}; that is, consider them to be zero. (It may be convenient to change the format statement so that these values are represented as zero.)

10. Plot the magnitude of q and the zero line (ordinate) vs. the index (abscissa), that is $|q_j|$, 0 vs. j. Change the plot type to e. Let the abscissa range from -1 to $M + 1$ so that all the elements can be seen within the plot region. Let the lower limit of the ordinate be a small negative number, -0.1 or -1 (adjust to please the eye) so that the bar lines do not start from the lower edge of the graph. Because it is the magnitude of q that is plotted, q's complex nature is obscured.

11. Take the inverse transform of q, $iq := \mathrm{ifft}(q)$, and plot iq_i vs. i. Compare it with the plot y_i vs. i.

• • Load FFT1, FFT of a sine wave (see Fig. 18.2).

Be aware that starting with a wave of a single frequency does not necessarily mean that the transform will have only one nonzero element. A transform will result in a single nonzero element only if the frequency of the signal is an integral multiple of the frequency interval δf.

For example, if the frequencies, as represented in the frequency domain, range from 0 to 8 in integral steps, and if the frequency of the input signal is some nonintegral value such as 2.4, then to represent this signal all the frequencies will be required. On the other hand, if the frequency has an integral value, only the corresponding component of the transform will be nonzero. (Integers are represented by a single line in this case because they are multiples of δf, not because they are integers.)

• • The file FFT1 contains the listings for the above example. Load and run it and gain some feeling for the role of the quantities δf, f, N, s_c, and n_c in the FFT process. Take the transform of a sinusoidal wave, change

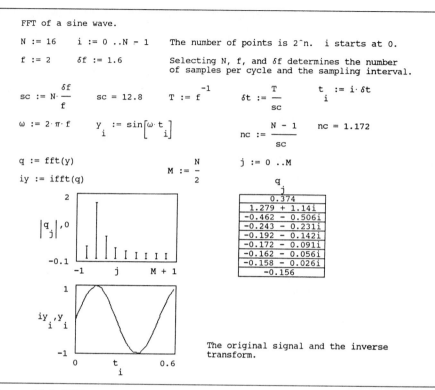

Figure 18.2 FFT1, FFT of a sine wave.

the value of δf, take the transform again, and observe the change in the transform. Follow the steps outlined below first, then try some variations of your own, and learn to control the process.

How many samples per cycle are there in this case?

Is there more than one nonzero element in the transform? Is f an integral multiple of δf?

Note that the plot of the transform contains nine elements. What are the frequencies that are associated with these lines? Write an expression for f_j and evaluate it.

Leaving f and N unchanged, find the value of δf that will yield the number of samples per cycle to be two. (Try different values of δf.)

What has happened to the transform in this case? Look at the numerical values as well as the plot. Is the problem because $s_c = 2$, or is there some other reason?

cies. By removing those elements in the transform whose magnitude is less than some arbitrary value, some of the noise can be removed. Of course, some signal information will be lost as well. Taking the inverse transform of the filtered frequency information, we can observe the results of the filtering process in the time domain.

● ●   Load FFT3, digital filtering (see Figs. 18.4 and 18.5).

The signal y is the sum of two sinusoidal signals of different frequency. Noise is generated and added to the signal. We then observe the transform of the original signal, t_y, the noise, t_n, and the signal plus noise, t_z.

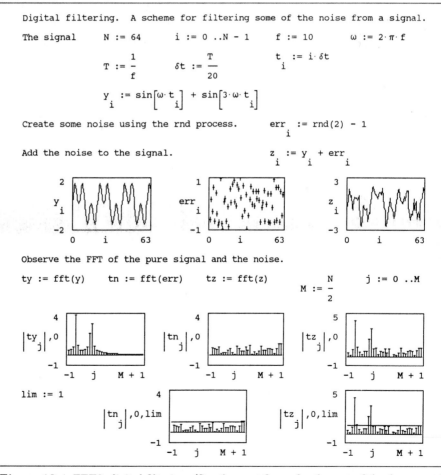

Digital filtering. A scheme for filtering some of the noise from a signal.

The signal $N := 64$ $i := 0 .. N - 1$ $f := 10$ $\omega := 2 \cdot \pi \cdot f$

$$T := \frac{1}{f} \qquad \delta t := \frac{T}{20} \qquad t_i := i \cdot \delta t$$

$$y_i := \sin\left[\omega \cdot t_i\right] + \sin\left[3 \cdot \omega \cdot t_i\right]$$

Create some noise using the rnd process. $err_i := rnd(2) - 1$

Add the noise to the signal. $z_i := y_i + err_i$

Observe the FFT of the pure signal and the noise.

ty := fft(y)   tn := fft(err)   tz := fft(z) $M := \frac{N}{2}$ j := 0 ..M

lim := 1

Figure 18.4 FFT3, digital filtering. (See the next figure for the rest of the document.)

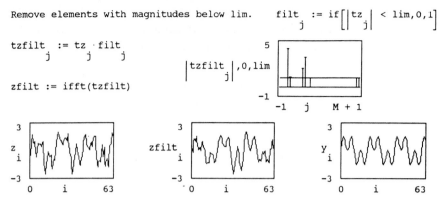

Recall that the amplitude of the noise was equal to that of either component.

Figure 18.5 FFT3 *continued.*

The transform of the noise and the data plus noise are shown together with a horizontal line at the level *lim*. By arbitrarily adjusting the value of *lim*, we can see the number of frequency components that would be eliminated. $t_{z\text{filt}}$ shows what remains.

A final set of plots show the corrupted signal, the filtered signal, and the ideal data. The result is not perfection, but it is a significant improvement. The high-frequency noise has been removed; some low-frequency noise and some of the signal have been removed as well.

Try this with different levels of noise, different frequency values, and different values of *lim*.

Can you think of other filtering schemes that you could implement? Try them.

18.5 Aliasing

Earlier, the notion was introduced that the number of samples per cycle (of the highest frequency component of the signal) must be at least two. To see why this must be so, violate the rule and see what happens. If the signal is of one frequency, the result is easier to observe.

The following provides one means of examining the problem.

• •   Set up a document for performing a FFT with 17 elements in the frequency domain. (Define N, i, M, j, δf, f, ω, s_c, T, δt, and t_i — or load FFT2, which has these quantities at the beginning.) Let δf equal one. Given this set of parameters, what would be the maximum allowed value of f that the FFT can correctly process? (The frequency value when the number of samples per cycle is two.)

Take the FFT of the function $y_i = \cos(\omega t_i)$. Plot the FFT in the standard way. Let f take on the values 14, 15, ... , 18. What happens to the frequency (as interpreted in the frequency domain) when it exceeds the maximum allowed value? Continue taking transforms with $f = 30$, $31, \ldots, 34$.

These procedures clearly demonstrate that high-frequency signals — that is, signals of sufficiently high frequencies that less than two samples per cycle occur — fold back into the legitimate frequency domain. They do so in a way that renders them indistinguishable from the lower frequency signals that they masquerade.

• •   Write an expression for f' (a frequency outside the proper range) in terms of f, N, and some arbitrary constants that will give the same transform as some f within the range. Try not to look at the answer that follows. Base your expression on the empirical observations that you just made when looking at the different frequencies and observing the foldover.

Test your result with several different sets of values. Let y represent the properly sampled signal; let z represent the high-frequency signal that appears to have the same frequency as y. Test them by calculating the expression $z_i := \cos(\omega' t_i)$, where $\omega' := 2\pi f'$, and taking the transform $q' := \text{fft}(z)$. Plot y_i, z_i vs. t_i and plot the magnitude of the transforms q and q'. If your expression is correct for f', the plot of z_i should lie exactly over the curve for y_i. The transforms should be identical as well. Does your expression work for nonintegral values? (The answer is $f' = k N \pm f$ where $k = 1, 2, \ldots$.)

It is especially interesting to plot the two functions, $y(f)$ and $z(f')$, which appear the same when using the δt above, with a new value $\delta t'$ that is much smaller. When this is done, both signals are adequately defined and can be seen as distinct. By observing where the curves intersect, it becomes intuitively clear as to how these two signals that are so different could be interpreted as the same.

• •   Load file FFT4 (see Fig. 18.6). This file demonstrates all the points outlined above in the exercises. Define $q' = \text{fft}(z)$ and plot q' in the same way as q is plotted. How do q and q' compare? What are the associated frequencies? How do the plots of y_i and z_i compare? Why? Try some other frequency combinations that result in similar transforms and y, z curves.

Aliasing. Two different frequency signals appear the same in both the time and frequency domains.

$N \equiv 16$ $i := 0 .. N - 1$ $\delta f := 1$

$f \equiv 2.5$ $\omega := 2 \cdot \pi \cdot f$ $sc := N \cdot \delta f \cdot f^{-1}$ $sc = 6.4$

$f' \equiv 18.5$ $\omega' := 2 \cdot \pi \cdot f'$ $sc' := N \cdot \delta f \cdot f'^{-1}$ $sc' = 0.865$

$T := f^{-1}$ $\delta t := T \cdot sc^{-1}$ $t_i := i \cdot \delta t$

$y_i := \cos\left[\omega \cdot t_i\right]$ $z_i := \cos\left[\omega' \cdot t_i\right]$ $q := \text{fft}(y)$ $q' := \text{fft}(z)$

Plots of the two signals, showing the points at which the signal was sampled.

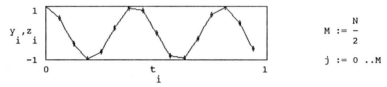

$M := \dfrac{N}{2}$

$j := 0 .. M$

The transforms of the two signals are also identical.

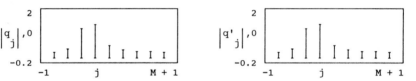

Let the period between samples be reduced by a factor of 16. Plot one fourth of the time range shown above.

$k := 0 .. 4 \cdot N - 1$ $\delta t' := \dfrac{\delta t}{16}$ $tt_k := k \cdot \delta t'$

$yy_k := \cos\left[\omega \cdot tt_k\right]$ $zz_k := \cos\left[\omega' \cdot tt_k\right]$

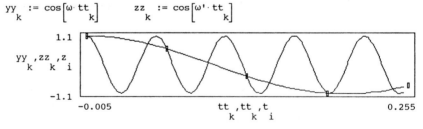

The separation between points, as sampled initially.

$\dfrac{f'}{f} = 7.4$ $angsep := \dfrac{2 \cdot \pi}{sc}$ $angsep' := \dfrac{2 \cdot \pi}{sc'}$ $angsep = 56.25 \cdot \text{deg}$

          $angsep' = 416.25 \cdot \text{deg}$

$rad \equiv 1$ $\deg \equiv \pi \cdot 180^{-1} \cdot rad$

Figure 18.6 FFT4, an example of aliasing. These frequencies are indistinguishable given the sampling limitations.

At the end of the document is a graph displaying the original function, y_i, sampled at δt (the points are shown as open rectangles), and the functions y and z, called yy and zz, sampled at 16 times the previous rate. Explore this a bit, for example; change f and f' and observe the results.

The equation for f' above indicates which frequencies will have common intersection points as shown in the graph. Because the FFT process assumes that frequencies above the maximum ($s_c < 2$) are not present, the lower frequency is the only choice consistent with the data.

In audio recording, if any frequencies above 20 kHz are first removed from the signal by a filtering process, and if the signal is then sampled at a frequency greater than 40 kHz, then precise information about all frequencies below 20 kHz is theoretically possible. Other practical factors, like noise, may get in the way of a perfect recording, but they are not directly associated with the value of the sampling rate.

18.6   Spatial Frequency and Period

The FFT provides a mechanism for passing back and forth between the frequency and time domains. Similarly, the FFT provides a mechanism for treating a different pair of domains, spatial frequency and spatial period. Whereas the first example was that of frequency f (or ω) and period T, the second is that of wavenumber k and wavelength λ. For example, the amplitude and phase of a diffraction pattern is related to the amplitude and phase of the spatial measure of the diffracting object.

We explore this relationship by considering the rectangular slits of, for example, the Young's two-slit interference experiment, as having a certain amplitude x where the slit openings are located, and zero amplitude elsewhere.

• •   Load FFT5, two-slit pattern (see Fig. 18.7).

The slit openings are defined by setting x-values over the entire range to be zero. Then a few specific values representing the slits are set equal to one. See the k and l indices. In the document, we consider the cases of single-slit and double-slit simultaneously; y represents the single-slit and is defined in the same manner as x.

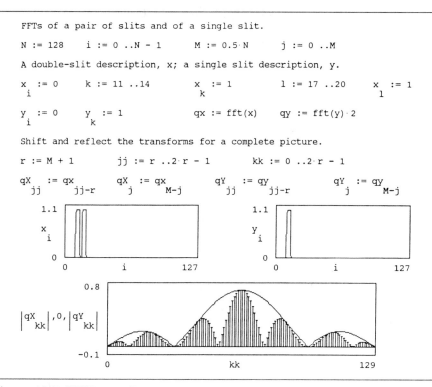

Figure 18.7 FFT5, two-slit pattern

A plot of the FFTs, qx and qy, would show only one half of the pattern. To see the pattern in more typical form, we do two things. First take the transform and then shift the entire set of elements to the right. If M is the number of elements in the transform, each element is shifted over M steps. Then we reverse the order of the elements in the transform and put these reversed elements in place of the original, in the first M spaces. The entire pattern is shown by qX. Since qX represents two slits, and qY only one, the intensity of qY is doubled for purposes of comparison.

We see the combined interference and diffraction effects that one would expect in a double-slit combination. The single-slit pattern is plotted in the same space, showing the modulating envelope expected from the diffraction effects.

Let $k := 14 \ldots 14$ and $l := 18 \ldots 18$. Observe now the interference and diffraction effects. Explain the results and the differences from the previous example in terms of interference and diffraction.

Observe the following cases:

$$k := 13 \ldots 14; \quad l := 17 \ldots 18;$$
$$k := 6 \ldots 14; \quad l := 17 \ldots 25;$$
$$k := 13 \ldots 14; \quad l := 27 \ldots 28;$$
$$k := 11 \ldots 14; \quad l := 27 \ldots 30.$$

Explore the behavior in a general way. Make the slits narrower or wider. Change the separation between the slits. In each case provide a qualitative explanation for each observed change.

What happens if the two slits are not assigned the same x-magnitude? Compare your results with those of the previous chapter.

Increase the number of slits to three and compare your results with those of the previous chapter. Try four slits.

• • Examine single-slit diffraction; remove all double-slit related regions from the previous document. First let there be one slit one unit in width. Observe the width of the transform pattern. At what point does the first zero occur in the pattern? Now let the slit width be two units. At what point does the first zero in the pattern now occur? Let the width be four, and again find the zero.

In each case, what is the product of slit width and distance to the first zero in the pattern?

These observations are related to the uncertainty principle. The slit width is a measure of the uncertainty in position as a photon (or an electron in an analogous experiment) passes through the slit. Call this value Δx. After the photon (electron) passes through the slit, the spread of the diffraction pattern is a measure of the momentum of the photon in the x-direction.

When the slit is narrow, the diffraction pattern is wide. As the slit width increases, the pattern width decreases. The quantum mechanical relation for the uncertainty principle in terms of position and momentum is

$$\Delta x \cdot \Delta p \geq \hbar. \tag{18.14}$$

The products you observe of slit width and distance to the first zero should be approximately constant.

• • Another method for gaining experience with this general concept is to add a set of waves with different frequencies and make some observations about the envelope associated with the sum.

Plot two cosine waves as a function of x, $\cos(kx)$. Let one wave number be 10 and the other 14. Let the plot range include about eight cycles of the slower frequency. Plot the individual waves in one region and plot the sum separately. Observe the modulation resulting from the sum of two similar frequency waves. If the k-values are described as spanning a range from $k+\Delta k$ to $k-\Delta k$, what is k? What is Δk? If the width Δx is defined as the distance from the maximum to the half maximum, what is Δx? What is the product $\Delta k \cdot \Delta x$?

Next plot the sum of four waves with wavenumbers 10, 11.33, 12.67, and 14. What are the values for Δk, Δx, and the product $\Delta k \cdot \Delta x$?

Repeat for eight waves with wavenumbers uniformly spaced between 10 and 14.

The product $\Delta x \cdot \Delta p$ and the product $\Delta x \cdot \Delta k$ can be related through the deBroglie relation

$$\lambda = \frac{h}{p}. \tag{18.15}$$

$$\Delta x \cdot \Delta k = \Delta x 2\pi \frac{1}{\Delta \lambda} = \Delta x 2\pi \frac{\Delta p}{h} \approx 1. \tag{18.16}$$

An alternate form of the uncertainty principle is

$$\Delta E \cdot \Delta t \geq \hbar, \tag{18.17}$$

where ΔE refers to the uncertainty in energy and Δt the uncertainty in time.

• •   Create a vector, y_i, to describe a sine wave with eight cycles. Display the FFT of this single-frequency sine wave. Let f be an integral multiple of δf. The transform should have only one nonzero element.

In a region after the definition of the vector y, let $y_i = 0$ for the first two and last two cycles of the display. (Define two indices j, k; let j have the same values as i for the first two cycles and let k have the values for i for the last two cycles; let y_j and y_k equal zero.) In the plot, there should be four cycles left in the center. Take the FFT of y and observe the width of the transform.

Repeat this process, reducing the number of nonzero cycles first to two and then to one. Each time, observe the shape of the transform and estimate the width.

As the observation time of the signal decreases, we observe an increase in the number of harmonics needed to describe the signal. This spread in frequencies is directly related to energy: $E = hf$ and $\Delta E \propto \Delta f$. Again,

---

FFT of a Gaussian waveform.

N := 32   i := 0 ..N - 1 δt := 0.5   a := 0.8

$$t_i := i \cdot \delta t - \frac{N \cdot \delta t}{2} \qquad\qquad y_i := \exp\left[-a \cdot t_i^2\right]$$

q := fft(y)   M := last(q)   j := 0 ..M

Shift and reflect the transform for a complete picture.

k := 0 ..2·M q'_{j+M} := q_j q'_j := q_{M-j}

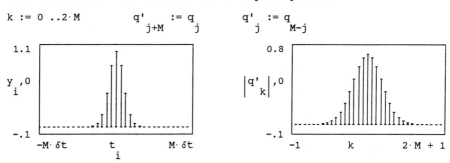

Figure 18.8 FFT6, FFT of a Gaussian waveform.

the product of time expressed in cycles and the width of the transform should be approximately constant.

● ●   Load FFT6, FFT of a Gaussian waveform (see Fig. 18.8).

A related phenomenon is that of the transform of a Gaussian wave-shape. Observe the width of the original shape and that of its transform. Alter the width of the original shape by changing the constant in the exponential of y. What can you conclude about the product of the widths in each case?

That test of yours. The Voight-Kampff test ...
did you ever take it yourself?

Blade Runner

C H A P T E R

19

Special Topics

In this final chapter, several unrelated topics are discussed. The first is an introduction to the concept of chaos using the logistic equation. The second is an introduction to a classic problem in quantum mechanics, the problem of a finite square well, sometimes referred to as a particle in a box. We also examine several details of nuclear decay series and finally look at the problem of extracting a signal buried in noise.

19.1 Chaos: The Logistic Equation

The logistic equation has a deceptively simple appearance. The difference form of the equation is

$$x_{i+1} = 4\,r\,x_i\,(1 - x_i), \tag{19.1}$$

where

$$0 < x < 1, \qquad \text{and} \qquad 0 < r \le 1.$$

To start enter a value for x_0 and return the value for x_1. This value is substituted back into the equation to generate the next value, x_2, and so on. We have examined difference equations earlier, but the values changed gradually. That is not always the case here. An examination of this non-linear equation provides a quick introduction to some features of chaotic phenomena. (For a more detailed study of these concepts see, for example, H. Gould and J. Tobochnik, *An Introduction to Computer Simulation Methods*.)

●●   Before we ask the computer to perform the substitutions, it is a good idea to generate one sequence of values by hand (given that we are dealing with an unfamiliar nonlinear equation).

Write the equation as $f(x) = 4\,r\,x\,(1-x)$. Let $r = 0.5$. Let the initial value, $x_0 = 0.3$. Generate the first six values for this sequence. Start by taking $f(0.3)$ and then substituting that result in f, for example, $f(0.462)$. Continue.

Given this specific value for r, the x-values converge to a stable value very quickly.

●●   Load ST1, logistic equation (see Fig. 19.1).

The logistic equation and a value for r are defined and a series of values for x are computed. Next $f(x)$ is plotted vs. x. A particular value, x_a, is marked on the $f(x)$ curve and on the x vs. x curve.

Change the x_a value until $f(x_a) = x_a$.

Simplify the plot region to $f(x)$, x vs. x. Change r values. Note the change in amplitude of the f-curve and the shift in the intersection point between the two curves.

Next, the values are shown that you should have obtained in the exercise above that you did by hand. A plot of these values is shown, computed using the function f and a subscripted variable.

Pay careful attention to the recursive nature of the function; note the values of the functional forms $f(f(x))$, $f(f(f(x)))$, and so on.

When observing the overall behavior in the following examples, distinguish between the initial transient behavior and the behavior after many iterations. Also after many iterations, note whether the x-values converge to a single value (a stable fixed point), to a pair of values (attractors of period two), or to something more complex.

●●   Examine values of x_i vs. i for different values of r. Let $N = 15$,

---

Logistic equation.

r := 0.75   x := 0,0.05 ..1   f(x) := 4·r·x·(1 - x)   xa := 0.3

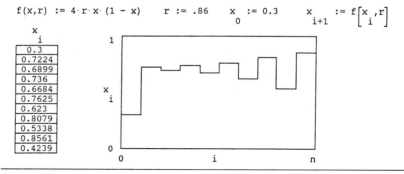

A sequence of values is generated. The function is written a second time
in order that the new r value be used.

r := 0.5   n := 10   i := 0 ..n   f(x) := 4·r·x·(1 - x)

Note the recursive nature of the function.

f(0.3) = 0.42

f(0.42) = 0.4872   f(f(.3)) = 0.4872

f(0.4872) = 0.4997   f(f(.42)) = 0.4997   f(f(f(.3))) = 0.4997

f(0.4997) = 0.5   f(f(f(f(.3)))) = 0.5

f(0.5) = 0.5

Plot a sequence of values. For a general statement of x, it is useful to
include r in the argument list. Above, it would have made for more clutter.

$f(x,r) := 4 \cdot r \cdot x \cdot (1 - x)$ $r := .86$ $x_0 := 0.3$ $x_{i+1} := f\begin{bmatrix} x_i, r \end{bmatrix}$

x_i
0.3
0.7224
0.6899
0.736
0.6684
0.7625
0.623
0.8079
0.5338
0.8561
0.4239

Figure 19.1 ST1, logistic equation.

$i = 0, \ldots, N$, $x_0 = 0.4$, and $r = 0.1$. Enter the difference equation in
subscripted form. Plot x_i vs. i; set ordinate limits at 0 and 1.

To get a rough sense of the behavior of equation 19.1, let r take on
values from 0.1 to 0.9 in steps of 0.1.

Repeat the sequence of r values for $x_0 = 0.8$. How does the final
behavior depend on the initial value of x_0? On the value of r?

For what range of r-values do the values of x converge to zero? Let
$N = 100$ and change the ordinate plot limits to 0 and 0.1. (For these

r-values $x = 0$ is a stable fixed point.) When $r > 0.2$ consider increments as small as 0.02 (for example, let $r = 0.20, 0.22, ...$).

Increase r to 0.7. There is a transitory behavior that dies out. Let $N = 200$ and increase r in steps of 0.01. (Two hundred iterations is not really adequate to decide whether the oscillations will die out or persist, but we will not be too fussy.) What is the approximate upper limit of r for which the iterated values of x converge to a single value?

Let r increase still further, in steps of 0.02. Note that the quantity x oscillates between two values (after the transients at the outset). How large can r be before this stable period two behavior changes?

Let $k := N - 10 ... N$, and display x_k in tabular form and in a plot vs. k.

For $r = 0.8$, what are the x-values of the attractors? Repeat for $r = 0.82$.

Repeat for $r = 0.87$. How many attractors are there now?

Repeat for $r = 0.89$. How many attractors are there now?

• •   Load ST2, logistic map (see Fig. 19.2).

In the previous study we examined x for various r. In this document, we plot a summary of all the information obtained in the previous example. (This document requires lots of memory. Exit MathCAD and reload.) This document takes a minute to process, so study the code while Math-CAD is doing the work.

In studying the behavior of the logistic equation, we are interested in values of x as a function of r. Usually, the transient behavior is not of interest. The equation is iterated $(N + P)$ times. Only the last P values are plotted (k index). If the last P values of x are the same, as they are for small r, then only one point appears. The values of L and M and the index j are used to specify the values of r. (The map is pale because of MathCAD's limitations, including, for example, the need to store everything. To pursue these calculations in more detail, it is necessary to use a standard programming language.)

Correlate the x vs. r plot with the results from the previous example. Recall, for example, the values for x as r increased above 0.7, and recall those values of r where the number of attractors increased.

• •   Load ST3, graphical display of the iteration process (see Fig. 19.3).

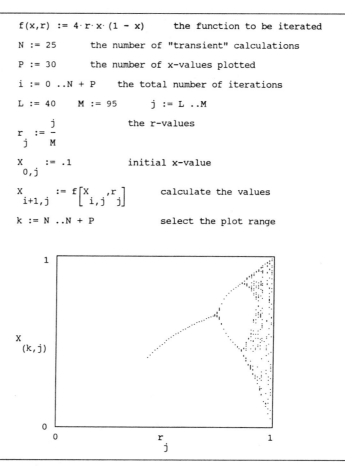

```
f(x,r)  := 4·r·x·(1 - x)        the function to be iterated

N := 25        the number of "transient" calculations

P := 30        the number of x-values plotted

i := 0 ..N + P     the total number of iterations

L := 40     M := 95        j := L ..M

      j                    the r-values
r  := -
 j    M

X      := .1            initial x-value
 0,j

X      := f⌈X   ,r ⌉     calculate the values
 i+1,j    ⌊ i,j  j⌋

k := N ..N + P             select the plot range
```

Figure 19.2 ST2, logistic map.

The iteration process starts with a particular x-value. The corresponding $f(x)$ is determined. This f-value is the next x-value, and so on. A line on the graph following this process gives a visually appealing view that should aid your intuition.

Start with a point on the x vs. x line. Draw a vertical line to the f-curve. Draw a horizontal line to the x vs. x line. Draw another vertical line to the f-curve and continue in this way. The process may lead to a stable point, to a pair of attractors, or to something more complex.

In the first plot, we show x_i vs. i, something that should now be familiar. In the second plot, we show $f(x)$ vs. x, x vs. x, and the line we described following the iterative path. Each x-value leads to an f-value that is the next x-value. The vectors XX and YY contain the coordinates

Graphical solution of logistic equation.

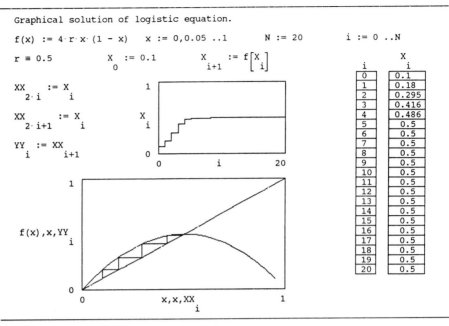

Figure 19.3 ST3, graphical display of the iteration process.

of these points. XX_0 and XX_1 contain the same x-value, x_0; XX_2 and XX_3 contain the x_1 value. The XX-values continue in this sequence. The YY values are offset by one. YY_0 has the x_0 value; YY_1 and YY_2 have the x_1 value. Thus we get the sequence of horizontal and vertical shifts.

Let $r = 0.6$. Change the limits of the second plot region to 0.45 and 0.7 for both axes. This is an easily visualized example of the iteration process "homing in" on a stable attractor.

Change the plot limits back to 0 and 1 and let $r = 0.2$ and process. Let $r = 0.7$ and process. Here, we see the oscillations diminishing in size.

Let $r = 0.75$. Now we see a stable attractor of period two.

Let $r = 0.82$. What do we have here?

Let $r = 0.84$. The values are not yet stabilized.

Let $r = 0.88$. The graph seems to be closing on itself. Examine this case carefully. What is happening?

• •   Let $r = 0.88$ and $x_0 = 0.512$. Define $g(x) = f(f(x))$. Change the iterative equation to $x_{i+1} = g(x_i)$. Change the plot region label to $g(x)$ and process. Explain the results. Where are the other two values that were present when we used $f(x)$? Find them.

• •   Another function that can be explored is:

$$f(x) = r \cdot \sin(\pi\, x). \tag{19.2}$$

Look at graphical solutions for values 0.82, 0.84, 0.8637, 0.8651, and 0.9 to see examples of period 2, 4, 8, 16, and chaotic behavior. (Professor Philip Pennance suggested these values.) Another function is

$$f(x) = x \cdot e^{r\,(1-x)}. \tag{19.3}$$

Examine r-values around and between 2 and 2.8.

19.2   Finite Square Well

A classic problem of quantum mechanics is that of a particle in a one-dimensional box. The particle's behavior is described, not by Newton's second law, but by the time-dependent form of Schrödinger's equation. We solve the time-independent equation form of the equation to determine the wave function and energy states of the system. This equation is

$$-\frac{\hbar^2}{2\,m}\frac{d^2\Psi}{dx^2} + V\,\Psi = E\,\Psi, \tag{19.4}$$

where $\hbar$ is Planck's constant divided by $2\,\pi$; m is the mass of the particle; x is position; V is the potential energy; E is the total energy; and Ψ is the wave function. This equation is quite different from Newton's second law, where x describes the position of the particle. Here, $|\Psi|^2$ is the probability of locating the particle within a given region. (See, for example, A. P. French and Edwin F. Taylor, *Introduction to Quantum Physics*, for a clear and lucid introduction to this equation.)

Imagine a particle contained within a one-dimensional box, free to move in the interior, colliding with the walls and rebounding, or possibly escaping. In a classical picture, one might visualize a marble rolling on a track that is flat in the interior and inclined at the edges. As the particle moves into the edge regions, the potential energy increases, and if the marble's energy is not too great, it will stop and return along the track. An analogous example with an electrically charged particle can be described. The particle is placed in a field-free region surrounded by a region with an electric field. The particle moves freely in the interior but is restrained from leaving the area by the forces associated with the field at the edges.

In quantum mechanics, we treat a similar problem but cannot describe the precise location of the particle within the box. It is fundamentally impossible to specify both the position and velocity of the particle simultaneously, as we noted in our examination of the FFT. The wave function provides information about the probability that the particle is located within a certain region. A series of measurements on similar systems would indicate the particle's location in correspondence with the square of the magnitude of the wave function.

One feature of these quantum systems is that not all energies are possible. Only certain values can be consistent with a given potential; the possible energy values of an electron in the field of a proton, the hydrogen atom, is an example. The problem of a particle in a one-dimensional box is much simpler than that of an atom but it does exhibit certain fundamental properties of quantum mechanical systems.

Our goal is to find the possible energies of a particle in a box, a potential well. If the energy is properly selected, the wave function trails off to zero as the distance from the well increases, indicating that the probability of locating the particle far from the well must go to zero. In fact, the wave function must go to zero as the distance from the well approaches inifinity. If an energy is selected which is not the correct one for the system, the value of ψ quickly diverges as the distance from the box increases.

The goal is to achieve some qualitative understanding, not to determine specific numerical values for a particular system. Consequently, we simplify the equation by letting the constants equal unity (but we do not lose the minus sign). The scaled equation is

$$\frac{d^2\Psi}{dx^2} = (V - E)\,\Psi. \tag{19.5}$$

This equation can be converted to a difference equation, in the same manner as the differential equation expressing Newton's second law was converted. The Euler algorithm is sufficient to provide the necessary qualitative behavior:

$$\frac{d^2\Psi}{dx^2} = (V - E)\,\Psi \qquad\qquad \frac{d^2x}{dt^2} = -\frac{k}{m}\,x \tag{19.6}$$

$$\delta^2\Psi_{i+1} = (V - E)\,\Psi_i \qquad\qquad a_{i+1} = -\frac{k}{m}\,x_i \tag{19.7}$$

$$\delta\Psi_{i+1} = \delta\Psi_i + \delta^2\Psi_i\,\delta x \qquad\qquad v_{i+1} = v_i + a_i\,\delta t \tag{19.8}$$

$$\Psi_{i+1} = \Psi_i + \delta\Psi_i\,\delta x \qquad\qquad x_{i+1} = x_i + v_i\,\delta t. \tag{19.9}$$

• •   Load ST4, particle in a box: finite square well (see Fig. 19.4).

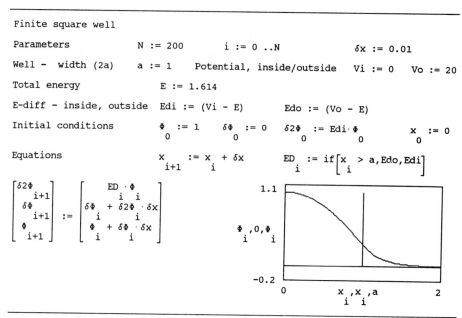

Finite square well

Parameters   N := 200   i := 0 ..N δx := 0.01

Well - width (2a)   a := 1   Potential, inside/outside   Vi := 0   Vo := 20

Total energy   E := 1.614

E-diff - inside, outside   Edi := (Vi - E)   Edo := (Vo - E)

Initial conditions Φ_0 := 1 $\delta\Phi_0$:= 0 $\delta2\Phi_0$:= Edi·Φ_0 x_0 := 0

Equations x_{i+1} := $x_i + \delta x$ ED_i := if$\left[x_i > a, Edo, Edi\right]$

$$\begin{bmatrix} \delta2\Phi_{i+1} \\ \delta\Phi_{i+1} \\ \Phi_{i+1} \end{bmatrix} := \begin{bmatrix} ED_i \cdot \Phi_i \\ \delta\Phi_i + \delta2\Phi_i \cdot \delta x \\ \Phi_i + \delta\Phi_i \cdot \delta x \end{bmatrix}$$

Figure 19.4 ST4, particle in a box: finite square well.

(If the response is too slow, let $N = 160$ and $\delta x = 0.0125$.)

The width and depth of the well are specified. Inside the well the particle is free; at the edges of the well the potential energy increases by V. The energies, E, that we consider are less than V. Consequently, the wave function should show that the probability of finding the particle inside the well is significantly greater than that of finding it outside the well. On the other hand, the wave function must make a smooth transition as it passes from within the well to outside the well. Consequently, there is some possibility of finding the particle outside the well even though this is energetically not allowed in classical terms.

The initial values of E and V_0 are 1.39 and 10. Process. As noted above, if a possible energy is selected, the wave function must go to zero as the distance from the well increases. Note that within the well, the curvature of the wave function is toward the axis (abscissa), and outside the well, the curvature is away from the axis. This sense of curvature (and its energy dependence) is a powerful tool for qualitative descriptions of wave functions for various wells.

Increase E to 1.45 and observe the shape of the wave function.

Decrease E to 1.3 and observe the shape of the wave function.

For the previous two examples, describe where there was too much curvature and where there was not enough. Devise a rule so that by looking at a wave function for this problem you can tell if the energy is too high or too low.

Let $V = 20$ and $E = 1.39$. Process. Is the energy 1.39 too large or too small? Clearly, the energy levels depend on the potential, V_0. Find the appropriate E-value for the lowest energy state.

With V_0 still 20, let $E = 13.75$. How is this shape different? Be specific. What can you say about the difference in curvature between this case and the previous one, the lowest energy state?

Let $V_0 = 20$, $\Phi_0 = 0$, $\delta\Phi_0 = 1$, and $E = 6.4$. How is this state different from the other examples? How many energy levels are there for this $V_0 = 20$, $a = 1$ well?

If the well width is increased, does the number of possible states change? Keep $V_0 = 20$.

For the original width, what happens if V is increased to 100? Find the lowest energy level and compare this wave function with the $V = 20$ case. What is happening to the wave function near the edges of the well? Let $V = 1000$ and compare again.

19.3   Radioactive Decay Series

Nuclei that are radioactive transform by emitting alpha particles (^{4}He nuclei) or beta particles (electrons or positrons). The probability that any particular nucleus will decay per unit time is given by the probability constant, λ. This constant is known as the radioactive decay constant. The decay rate, the number of decays per unit time, is proportional to the number of nuclei in the sample:

$$\frac{dN}{dt} \propto -N.$$

If N is doubled, then there are twice as many decays per unit time. The minus sign indicates that as the nuclei decay, fewer nuclei remain (in the original state). The decay rate, dN/dt, is the number of radioactive nuclei present multiplied by the probability that a nucleus will decay per unit time. This product, $N\lambda$, is referred to as the activity of the sample. The equation

$$\frac{dN}{dt} = -\lambda N \tag{19.10}$$

is easily integrated and yields

$$N = N_0\, e^{-\lambda t}, \tag{19.11}$$

where N_0 is the number of nuclei present at $t = 0$. The units of λ are time^{-1}.

The half-life of a sample is that time period during which one-half of the radioactive nuclei have decayed. We obtain the half-life by substituting for N the value $N_0/2$ in equation 19.11 and rearranging. The half-life is specified by

$$t_{1/2} = \frac{\ln(2)}{\lambda}. \tag{19.12}$$

● ●   Find the times at which one-tenth and nine-tenths of the nuclei have decayed. Specify your answers in terms of the half-life.

● ●   For $N_0 = 10^6$ and $\lambda = 1$, plot N vs. t for three half-lives. Plot the activity for three half-lives. What does the area under the activity curve represent? Integrate the activity as time goes from zero to infinity (of course, if you are using MathCAD to perform the integration, simply let the upper limit be a large value compared to the half-life, 10×, for example).

Not infrequently, a radioactive nuclide decays into a nuclide that is also radioactive. Some decays may involve a sequence of ten or more decays. It is not difficult to keep track of such decays.

● ●   Load ST5, alpha decay series (see Fig. 19.5).

In radioactive decays or in low-energy nuclear reactions, the particles emitted or exchanged include β^-, β^+, α, γ, p, and n, where p and n represent the proton and the neutron. If any one of these particles is emitted, the change in the number of neutrons and protons in the nucleus is listed in the matrix, X. Given the number of neutrons and protons and the particle emitted, the function A' returns the number of neutrons and protons remaining. This is a rather minor bookkeeping matter. However, it is more useful when a series of decays occur, such as in the α decay series.

For example, in d_i are listed one sequence of decays associated with the decay of ^{232}Th. The amusing feature is that the decays are listed in terms of the names, for example, α, and not in a direct numeric form. In the matrix B is listed a summary of the decay sequence in terms of the total number of nuclides, A, the atomic number, Z, and the neutron

An examination of some elementary features of nuclear decay or reaction.

In the matrix X, are the changes in value of proton and neutron number associated with a particular decay.
ß – m or p for negatron or positron decay. τ is gamma decay. ec is electron capture. The numerical values refer to columns in the X matrix.

$$X := \begin{bmatrix} 1 & -1 & -2 & 0 & -1 & 0 \\ -1 & 1 & -2 & 0 & 0 & -1 \end{bmatrix}$$

$$\beta_m := 0 \qquad \beta_p := 1 \qquad \alpha := 2 \qquad \tau := 3$$

$$p := 4 \qquad n := 5 \qquad ec := 1$$

A single decay is obtained with the function A'. Give initial P,N values and specify decay mode.

For example, carbon 14 beta minus decay would be

$$A'(P,N,d) := \begin{bmatrix} P \\ N \end{bmatrix} + X^{<d>}$$

$$A'\begin{bmatrix} 6,8,\beta_m \end{bmatrix} = \begin{bmatrix} 7 \\ 7 \end{bmatrix}$$

A sequence is similarly obtained. m is the number of decays in the sequence. P[0 and N[0 are the initial values for the proton and neutron number. d[i lists the decay sequence. B lists the sequence for A, Z, and N.

$$m := 10 \qquad i := 0 \,..m-1 \qquad j := 0 \,..m \qquad P_0 := 90 \qquad N_0 := 142$$

$$d := \begin{array}{|c|}
\hline i \\
\hline \alpha \\
\hline \beta_m \\
\hline \beta_m \\
\hline \alpha \\
\hline \alpha \\
\hline \alpha \\
\hline \alpha \\
\hline \beta_m \\
\hline \beta_m \\
\hline \alpha \\
\hline \end{array}$$

$$P_{i+1} := P_i + X^{<d_i>}_0 \qquad N_{i+1} := N_i + X^{<d_i>}_1 \qquad Z := P$$

$$A_j := N_j + P_j \qquad B^{<0>} := A \qquad B^{<1>} := P \qquad B^{<2>} := N \qquad z := 82 \,..90$$

$$El :=$$

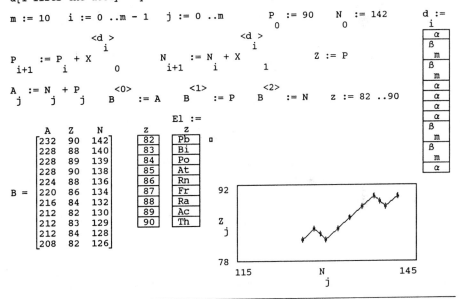

$$B = \begin{bmatrix} A & Z & N \\ 232 & 90 & 142 \\ 228 & 88 & 140 \\ 228 & 89 & 139 \\ 228 & 90 & 138 \\ 224 & 88 & 136 \\ 220 & 86 & 134 \\ 216 & 84 & 132 \\ 212 & 82 & 130 \\ 212 & 83 & 129 \\ 212 & 84 & 128 \\ 208 & 82 & 126 \end{bmatrix}$$

z	z
82	Pb
83	Bi
84	Po
85	At
86	Rn
87	Fr
88	Ra
89	Ac
90	Th

Figure 19.5 ST5, alpha decay series.

number, N. Once the decay is specified, it can be plotted and examined easily.

The plot shown is in the Z vs. N form. Plot the corresponding data as A vs. N.

Plot two of the decay sequences associated with ^{235}U. All these β are β minus:

$$\alpha, \beta, \alpha, \alpha, \alpha, \alpha, \beta, \alpha, \beta, \alpha \qquad (19.13)$$

and

$$\alpha, \ \beta, \ \alpha, \ \beta, \ \alpha, \ \alpha, \ \alpha, \ \alpha, \ \beta, \ \beta, \ \alpha. \qquad (19.14)$$

Plot separately and together in the same region. Look at both Z vs. N and A vs. N forms.

Such sequences provide information about the path but not about numbers of a specific nuclide as a function of time. Consider a decay such as that of

$$^{131}\text{Te} \rightarrow \ ^{131}\text{I} \rightarrow \ ^{131}\text{Xe}. \qquad (19.15)$$

If at $t = 0$ there are N_0 Te nuclei, what are the numbers of Te, I, and Xe nuclei as a function of time?

In general, for a decay $A \rightarrow B \rightarrow C$, the activities of A and B are $A \lambda_A$ and $B \lambda_B$. The rate of change of the number of B nuclides is supply minus loss, that is

$$\frac{dB}{dt} = A \lambda_A - B \lambda_B. \qquad (19.16)$$

If $A = A_0 e^{-\lambda_A t}$, then B can be expressed as

$$B = A_0 \left(\frac{\lambda_A}{\lambda_B - \lambda_A} \right) \left(e^{-\lambda_A t} - e^{-\lambda_B t} \right). \qquad (19.17)$$

The activities of A and B are given by

$$A \lambda_A = A_0 \lambda_A e^{-\lambda_A t}$$

$$B \lambda_B = A_0 \left(\frac{\lambda_A \lambda_B}{\lambda_B - \lambda_A} \right) \left(e^{-\lambda_A t} - e^{-\lambda_B t} \right)$$

$$= A \left(\frac{\lambda_A \lambda_B}{\lambda_B - \lambda_A} \right) \left(1 - e^{-(\lambda_A - \lambda_B)t} \right). \qquad (19.18)$$

• •   Load ST6, nuclear decay series (see Fig. 19.6).

Enter values for the half-lives of A and B. Display in linear and semilog form the values of A and B as a function of time. Examine the shapes of the two curves as T_A ranges from 10 T_B to 0.1 T_B. For what relative values of half-life does B reach its maximum at the largest time? Do the A and B curves always intersect? If the answer is no, find the ratio of T_A/T_B, which corresponds to the limiting case. Viewing the curves on a semilog graph is useful here. Let $\delta t := T \cdot 0.5$, and let t go to $15 \cdot T$. Change the lower limit of the semilog plot to 10^1. Increase the vertical size if you like.

The accumulation of stable end products can be obtained by equating the sum of all nuclide decays at time t plus the active nuclei remaining to the initial value. For example, if B is stable, $A + B = A_0$, where A is

---

```
Nuclear Decay Series

Enter values for the half-lives  and initial numbers of A nuclides.  Do not
give the two half-lives identical numerical values.
```

$TA_{hlf} := .5 \qquad TB_{hlf} := 1 \qquad\qquad A_o := 10^6$

Decay constants $\quad \lambda A := \ln(2) \cdot TA_{hlf}^{-1} \qquad \lambda B := \ln(2) \cdot TB_{hlf}^{-1}$

$$A(t) := A_o \cdot \exp(-\lambda A \cdot t) \qquad B(t) := A_o \cdot \left[\frac{\lambda A}{\lambda B - \lambda A}\right] \cdot (\exp(-\lambda A \cdot t) - \exp(-\lambda B \cdot t))$$

$$T := \text{if}\left[TA_{hlf} > TB_{hlf}, TA_{hlf}, TB_{hlf}\right] \qquad \delta t := T \cdot 0.1 \qquad t := 0.1 \cdot \delta t, \delta t \,..3 \cdot T$$

linear   semi-log

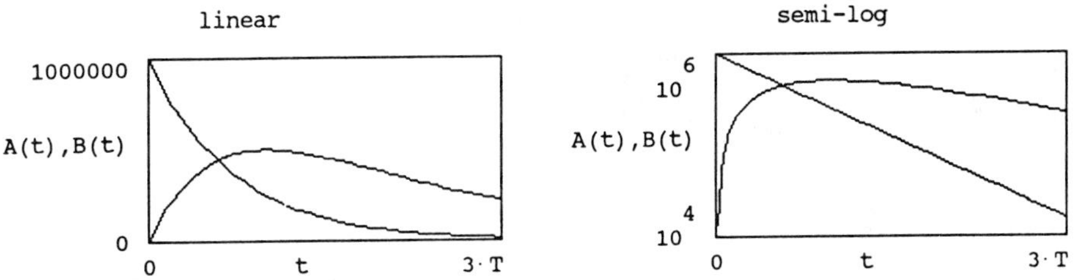

Figure 19.6 ST6, nuclear decay series.

a function of time:

$$A_0 e^{-\lambda_A t} + B = A_0 \qquad \text{or} \qquad B = A_0\left(1 - e^{-\lambda_A t}\right). \qquad (19.19)$$

To express in terms of A, we note that $A = A_0 e^{-\lambda_a t}$ or $A_0 = A e^{\lambda_a t}$:

$$A + B = A e^{\lambda_a t} \qquad \text{or} \qquad B = A\left(e^{\lambda_A t} - 1\right). \qquad (19.20)$$

Similarly, in the decay $A \to B \to C$, $C = A_0 - A - B$. The time dependence of C can be expressed as

$$C = A_0\left(1 - e^{-\lambda_A t} - \frac{\lambda_A}{\lambda_B - \lambda_A}\left(e^{-\lambda_A t} - e^{-\lambda_B t}\right)\right). \qquad (19.21)$$

• •   For the case $A \to B$, plot A, B vs. t.

• •   For the case $A \to B \to C$, plot A, B, C vs. t.

19.4   Signal Averaging

Noise is the bane of all experimental scientists. Many experimental techniques have been developed to improve the signal to noise ratio. Here we provide a glimpse at one process, signal averaging.

If a signal's amplitude is small compared to the noise surrounding it, observing a clean signal may be impossible. However, if the signal repeats or is periodic, then we have a handle by which we can greatly improve our ability to distinguish the signal from the noise which masks it. If there is a means to latch onto the periodicity or repeatability of the signal, we can sample the data starting at the same point in the data sequence over and over again. If we can do this, the signal will tend to make essentially the same contribution with each new sweep, while the noise will tend to cancel itself out. If many samples are taken and the entire set is averaged, the signal is enhanced and the noise is significantly reduced. The effects of the noise cannot be removed completely, but in some cases, a signal can be extracted from data where in any given sampling, the signal is completely obscured. This technique can improve enormously the signal-to-noise ratio, the ratio of the signal amplitude to a measure of the effective noise amplitude.

• • Load ST7, signal averaging (see Fig. 19.7).

A data sequence of N points, consisting of signal plus noise, is sampled M times. The signal has amplitude B, which may be more or less (you adjust) than the noise amplitude, C. To improve the signal-to-noise ratio, we sum the different data sequences and take the average. The signal emerges gradually from the noise.

In the example, the noise is generated using the rnd function, and the data are generated using the sine function. (Note the subscript on t.) With each pass of N data points, the signal repeats. The summation is the means by which we perform the averaging. The data index b corresponds to the different data values in a particular sample; a is the sampling index which specifies which data set we are considering. For each given value of b in the data sequence, we sum each of the sampled values of a.

It is instructive to see the progression of the averaged signal as the number of samples increases. Start out by observing one sample; let $X = 0$. Then gradually increase the limit on the sum; let X take on successively larger values, from 0 to $M - 1$, and observe the signal as it emerges from the noise.

Increase M to 64, and see what effect this change has on the quality of the average. In actual experiments, signals may be determined after thousands or even millions of sequences.

With a modest M, how large can the ratio C/B be and still expect to extract some signal information?

Store the data and take the FFT of S_0 and S_{M-1}. Compare the frequency compositions for an impressive view of the noise reduction. Would digital filtering be as effective as the averaging process?

Signal averaging

Sixteen sweeps of a data set containing 64 samples.

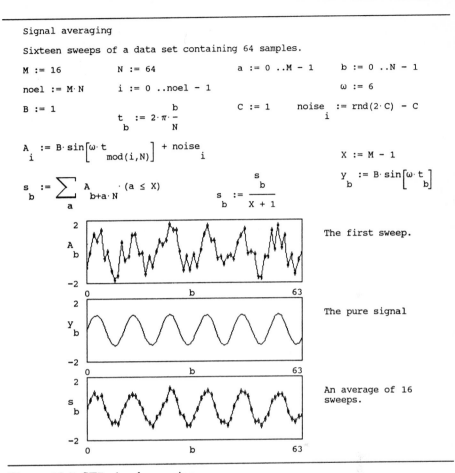

$M := 16$ $N := 64$ $a := 0 .. M - 1$ $b := 0 .. N - 1$

$noel := M \cdot N$ $i := 0 .. noel - 1$ $\omega := 6$

$B := 1$ $C := 1$ $noise_i := rnd(2 \cdot C) - C$

$$t_b := 2 \cdot \pi \cdot \frac{b}{N}$$

$$A_i := B \cdot \sin\left[\omega \cdot t_{mod(i,N)}\right] + noise_i$$ $X := M - 1$

$$s_b := \sum_a A_{b+a \cdot N} \cdot (a \le X)$$ $$s_b := \frac{s_b}{X+1}$$ $$y_b := B \cdot \sin\left[\omega \cdot t_b\right]$$

The first sweep.

The pure signal

An average of 16 sweeps.

Figure 19.7 ST7, signal averaging.

"...because you don't take care of me, or want me,
or want to make any kind of commitment to me.
I'm completely finished with you, Zack.
You just find some other girl to be your little pet
— that doesn't seem to be too difficult for you anyway
— because I've had it with you, and with your stupid music,
and this is just really boring... ."

"I guess it's over between us."

Down by Law

Index